1020

SUDOKU PUZZLES

5 Difficulty Levels with Full Solutions

INDEX

HOW TO PLAY SUDOKU

Sudoku is a puzzle based on a small number of very simple rules:

- Every square has to contain a single number.
- Only the numbers from 1 through to 9 can be used.
- Each (3x3 box) can only contain each number from 1 to 9 once.
- Each (Vertical column) can only contain each number from 1 to 9 once.
- Each (Horizontal row) can only contain each number from 1 to 9 once.

At the start, you will find that some squares are already filled. Our job is to use logic to fill in the missing numbers so that the grid is complete.

The easiest way to start is to choose a number and try to fill in all the places in the puzzle where it should appear.

When you finish dealing with the first number, choose another number.

Repeat these steps for all the numbers from 1 through 9.

In the end, there will be no empty boxes left, and the puzzle will be complete.

Example:

	?							
7	9	5	4	8	3	1	6	?
	6							
9	1	4						
8	3	7						
2	5	?						
	7							
	2							
	4							

The three question marks are in places where there is a missing value. Because the rest of the line or box is complete, it's easy to find which value must be left:

- The only number missing from the horizontal row is a 2.
- The number missing from the vertical column is 8.
- The value missing from the box is a 6.

EASY - 1

	3				9			2
		2		3		4	5	
5	8	6		2		1		3
1	4		3		2		6	9
2	5	7	9	6	1			
	9		8			7	2	1
8	1	9	2		7	3		6
	6	5	4		8		1	7
		4		1	3	9	8	

EASY - 2

				9	6	4		
4			5	1	2	3	8	9
	9				8	6	7	1
	1		2			9		8
6	2	8	1	4				7
9		4		7	3		6	2
8	6	2	9	5	4	7	1	3
7	4					8	5	
		1	6	8			9	

EASY - 3

	2			1	5	7	3	4
		3	2		4		8	5
5	1	4	8					
2	8						9	6
		5		8	9			
		9	6		1	5	7	8
9	5	2	4	3	6	8	1	
		8	7	9	2	4	5	3
	3			5	8	9	6	

EASY - 4

6	5		7		2	4	9	3
		2	5	9	4		6	1
9	4	8	3				2	
1				2				4
	9	4			7	5		
3	8	5	6	4	9	2		
				6	3	1	5	9
5	1		9	7		6	4	2
4		9				3	7	

EASY - 5

	7	9		2	1		5	4
5			8	6	9		2	
2		1					3	8
	2		7	8		1	9	3
3				9	5	4	7	2
	1	7	4		2	8		5
6	5	2	9	1	4		8	
	9			7	3		4	
	3	4	6					9

EASY - 6

4	2	1		3	7	6	5	
3		5				7	2	
7	6			1	5		9	4
	8		4	5	2	1	3	
1	3	4	7	8			6	
2	5	7		6		4	8	
		6	1	2	3			
5	4	2	6				1	
8	1			9		2		

EASY - 7

9	5				1	4	8	2
	2		8	9			3	7
4		3	5		7			
5	9	6		8	2	7		3
7	3	8	9	1	5	6	2	4
	4	2	3				9	5
			1	5	3			9
				6			4	
2	1		7	4			5	6

EASY - 8

8	3		9	5	7		2	4
7			8	1	3	9		6
9	5	1	4			7	3	
6	7		1		8	3		5
		9		6				
4	8	5			9			1
1			3					7
2	6		5	7	1	8		9
	9	7	6		4	2		3

EASY - 9

1	4	2			9		8	5
5		8					6	3
	6	9	8		5	2	1	4
2	3	7	5	4	6	1	9	8
				2	1		7	6
	5	1			8			
			1					9
	9	3			7	6	2	1
	1		9	6	2	8	3	7

EASY - 10

5	8		6		9		2	3
3		9	1		2		6	
		6			3	8	9	1
8	6	5		2		1		9
2				9			7	6
9	1	7		3	6			
6		2	3					7
4	7	8			5	3	1	2
			2	4	7	6	5	8

EASY - 11

7	2	9	1	3	4			8
4	3	1	6		5	9		
6		8			2	4	1	3
5				2	9	3		7
2		7		6	3	5	8	1
3	8	6	7	5	1	2	9	
		5	2	1		7		9
			3				4	5
					8			

EASY - 12

		6		2	4	9	8	
	8			9	1	5		
2	3	9		8	7	4		6
8		3	7	1		2	6	
6	4		2	5	8			9
9		2			6		5	8
7		8	9		5	6		1
	6	4			2	8		5
1	9				3		4	

Solutions on Page: 90

EASY - 13

	5	9	8	4	1	2		6
8	1		6	2	7	5	9	
7	2			9	3			8
	8					9	4	7
1		5	9	7				2
9	4	7	2		8	3		
	6		3				8	4
	9	8	7	1	2	6		5
	7			8	6		2	

EASY - 14

			3		4	6	7	2
6		5			2	3	9	
7	2				6		1	
		4	5		9	2	3	6
	7	6			3	8		
3	5	9	6		8	7		1
5	6	2	4	3	1	9	8	
	9		2	8	5	4		3
4	3	8				1		5

EASY - 15

3			6	5	8	2	1	9
	2	5	1	3	4	7		6
1	8		2	9	7	3	4	5
	1	8	4		5		2	
4					3	6	7	
2	6							4
	9						6	8
8	3	2	5	1	6		9	7
	4	1	7		9	5		

EASY - 16

3	4		9	1	2	8	6	7
			5			2		1
8		1	6	7	3		5	
9		6	4		7			
4					1	6	9	5
			3		6	7	8	
	7	4	1		9		2	8
5	9	8	2	6	4	1		
	1	3	7	8		9		

EASY - 17

8	1				7	2	3	6
			5	3		9		1
	3	2	1		6	7	4	5
	4		8	7		1	5	
3	5	8	6		9		2	
	9	7	2	5	4	3	6	
7		1	3			6	9	
				4	1		7	
			7	6	8	5		3

EASY - 18

						3		
6	7	5		2	3			
9	2	3	8	4	6	5	7	
2				8	9	1	4	
8	5	4	3			9	2	7
	1	9			2	6	8	
3	6	2	5	9	8	7	1	4
			6	1	4	2	3	
4		1	2					6

EASY - 19

	9	1	8		4		2	
			2	5	1	7		9
8	7	2	3	6			1	5
	6					5	4	8
2	3	8	5			1		7
	5	4	7	9	8			2
4	8	9			7		5	1
3	2		9		5	8		4
	1			8			3	

EASY - 20

3	8	9		5	7	6		1
5	2		9	3	6	4	7	8
	7	6		1			9	
		4		8	5	9		
6		2	1	9		7		4
9	3	8	7					5
	6			2	8	3		9
2	9		3				6	7
8		3		7			5	2

EASY - 21

9		6	8	2	3		4	
3	8	7		5	4		2	1
4		5		6		9		3
6	9	3	5	1	2	4	7	
	4	8		3				
5		2					3	6
2	6		3					
8			2	7	6	3	1	4
		4	1	8		5		2

EASY - 22

			4		8	3	1	9
	4	1			6	7	8	5
3	7	8	9	5	1		2	
		9	7				4	8
1		7		8	4	9	6	2
				6		1		3
8	9	4		3			5	
7		3	1				9	
2	1	6	8		5	4		7

EASY - 23

8						1	5	6
5	6	3	9		1	4	8	7
7		4	8	6	5		3	2
2	3			4			9	5
4	5	7		9		3		1
	9	6				8		
3		5	4	8		6		
	8			7	9			
9	4	2	6		3		7	8

EASY - 24

9	5	2	4			7		8
6	8		9				2	5
		4			2		1	
			1	5	4	3		2
2	4	9	3	6	8	1	5	
5		3	7	2	9	8	4	6
	3	6	2		5	9		4
					1	2	7	3
			8				6	1

Solutions on Page: 90

EASY - 25

7	2			8	9		4	3
4	5	1	2		3	8	6	
3	9			1			5	
5	6	3		4			2	7
	7	2			5	4		6
			7	6	2	5		
	8	9	1	2		3		
			3	5			9	1
1		5		9	7	2	8	4

EASY - 26

6	7		3	5	4		8	
4			9	6		7		3
2	3	9	8	1	7			
1	4	7		8	9		3	6
	6		1	7	3	8	9	4
8	9		6	4		1	2	
				9	8		4	1
9				2	6		7	
7	2		4		1			

EASY - 27

4	7	6			3		8	
1	2		7	8		4	6	3
		9	6	4	5			1
3	9	1	4		8			2
	8		9			6		4
5	6	4	3	7				8
	5	3		9	4	8		7
9	4				7			6
7			5	3	6			9

EASY - 28

6			1		9			
		4	2	7	3		1	6
1	7		5			9	2	4
2		6	7	9		4	3	1
	1		3	2	4		6	
			6	1	5	2	7	9
4	6		8		2	3		
			9	6	7		4	
7		9	4	3		6		2

EASY - 29

	6		2		9	5	7	1
	2	7	5	3		8		
	5	9	6	8	7		2	
8	7	6			5	4	3	9
2		4	7	9	3	6		8
	9		4	6	8		1	
	4				6	2		
	8				4	1	6	3
6	3		8					7

EASY - 30

5	9	6			4	7	3	2
		8		2	9	4	1	
4	1	2	3	7				9
1	3		9		6		2	8
	2	4	5					
6	5		7				4	3
7	6	1	2			9	8	
	4			9	8		7	6
9		3		6	7	2		

EASY - 31

3		9	2	1	6	8		4
6	2	8		3				1
					9	3	6	2
				6		4		
	5	6					2	7
8	4	3	9		7	6	1	5
	8	2	6	9		5		3
5			4	8	3	2	7	9
	3		7	5			8	6

EASY - 32

1	4	5		2	6	3		9
	2	3			9		1	6
		9	3		1			7
	9	4	6		8		3	
	8		9	1			5	4
	7	1	5		3	6		8
		8		9	7		6	
	3	2	1			8	7	5
6	1		8		5		4	2

EASY - 33

3	9	4		5	2	6		
	7			6	8	3		1
	6	1	3	7		5		
1	3		9	4	6	8	7	
5				1	7	9	4	3
		9	8		5		6	
4	1		6	2	3	7	5	
		3		8	4	2		
	2	7	5			4		8

EASY - 34

		8	2		5			
				4				2
2		5	9	6		8	7	4
1		7		8	9	4	2	6
5	9		6			7		8
8	2	6	7	3				5
9		3		7	6			1
6	7	2	1	5	8	3	4	9
4				9	2	5	6	

EASY - 35

8	2	7			3	4	6	9
	6		7		4	5		8
	4	5	2					
				2	9		3	
		9		3	7	6	8	2
6			1	5	8	9	7	
4	9	8			2	7		
3	7						9	6
2	5	6	9	7	1	8	4	3

EASY - 36

					4	1		
		5	7				8	9
8	7	1				3	5	4
5	4	3	1	9	6			
7	1		4	2	8	5	9	3
			3	7		4		1
3		9	6	1	2	7	4	
6	8	7		4	3			2
1				8		6	3	5

Solutions on Page: 90 & 91

EASY - 37

9	1	2	5	3		7	4	
	6	4		9				
3	7				4	6	5	9
2			7			4		1
	8	3		4	1	5	9	
			3		9	8	2	
	5	1	6	7		9	8	
	3	7	9	5	2		6	
6		9	4		8	3	7	

EASY - 38

			9	4		6		2
5	2		1	7	8	3	9	
4				2	5		8	
		2	7		6	4	1	
3	4	8	2	5	1		7	
6		7			4	5		
2	8	9	4	3	7	1	6	
	3	5		6		2		
7	6				2	8		9

EASY - 39

	2	8		9	6		5	
3	6	9	7	5	4		1	2
1	4	5	3	2			9	6
	8	3		7	1	6	2	
4				8	9		3	
		1	4		3	9		
	1	4	6		7			
			9	4			6	5
	9	2	8	1		3		7

EASY - 40

3		1	4		8			5
	4			7	2	3	1	9
2	7	5		1		4	8	
	3	8		5	1	2	9	7
5	1		7	3	9		6	
9	6				4	5		
	8		3		7	6		2
6			8	4		1	7	
7	5	3	1	2	6			

EASY - 41

	7				9	5		2
	3		5		1			9
6		9	8	7	2	1		4
5	4				3	8	7	1
1	8		2	5		4		
3	9	7	1	8	4	2		6
	6	3			5	9		8
9	2		4	3				
7		8	9			3		5

EASY - 42

5			8	1	7		4	3
4			9	6	3	8		5
3	1		4			9		
6	3	7	1		8	4		2
	5	1	2		4			8
8	2	4	3			1		9
2			5		1			6
	8		6		2	3		1
	6		7		9	5		4

EASY - 43

		7	6	2	8			9
2	8				9	1	6	3
5	6		3	1		7		
9	5	8	1	4	2	3		6
	4	3			6	2		5
6	1		5	3		9		
4		1	9		5			
	9			7	3	6		1
	7		2		1	5		4

EASY - 44

2			5	6			3	
4		6	7		2	9		1
3		9			8		5	6
8		7	2		6	3	9	5
	2	5		8	3	1		7
9	4	3			7		6	
		8		2		4	1	
	3			7	1	6		
1	6	2	3	9	4			8

EASY - 45

8	7	5	9		6			
	4			2		8		
		2	5			7		6
9		1		7	5	3	8	
	5		4		8	1	6	9
4	8	3	6		1		7	2
7		4	1	5			3	
3	2		8		7	9	5	1
5				6	9		2	7

EASY - 46

4		1	2	5	8		9	
	8	3		1		5		2
		6		9	4	7	1	8
8		7	5		1			4
2	1	5		6			7	3
	3	4		2	7		5	9
7	5		9				8	1
3		9		8	5			7
1				7	2		3	

EASY - 47

9	2			5	8		3	7
3			1		6			
8		5	7		3		4	6
1	8				7	3	2	9
5	3	9	2	8			6	1
	6				1	8	5	
6	7		3	4	2		8	5
4	9		8		5	6		2
2	5				9		7	

EASY - 48

4	3				2	5	1	7
8	7	9	1		5		6	2
5	1		3			8		9
	8		6	7	4		2	3
	4	7		2				
2	6	3	5	9	1	4	7	
7		8	4				9	5
			7	1	9	2	8	
6				5			3	

Solutions on Page: 91

EASY - 49

1				5			4	9
						7		
4	6	7	3	2	9			
	4	3	2	7	5		1	
2			9			5	6	4
5	9				6	2	3	
3	8		5		2	4	7	1
7		5	4	3	1	8	9	6
9	1		8	6	7		5	

EASY - 50

4	3	2			5	9		8
6	7			3		1		
8	5			2		3	4	7
	2			4	6	5		3
5	9		7	8		2		
		8	2			7	9	4
7	8	6		9	2		5	1
9	4	5		1		6		2
2		3				8	7	

EASY - 51

8	1	3	9	6		7	2	5
				2	7	1	4	3
2	4			3		6		8
6	2		4	9		3		7
4			5			2	8	1
5	7	1	2	8	3	9	6	
7				4	9			2
	5	2	6		8			
	9		7	5				

EASY - 52

		1	4	6	2		8	
3				8			7	2
	8					5	6	4
1	9	3	7		6	8	4	5
7	6	2	8	5	4	9	1	3
	5		9	1	3		2	6
6	1						5	7
					5			
	2	5	6	7	1	4		8

EASY - 53

5	6		7	8		2		4
	4	9				5	3	8
8		2	3	4		9	7	
	2		1	3	6	7		5
	5	8		7	4	1	6	2
6	7	1	8	5		3	4	
	3		4			8	9	
							5	
		7	5	1		6	2	3

EASY - 54

9	6	4	3			1	7	
5	3	7						
8		2		6	4	5	9	3
		1	6	5			3	
4	2	3	1	9	7	8	5	6
6	8	5	4	2		7		9
	4			3	6	9	2	
	7	9		4	1			5
2		6				3		

EASY - 55

3	1		8		4			
7	8	2	6	9		1		4
	6		3	2	1	7		8
2		4			8	9	1	6
9	3			5	6	4		
6		1		4	2	5		3
5	2	7	4	6	3	8	9	1
	9	3	5	1		6		
		6			9		7	5

EASY - 56

9			1	8	5	6	7	3
	6	5		2	9			8
7	1				3		5	
			8	1			3	
	5		9	7	6	4	2	
				5	2		9	
1		7	2	3			6	9
	8	2				3	1	4
5	3	9	6	4	1	2	8	7

EASY - 57

4	7	5	3		9		6	1
	2	1				7		
3	9						5	4
5	8		6	1				
		9	2	5	8	4	1	7
7	1	2	9	4		6		
		8	7	3	6	1	4	9
1						3		6
9	6		4	2	1	5	7	

EASY - 58

5		4	8		7	1		
3			4			7	8	
1		8			2	5	6	
	4	7	9		8		1	6
6			1		4	9	2	5
9		5				8		
	5	6	2		9	4	7	
7		2	5	4	1		9	8
4	9		7	8	6	2		

EASY - 59

		8			2	1		
6			4	5	1	2	8	
		2		6	3	5		4
9	6		7	4		3	5	2
2		5	3			6		8
3		7	5		6	9	4	1
	2		1	9	5		3	6
1		3	6	8	7			5
5	7		2					

EASY - 60

			9	1		8		3
9							2	
6	5			8	3	7	9	4
4					2	9	1	6
5		2	8	9			3	
3	1	9		7	6			
1	3	5		2	9	6	4	
	9	6	5	3		2	7	1
7		4		6	8		5	9

Solutions on Page: 91

EASY - 61

			1	5		4	9	
4			8	3		2		1
1		9		2	7			5
6		4	7			1	5	
		1	2		5	3		4
3	7	5	9			6		8
7	8	3		9		5	1	2
5			3	7	8	9	4	
9		6	5			7		3

EASY - 62

9		4	6	1	3	5		7
7	5	1	2	8	4	6	3	
	3	6		7	9	8		
					7	1	6	8
3	7						5	
	1		4			9	7	3
1	6	3	7	2		4		
8		2	9	3	5		1	
5	9	7			6	3	8	2

EASY - 63

			5			7	4	
4				2	8		3	
1	6	9			7	8	2	5
8		3			4		5	2
		6	3	5	1	4	7	
7	5	4	9		2			
2	4	7	8	1	5	3	9	
6	9			4			8	7
		8	7	6	9			4

EASY - 64

		7	6	9	3	2	1	4
4	3			1		9	7	5
9						8	6	3
6	4	5	9		8	1	2	
3	7		2	5	1		4	8
		1		4	6		5	9
1	2		4		7			
		8		6		4		
5	6			2		7	8	

EASY - 65

1	8						2	4
						1		7
		2	5	1			9	8
	9	1	4	6	8	7	3	
	2	6	3	7	1		4	9
7	3	4	2	5	9	8		6
9			1	8	6		7	
	5	8	7		4	9		1
6	1		9	2				

EASY - 66

2	4		7	3			5	6
	9	6				3	2	7
3	5	7	6	2		9	8	4
			2	8	4	7		9
7	3	2		9	6			5
4			5		3		6	
	7				2		1	
	1	3		5	7	6	9	2
		5		1	8		7	3

EASY - 67

3		5	8	7	6	9	2	
7			5	1	9		6	
1		6		2			5	8
5		2			8	1		
6	1		2	3				5
8	7		1			2		6
2	5	1			4	6		
9	6	3	7	8	1		4	2
4		7			2			9

EASY - 68

9	4		6					
2			9	1	7		4	8
		8	5	4		2	6	
	9	3	1	7	4	6	5	2
4	5	2		9		1	7	3
6				3	5	9	8	
	8		3	5	1			
5			7			8		
3	6	1		8		7	9	5

EASY - 69

4	7	6	1					8
2				7		1		
	1	3			2	5		
6		1	7	9	5	3	4	
				3	4	6	5	1
5	3	4	2	1		8	7	9
1		5				7		
3	6	8	9		7	4	1	5
		2		6	1	9	8	

EASY - 70

9			3	4			6	
	5	1				8	3	9
	2		5			7	1	4
7	8	2	4	9			5	3
6			2	1	3	4	8	7
1	3		8			9	2	
	4		1	7	8	6	9	5
	1	9	6		4		7	
8	6	7			5	2	4	

EASY - 71

1	4		3	6		9		7
				9	2	8	1	6
8		6	1	5	7			3
5		9	6	3		7	8	
4	3		2	7		5	6	
	7		5		9			4
2		7			5	3		
3	8	5		4		1		2
9	1		8		3			5

EASY - 72

	5	8		7	6		2	1
		9	5	2	4	3		8
	6	4	8		1	5	9	7
	8		1	9			7	3
	3	7	4		2		8	
6	9			8		2	4	
9	2	6	7				5	4
8	4		2		9	7		6
	1				8	9	3	2

Solutions on Page: 92

EASY - 73

		2	6		1		9	7
	6		9	8				4
7	1		5			3	8	6
	9	8		1	6	7	3	
1	2	3		5	8	4		9
6		4	3		2	8	5	
		6	1	7	5	9		
		1		4			7	3
	8	7	2	6		1		

EASY - 74

	2			5	3			7
7	1		8		6		3	5
5		6		4			1	9
6		1			5		9	
2			4	8	9		6	
	5	8	6	1			2	4
	6	7	9		8	4	5	3
4		5	3		1	9		2
3	9	2		7			8	

EASY - 75

5	8		9		7			4
9	1	6		8		7		2
2		7			6	5		9
				5	4	9		3
		2	8	7				
4				1			7	6
3	6	9			1	2	5	8
1	5		2	6	8		9	7
7	2	8		9		4	6	1

EASY - 76

9	7	4	5			8		1
8	2	1	7	9		3	5	6
	3	6	1				9	7
		5	6	4	3	9		8
				5			1	
3			2	1	7		4	
	8	9		7		5	6	4
6	5	2				7		9
	4	3	9	6	5	1	8	

EASY - 77

9			6	3			4	2
8	5	6	1		4		7	9
		4	9	7	5	6		
7	8			5	9	1	6	3
3			7	6	1		5	
5	6		2	8	3	4	9	7
6	7	8	5		2			
1					7	8	2	6
4		3						

EASY - 78

6	9					8	7	1
4		1	5		9	6	3	
2	7	3	8	1	6	4	9	5
				8	3		1	9
1	3		9		5			
		8	7	2				
	1	9	6	5	4	3	2	
3			2	9	8	1		7
	5			3		9		6

EASY - 79

					7	6	5	
6	4		8	1	5	9	2	
5	1	3	9	2		8	4	7
4	7	2	6		3		9	
1		8		5	4	3	6	2
3	6	5	1			4		
	5	1	4	6		7		9
	3							6
			5	3	9	2		4

EASY - 80

5		6		1	2			
4			6	5		2	1	7
		2		4			5	
8		1		6		5		2
2	4					6		8
6	7	5	2	8				4
3	5	7				4	2	1
1	2	8		7	4		6	9
9		4	3		1	7	8	5

EASY - 81

8	2		5		4	6	9	1
4	6		2	7	1	8		
3			9	6	8			
7			8	1				2
6		8			5	3		4
			6	4	9		8	
9	3	2	1				4	6
1	8	4		5	6	2	7	
	7		4	9		1	3	8

EASY - 82

			1	6	2	8	9	3
	2	9	4	3	8	5		
6		3	9	5		2		1
9	4		5	8			3	
5	3	8		4				7
	1	6	3	7				5
			8	2		3		9
	6	2		9		1	5	4
	9	4	6	1			2	

EASY - 83

1			5		2		8	
8		6	9		7	5		3
		5		8			6	1
	9	4	2	5		7		8
	1	7	8		6		9	
	8	2	7	9		6	4	5
9	7		6	2		8		4
	6	3			8	1	5	
4		8			9	2		6

EASY - 84

	7	1	2	8	5		3	
2		6	1	4	3	7		
3	5			6				1
6				9		1	7	8
		9	7	5		2		3
7	2	3	8	1		9	6	
	9	7	5	3	1		8	2
		8	4		9		1	
1			6	7	8			

Solutions on Page: 92

EASY - 85

	5	9	3		4	6		1
	1		9	6	5	4	3	8
	6	3	2			7		9
		4	6	5	9	3	8	7
			8			1	6	4
	8		1	4			9	5
	9	7			8		1	
5	3	1				8	4	6
			5	1	6			3

EASY - 86

6	7	9	3	2	5	8	1	4
4					6	3	9	
8	1	3	4		9		5	
7	8					9	6	
			6	3	7	4	2	
			8	9	4			
5	6				3			9
3	2	8		6		7	4	5
9		1	7		8		3	2

EASY - 87

	8	9			3			6
6		5				9	2	3
3	7							8
		1	5	8	4	3	6	7
	6	3	7		1	5		
	5	4		3	9	2	8	1
2	4				6		3	5
5			4	7	8	1		2
1	9	8	3	5		6	7	

EASY - 88

			9	5		1		8
2	1	4	6	8	7	5		9
9		8	3	2	1		7	6
		2	5		6	3		7
4	7		2			6		
	3			4		2	9	5
			4	6	9	7	5	
7	2		1		5			4
5		6	8	7	2		1	

EASY - 89

8	1	4	3	5		9	6	7
5	3	9	4			2		
2	7	6	9		8	4		
	5			4			9	3
			1		9	5	4	2
			5	6	3	8		
3	9				4			6
4	8		6	3			2	9
6	2	7	8		1		5	4

EASY - 90

	1	6		5	7		2	
8		7				9	4	
9			4	1		7		5
1	7			4		8	5	
6		2				4	1	
		5	1	8		2	7	9
	6	1	8	2	9	5	3	4
2		8				1	9	
5		4	3	7	1	6	8	

EASY - 91

	6		4	3				1
	8	5	2	7	1			
	9	1			6	7	2	3
	1		6			9	8	
6	2	9	5					
	7	4		2	3	5	1	6
2	3	6	7	9			4	8
	4	7	3		2			5
	5	8	1	6	4	2		

EASY - 92

1					2	7	6	
		4	5		8			9
	8		7		9	3	4	
6		8			7	5		4
5	4	1	9		6	2	3	
3			1	5	4		9	
4	7		8			6	5	3
9		3	2		5	4		1
8	1		6		3	9	7	

EASY - 93

3			2				6	8
						2	7	5
2		5	6	7				9
4			9	2		8		7
		8	3			6	9	2
9	2	6	7	5		1	3	
5	1	2	4		9	7	8	3
	9	7		3	2		4	6
6	4	3	5				2	

EASY - 94

9	1			4		8		
	8		1	5	6	7	3	9
3			8	9	7		2	
8	4		3	2	5	9		
7	6	3		8	9	2		
5				6		3	4	
4	3	5		7	8		9	
		8	5	3			7	4
6				1	4	5	8	

EASY - 95

2	3		5	8	6			
	1	5	3				2	8
4	9		1		2	6		5
5	8		7	1		9	6	3
	6				5	4	8	
9	4			6	8		1	7
3	5		6		1	8		9
	7		4	9	3		5	1
1	2			5			4	6

EASY - 96

4	8	2	3					9
3	7		4	8	6		5	1
1	5	6	2	7	9	3	8	4
7	6					4	3	8
		5	7			6	9	
2					4			
5			6		8	9	1	7
6				4	7	5	2	
9	2	7		5		8		

Solutions on Page: 92 & 93

EASY - 97

	5	6	3	2			1	
		1				2		5
	8	2	1	4		6		9
6	2		9	8			5	
			7	6	3	1	9	2
	1	9	2	5	4	8		6
1		8	4	9		5		3
5	7		8	3		9	6	
2		3		1				7

EASY - 98

	6	2			3	1		7
	1	9		6		8	5	
		3		4	5	6	9	2
	5			3	1		8	
3	2	6	8	9	4	5		1
	8		5	2			3	
1		8			2	3		5
	4	5	3		8		2	9
		7	6		9	4		8

EASY - 99

3	5	9	8		6	2	1	4
1		4	9		5		7	8
	8		4			9		5
2		3				1		6
	9	8	1		2		4	
4		1		9	7	8	5	
	4				9			3
	1	6		5	3	4	2	
9		5	2	8		7		1

EASY - 100

5	6		2	1		7		3
1	8		9	7		2	6	
4		2	3		5	9	8	
	2	7						9
3			8		7	4	1	2
		4			2	8	7	
6	4	1	7		9	3	5	
	9		6	5	3		2	4
2	3	5				6		

EASY - 101

4		5	3	8		7		2
3		9		5	7			8
		8	6			9	5	
1		6	7		2	8		5
8	4	7	9	6		3		
5	9	2		3		4	7	
7			4	2	3	6		9
6		4	1		8	5		7
			5	7		2	1	

EASY - 102

4	6	1		3	7	8	9	
8	3	2	1	5	9	4		7
	7		4			2	1	3
		9	3		8			
				1	2			
6			9	4	5		3	2
3	2				1	5	4	8
	9	8	5			3	7	6
	5	4	8		3			1

EASY - 103

9		8			3	6	7	
6				9	2	5		8
1		7		8	6	3	2	
3	7	1	8	2	9	4	5	6
		6	5	4		7		
4		5	3		7		9	1
		2		1	8		3	
7	1	9	2	3	4			
8		3	9					

EASY - 104

7	9	3		2				6
		8			6	7	1	9
5	1	6			8	3	2	4
8	7	1		4		6		
	2	5		1		9	7	8
3	6				2			
1	5		9	6	7			
		7	2	8		1	6	
6	8	2	1		5	4		7

EASY - 105

5	1		7			9	6	3
6	9		4				2	7
3			6	9	5	4		
		7	8	5		3	4	
9		5	3	4	6	7	1	
	4	3	9	1	7	6	5	8
8	5			6		1	7	
	2		1	7	9	8		
			5			2		

EASY - 106

	1	5	6	7	8	4	9	
3		9	1				6	
7				9	4	1		
	6		2	1	7	8		9
1	2					5	3	6
			5	3	6	2		1
		4	7		3		1	2
8	7		9	6	2	3	5	
6	3	2	4				8	

EASY - 107

3	8	4	7	5	6			
		1	3			4	5	7
7	2	5	4		9	8		6
8	1	6			5		2	3
		7			2	5	1	8
	5				7	6		
1	3	8		6		9	7	
		2		9		3	6	4
4	6		5		3			1

EASY - 108

	9	2	8	3	4	1		6
	1	4		9	5		7	8
	5	6	7		1	4		3
	3	9	1		6			
	4		9					1
	8	1	3			5		9
	6	5			8		2	
4	7	8	2		3		1	5
1	2		5		9	6	8	4

Solutions on Page: 93

EASY - 109

8		2		9	3	6	4	
6	5		2	4		1	7	
9	4	7			5		3	
	7	9	6	1	2	3	5	
2		1					9	
5				8	4	7	2	
	2		5	3	1	9		
3	6					5	1	7
		5	4	7	6		8	3

EASY - 110

2		6	5	1		8	4	3
7	5	4	6	3	8	9	1	2
	3		2		4	6		7
3	6	8	4					
		2						
	4	5	9		3	7	2	8
			1	8	6	5	7	9
6	8	9			5	4		1
5	1		3				8	6

EASY - 111

		8	5	6			3	9
5		6		8	3	1	2	7
3	7		2			6	8	5
	3	5	4	2	9	7		
7		4		1			5	3
8	9	1	3	7	5	2	4	6
1		7		5		3	9	
	8	3	7					
			1			8		

EASY - 112

	1			8	5		9	3
6	9	4			2	5		1
8	3		9	6	1	7		4
5							3	9
	2	8	6		3		4	7
3				4	9	2		6
	5			2	6	9		8
	8	2	7		4	3	6	
7	6	9	5	3	8	4	1	

EASY - 113

	9		6	1			5	2
5	8	6		3				
		1	9					
1			3	5	6	2	8	9
	5	2	7	8	9	4		3
8			1	2	4	6	7	5
	1	5	4		2		6	8
2		8			1	3		7
9	7	4	8	6		5		1

EASY - 114

2		1		4	8	3	7	9
3		8	6	9	1	4	5	
9	4			3		6	8	
		2	1		9			
8		7		2	5	9		
1		4	3		6	5	2	
4		6		1		2	9	
7	1	3		5	2	8		
5	2	9						7

EASY - 115

		4	7			2		5
	7	9	5			1	4	8
	3			8	4			6
	8	3			5	7	6	
		2				5	1	9
1			4	6	9			3
	2	6	9	5	1	4	8	7
	9				8		5	2
	4	8	6	7	2	3	9	1

EASY - 116

3	8		6	1	5		4	9
1	9				4	3	2	6
	7	4	2	3		5		1
	3	9	7		2			
2	1	7		5	6	8		4
8	5	6	1	4	3	9	7	2
	6		5				9	8
		1	3	6	8			
				9		6		

EASY - 117

6	2			1	9		4	
1		4	5					9
	3		2	4	6		1	7
3				6		9	7	4
7		6	9		4	1		
	4	1	3	2	7	5	8	6
		5		7	8		6	3
2		7	6		5			8
8		3		9			5	1

EASY - 118

9	3	4	5	1				8
	2	1	3		8	4	9	
7	8	5		4	9		3	6
1	6		9	8		5	2	4
5	9			3	4	8	1	
				5	2			
2	7				3			1
4	5		7	2	1	9	8	
3				6	5			

EASY - 119

		2		3			6	1
	4			9	8		5	7
7	5		6	4		8		
9	1	5	3		4	7		6
4		7	5		9			3
	2		1	7		9		5
6	3	4	8		7	5		9
5	7			6	2	1		8
		1		5		6		4

EASY - 120

1		3	9	6			5	2
8		2		5	1			4
		5	8	4			1	
3		1	2		5	7		9
9	8	4	6	7	3	1	2	5
	5		4	1		3		6
			1	3	4	5	6	8
			7		8			1
		8	5	9		2		

Solutions on Page: 93

EASY - 121

9		3		2	6		1	
7	2	5	4		1		6	
4	1		3	9	5			8
			8	1			4	
	7	1		4	9		3	
6		8	5	3	2		9	7
			2		8			3
2	5	4			3	9		
8	3	7	9	6	4		5	1

EASY - 122

2		9	3		7		5	
				8		9	4	
		5	1	9	6		8	
1		4		3	9			5
9	3	6	7		1	8	2	4
7	5	8			4	1		9
	9	7	4	6		5	1	2
		2			8			6
5	6	3		1	2			8

EASY - 123

7	1	6	5	8	4		3	
9	5			3	1		7	6
3				6			4	5
			6	2		3	1	7
1	6			4		5		
2		9	1				6	8
		3	7	5	2	6	8	
8		1		9	6			3
	7	5	8		3		2	4

EASY - 124

	4	7	5			2	1	6
6	1		4			5	7	
	8			6		4		3
2	6		8	4		7	5	
1			9		6	3		2
3		8	2	7	1	9		
			3	8				7
7		5	6	1	4			9
8	3	1		2	9	6	4	

EASY - 125

3	4		8		6		1	9
	8	6	1	5	2			
1	5		9				2	8
4	3	9	6		5		8	
2	6	1			8		9	5
		5	4		9	1	3	
6	9	4			1	3	7	2
				9		8	6	
7	2	8		6				1

EASY - 126

9	2		7					1
1	7	4	2	5	8	9		3
5		8	1		3	2		4
3	5				6	1	2	7
		1		3	2	6		9
6	8	2		7	1	3	4	5
	9	6	3	2		5		
		5				7	3	2
	3						9	

EASY - 127

1	4			9		7	3	
7		9				6	2	
		2	8			9	1	4
2	5	4	7	3	1	8	6	9
8		1				3	5	
3		6	5	2	8	1	4	
	1	8		4	5		7	
4	6	3			2	5		
5			1	8		4		

EASY - 128

	7		3	5	8	2		4
	6	4		9		5		3
2	3		6	4	7	8		
	9			2	4			8
	5	2	8		1			6
7	8		9	6	3	4		2
				3	6	1	4	5
	4	7	2			6		
5		6	4	8		3		7

EASY - 129

4		9			1	7	8	
				3			6	
5	8	3		4			9	2
7		6	8		2		1	5
8	5			1	4		7	
		4	5	7	6		3	8
			4	6	5	8	2	
2	7	5	1		3	9	4	6
	4	8			9		5	1

EASY - 130

				8	4		7	5
	5	7		3		8	6	
	6	4		5	9	3		1
2	7			1	6		4	
6	9		8	4	5	7		2
4	8	5	3	2	7	1		
5			4	9	3		1	7
	3	6					8	
7		9		6		2		3

EASY - 131

	7					4	3	9
		1	3			7	6	
4		3		9	7	2	8	
5					3		2	8
9				4	2	1	7	
1	3	2		7		6	5	
7	2	5	4	3	9			6
8	1		7	5		3	4	2
3	6	4	8	2	1	5		7

EASY - 132

4	8	1		7			5	
		5	8	1		7		
7	9	2	3			4	1	8
				3	2	9	8	
8	7	9		5	1	2	4	3
2	3	4	9	8		1	6	
1					6	8	9	4
6		7		9		5	3	2
9				4				

Solutions on Page: 94

EASY - 133

		1	9				8	
				4	1	9		2
2			3		6	1	4	7
	3			2	5	6	9	8
	6	2	7			4		3
8	1	9						5
4	8	6		9	7	3	5	1
1	2	7	5			8	6	9
	5		6	1		7	2	

EASY - 134

5			6	7		4	8	9
						5	6	7
	6	8	4	5	9	1		3
6	5	1		2				4
8	7	4	1	6	5		9	2
9	3		7	8	4	6	1	
	8	5	2	9			4	
	2	6			7		3	
			8			2		6

EASY - 135

	5	9		1	6	2	4	3
					2	8		
8	2	1		4		9		
		5		8			9	6
9	6		3		1	5		4
1	8	4		5	9			
		8	2			3	7	1
3	1	2	7		4	6	5	8
	7	6		3	8		2	9

EASY - 136

8	9	3	2	4		5	6	
			9	3	5		2	
5	4	2		8				1
	6		1	5		2		8
9	8	4	7	2		1	5	
2		5		6				3
	3	6	5			7	1	2
		9		7		4	8	5
	5	8	4	1		6		

EASY - 137

				2	5	3		7
	1		9			5	2	
	2		7	3	1	9	6	4
	5	9	1		7		3	
2	7	8		9	6	1	4	
3		1	4				8	
	4		6		3	8	5	2
	3		2	7	8	4	9	1
	8					6	7	3

EASY - 138

		7	3		9	1	5	8
					5	3	9	4
5		9	4		1	6	7	
1		6	5		4		2	9
	9			2	7	4	3	
3	2	4	1		8	7		5
4	1		7	5			8	3
7					3	2	1	
	6	3				5		7

EASY - 139

3	2	1		8				5
6		8	3			4	1	
4		5	6		9	2		8
7		2	8	9		3	5	
		9	7	6		8		
8		6		3	5			9
2		4	1	7		9	8	
		7	9		3	5		2
		3	5	2		7	4	1

EASY - 140

		3		6			5	2
7		5		8	3	4	6	9
		6	2	5				3
	1	9	8			7	3	6
4	7			3	9			1
6		2	7		5	9	4	8
3		1		2		6	9	7
8					6		1	5
2				9	1	3	8	

EASY - 141

	4	7	8	9	2	3		5
5	6	3			7			2
9	8		6		5	4	7	1
		6		8		5	2	
	2		1	6	4		8	
					3	7		
4	3	8	2	7		6		9
2		9				1	4	8
6	5	1				2	3	7

EASY - 142

1	3	7			6		4	
	6	4		1	7	9	3	2
		2	4	3				
6	9	1					7	8
	4	3		6				9
	5			9	1	4		
8			6	4	2		5	
			1	8	3	6	9	7
3	1	6	5	7	9	2	8	4

EASY - 143

1	4		7		2			6
8		3				1	7	4
9		7	1	8				3
2	5	1	4	7			6	
6		4	3	2	1	7		
	3	8		5	9	2	4	
4				1	3		8	
		9	2	6	5	4		7
5	1	6		4		9		

EASY - 144

	3	9	1		2		5	
5			8		4		2	3
4	8	2		3	5	6		
8	6	3	4	5	9	1		
1			6	2			8	
2		7	3	1		5	4	
6	4	1			3	2		5
9			5		6	7	3	1
		5				4	6	

Solutions on Page: 94

EASY - 145

	3			2			5	9
	6		4				3	1
9			7	3		6		2
	2	6		8		5	1	3
8	9		5				7	
		1	3	4		8	9	6
	4		8	7	3	1		5
7	1	5	2		9		4	8
3	8		1	5	4		6	

EASY - 146

5	6	9	4	1	2	8	3	7
2	7	8	3	9	6	1		4
3	1	4					6	
	2					7	8	
4		1	6	7				
8	5	7	2	3			4	
6	4	5	7	2		9		
	8	2		6	4	3		5
				8	5		2	

EASY - 147

	5			9	8	4		
6		7	3					
		4				3	7	
	7	8	9	2	3		4	6
2			8		4	9	5	7
1	4	9		7	6	2		3
7			4					2
4	2	1	6	5		7	3	8
9	3	6			7	5	1	4

EASY - 148

3	6	7		5		1		8
1	5		2		8			7
2	4	8	3	7	1	6	9	5
7	1	2			6	9	8	3
9		5	1	8	2			6
4			7			2	5	
6	2		8	9			3	
8					7			
5		4			3	8		

EASY - 149

4		1	7	3		5	8	
7							3	2
8			2	5	1	7		
2			8	9	3	4	6	1
9		6		4	2		7	3
3		4	1	6	7	2		9
			9			6		
6		7	3	2			1	8
1	9	8			4	3		5

EASY - 150

4				7	9	5		6
	6	3			1		8	9
2	7		6	5				4
			8	9	6	2		5
	2	6	4		5	9		
9	5	4	1	2		6		8
	9			1	4	8		3
1	3	5	9	8	2	4		7
7	4		5		3	1		

EASY - 151

	9	8				7		2
3		4	8	1	7	9	6	
6	5	7	4	9		1	3	
4					9	5	2	1
7		5	2		3		9	4
2	6				4	8	7	
8		2	9		1	4		6
		6	5		8		1	
		1	3		6	2	8	

EASY - 152

	2	9	5	7		3	8	4
4		7		1	8		9	
6	8	3	9	4	2	1	5	7
				6	3	4		1
8		1	2	9	4	5	7	
	6	4		5		9		
5		8			7	2	1	9
	1						6	
7			1	2			4	3

EASY - 153

4	5		8	1	6	7	9	2
6	9	2	3	5		1	4	8
7				4	9			6
1	3	5		6		4		7
	7	4				9		1
9				7	4	3		
		1	7	9		8		
5				2	1		7	3
		7			5	2	1	9

EASY - 154

1	7	2		4		9	6	5
8	9	4	6	5	2			1
5			9	1	7	4	8	
			4	2		6		3
2		3	7		6	1	4	
6	4	7	1	9	3		5	8
4	8	9		6		3		7
7			8		4	5		
	6		2				1	4

EASY - 155

2				5		4	8	3
	6	4	2	8	3	1		
8	3	5	1		7		9	
		1		7	2		6	
7	4			6	9			
6		2	3	1	8			4
3	5		7	2		9	4	
1		6		9		5		7
4	7	9	6	3			2	1

EASY - 156

	2	8		1	9	7	5	
		4			2	3		
6	9	7	3	4		2		1
2	8			9		5		
1		5			3	8	9	2
		9	2	5			1	6
9	7			3	6	1	4	
8		1	4	2	7	9	3	
		3	9		1		2	

Solutions on Page: 94 & 95

EASY - 157

8	9	4		1		6	7	
				4	9	1		
5	1		8	7	6			
	5	1	6		7	4	3	2
7		6	1	3	4		8	9
4	8	3		5		7	1	6
	4	8			5	3	9	
					1		6	4
1	3	9		6	8			

EASY - 158

6	9	3	2	7		4	1	
7		5	6	4	9	3	2	8
2	4	8						9
5	8			3	7	1	6	
	6	4	5		2		3	
	2	7		1				
9	7						4	3
8	3		7	6	4		9	
		1	9	2	3	7		6

EASY - 159

8		1		2	7		3	4
5	7		4	1	3		8	
3	4	2				9		
		5	2		8			
6		7			1	4	2	
	2			7	9	1		
7		6		3	4	2	9	1
2	1	4	7	9	5	8	6	
	3	8	1	6		7		5

EASY - 160

1			9		3			2
3		5	8	7	2		6	
		2	6	5				4
2		7		9				3
9		6		1				5
		1	2		6	8	9	7
	1	4	3	2	7	9	5	8
5	7	3	1	8		2	4	
			5	6	4	7		1

EASY - 161

4			3		7	9	8	2
	7	6	8	9	1			5
8	3	9				1		
3		7	5		4			
	4			7	9	2		3
9	6	2		8	3			4
7							4	8
1	8		7	3	5	6	2	9
6	9	5	2			3	7	

EASY - 162

	1		2			8	4	7
2		4	3			5	6	
			8		6		3	1
	6	8	1		3			5
5	2	1		7	8	6	9	
7			5	6	2	4		
	9	6		2			8	4
8	4		6			9	5	2
1		2	9	8		3		6

EASY - 163

7			3		4	2	1	
	3	4	1		2			7
			5	6	7		4	9
			2	5	3	6		1
	5		9		1	8	2	3
	2		6		8			4
1		3		2			8	5
8	4	2	7		5			6
5	7			1	9	4	3	2

EASY - 164

4	5	6	2	8				
9		3	6	5	4	7	2	
7	8	2		9				6
				1		3		5
							1	2
	6	1	4			9	8	7
6	9	4	5	3				1
8	2	7		6	1	5	3	
		5	7	4	2	8	6	9

EASY - 165

	7		8		4	1		
3	1	5	6			8	4	
8	4	6						
9	3	2	4	1	8	5	6	
4	8		9	6	5	2	1	3
6	5	1	3		2			8
	9		7	4	3	6		
	6	3	5	2				4
7	2	4		8		9	3	

EASY - 166

	8				2			5
2	4	6		1	5	9		
	5	7		9			4	6
6		9		3	4			1
8	3	1		7				
		4			6	3	9	7
4	1	5		2	7	6	3	8
7	6	8		5		1	2	
		2		6	1	5	7	4

EASY - 167

1	7	3		8				5
4	8	6	5	7	9		1	2
5	2	9		1	3	8	7	6
9	5		3	6		4	8	
	4	1	9		8			
				5				7
8			2	9	1	5	3	4
	9			3	5		6	8
	1	5	8			7		9

EASY - 168

	7	8		2	4	9		5
4	5	2	8	9	1	7	3	6
		3	5	6		4		
1		6		7				4
	3	5		1				
2	4	9	6	5	8	3		
		7	2	4	6			
8			1		5		9	7
3	6				9	5	4	2

Solutions on Page: 95

EASY - 169

8	2		6		3	5		
9		7	1		8	2	3	
6	3	1		2				
2	8		3		5	6		7
	6	5				3		1
1	7		4	6	2		8	5
5		6	2	3	1	7	9	8
3			7			1	6	
	1		8	9			5	

EASY - 170

2		8	5		6	4	9	7
	5	4	1	9		2	6	3
	9	6		2	7		5	8
8		7			5			2
1	3		7	8			4	6
5	6	9		4	3		7	
4				7			8	
	8		3	6		7	1	4
6				5		3		

EASY - 171

7	3	6		8		9	1	5
			3	6	9		2	
2		8		5	1	6	3	
5		3	6		8	1	9	
	6	1		9	2		7	
	2	7	1	4	3	8	5	
1		4				2		7
3		2		1	6	5		9
6			8	2	7		4	

EASY - 172

9		5		1	6	3	2	8
7	8	2			5	1		
	3	6		2	9			
	1		6	3			8	2
	2			9	8	4		
	6	8		7	2		5	
	7	3	4			2	9	5
6	9	4	2		7	8		1
	5	1		8		6	7	

EASY - 173

			7	5			9	3
7	2	9	6			1		
			1		9		7	2
	7	1	5		8		6	4
5	9	4	2			8		1
	8	6	9	1	4	5	2	
8		2		9	5			
	3	5	8	7		2		9
9	1	7	4			3		8

EASY - 174

		9	2	4	7	5	1	
7	6			3	5	8		
	5	4	6	9	8		7	2
2		8	9	5				
4			8	2			5	1
6	9	5		7			3	8
5		7	3	1		6		9
9	2		5	6		7		3
	4						2	5

EASY - 175

3				7	9	4	2	8
	8							6
			4			9	1	5
8	2	3	7	9		5		1
9	4	5	8		1			
1			5	3	2			4
		4	3	5		6		9
5	9	6	2	1	8		4	7
7	3	8	9	4				2

EASY - 176

4	9	6	8			7		3
5	7	3			2			1
1		8	3		7		6	5
8	4	1	2			5	9	
3	6			7	8	2	1	
		2	1		9		3	
				8	4			2
6		5	7			1	4	9
2	3			1	5	8		6

EASY - 177

		2	7	3		1		4
3		1	2	8	4		7	9
4	7	5		6	9		2	
			6	2		9	1	7
2	8		4		1	6	3	
	1			5		8		2
7				1		4	9	
9	5		3		2			1
1		4	8	9	7		5	

EASY - 178

		6	5			7		1
7	2	1	3	9		4		6
	4		6			8	2	9
1	3	2	8		6	9	7	4
			7		2		1	8
	7	8		3		2		5
2	8	7	9	6		1		
		4	2	1		5		
3	1	5		8				2

EASY - 179

	1		7			3	8	5
3	7	8	9		4			
6	5	2		8		9		
	4	5	6	9	7	1	3	
7	3	9	1			4	5	
1	2		5	4	3	8		
5				1		6	2	3
	6	1		3	5	7		9
	9	3	2	7	6			

EASY - 180

2		5				1	9	8
9	6			2	1	5	4	
1	8	4	5		7	3	6	2
		7	2	5	3		1	
8	3		7					4
	2	1	6			9		
	1	2	9	6	8	4		
		9	3	4		7		
3	4		1	7		6		9

Solutions on Page: 95

EASY - 181

3	5		8	4	1	2	7	
	4	8		7		5	3	9
2	6					1	4	8
4			6	1		7		3
	9		4	5			1	2
6		5	3			8	9	
			1		4	9		7
	7	4	9		5			1
9	3	1			2	4	6	

EASY - 182

2		8						5
	3				2	4	9	8
	5	9			8		7	1
3	9	1	6	2	4	5		
8	2	7	9	1	5		4	6
5	6	4		7	3	9	1	
		5	2			8		3
9		2	1	3	6		5	
	4		5		7		2	9

EASY - 183

			7	2	1	4		9
7	2	1	6	9	4	8	3	5
	9	4	5	8	3	2		
	7	3				1	9	8
			8			3	2	
			3					6
3	6	2	9	4	7		8	1
	1	7	2		8			3
9		8			5	6	7	

EASY - 184

1				3	4	9		
9			1		5	8	6	
		8		9			1	2
2		4	5	1	3			9
5		7	2	8		4	3	1
	1	6	7				2	8
8	3			5	2	7		6
6		5	4	7		1	9	
7		9			1	2	8	5

EASY - 185

7	2	5		1		8	9	4
	4		2	7		1	5	
9	1	3			5	6		
5	6	2	9	3	4		8	
3	9		7	6	8	2	4	5
4		8				9	3	6
		7	4					
					1	4		7
2				9	7	3	1	8

EASY - 186

5	9			2	7	6		
			9	1		5	3	2
2	1			8	5	9		
3	5	8	6	7		1	2	9
	2	4	8			7		3
6		1		9			5	8
4	8	2				3		5
7	3	5		4		2	9	
1		9				8		4

EASY - 187

	5			9	4	6	8	3
4			5		6	9	2	7
9	2	6			3	5		
6	4			5				2
2	7	5		3		4	1	8
8	1	9			7	3		5
1	6			4	2	8		9
	9	4	1			2		
		2		6	5		7	

EASY - 188

		2	7	8			1	5
7		1	5		9	8	6	2
8	9		6	2	1	4	3	
	2		4	9		6	5	
			3	1	7	2	9	4
4			2		5			8
		4	9	5	6			
9		3	1	7			4	6
5				4	3	7		

EASY - 189

				2		6		4
		8	7	6			5	1
3		9	4	5	1	2		8
8	7					4	3	
5	1	6	2	3	4			9
9		4		7			6	2
2	8	5	6	4	3		1	7
6				1	2	8	4	
1	4	3	9		7			

EASY - 190

5					3	7	6	
6	8	7		1	9	3	4	5
3		4	6		5		1	2
		8	9	3	6	2	7	1
1	3	6	7					
	7	9	1	5	8	4	3	6
8	4		3				5	7
			5	9	1			4
			8				2	

EASY - 191

	9	6	7	8	5	4		3
7	8	1	4			6	2	5
5	3		6		2			8
6				4	7			9
		5						
3	4	9			1	8	6	7
9	6		1			2	8	4
		3	2	6	4	9		1
	1	2	9		8	5		

EASY - 192

5		4		1	8	2	3	
2		9	3			6	4	1
		6		2	4	8	5	7
7	9	8	5	4	6	3	1	2
			2	8			9	5
	5						8	
9		5	4					
4	6		8		5	9		3
			1	9	2	5	6	4

Solutions on Page: 96

EASY - 193

	4	5	7	2		3	8	9
	3	9			8	2		6
2		6	3	9	5	7	1	4
			4	6	1			5
4	1	8		7				3
			9		3			
8	2	3			9	4	7	
	7		8	3	4	5		2
6				1	7	9	3	

EASY - 194

8	2	4	3					
7		5	8		2	4		
3	1			9	7	8	5	2
		7	1	2	3	9		8
1		9		4	6	3	2	5
	3	2	9	5		7		
4		8		7		1	3	6
2	7	1	6	3	4		8	
			5	8				4

EASY - 195

9		6	4	1		2		
2	4		9	8				1
7	8	1		6				9
		9	8	2		5		7
3	1	8	5	7		6	9	2
	2			9			8	4
			7	4		9		
1	7		6	3	9	8		
	9	3	1	5		7	4	6

EASY - 196

5	6	3			7	1		8
1	8	2	6	5	3	7	4	9
	7		1				3	5
			5	6	1			
				3			8	
6	3		9	2	8			7
3	9	6	2		5	8	7	1
7	2	8		1				4
		1	8	7	9		6	3

EASY - 197

3		2	1		8		9	6
	7	1		5			4	
	6	8	7	9	3			2
		3	6				2	
	8	7	4			9	6	1
	2		9	8	1	5		3
7	1	5			4	2	3	
2		6			9	4	8	7
8			3	2	7			5

EASY - 198

	1		6	4	9	8		2
	2	5		3		1		
6	8	4	1		5	9		
	6			7	1			4
2	7	1	4		3	5		8
4	5		8	6	2			9
					4		8	7
	4	8	9		6	2	3	
5	3		2	8	7	4		1

EASY - 199

2			1	9	8	7	5	3
		3	6	2			9	
9	5	8	3	4	7	6		1
6	8		9	1	3			
	3				4	9	1	7
4		9		5				
3		1		8		5	7	4
		5	4	3	1		6	
	9	4		7		1		2

EASY - 200

	8			3				4
	4		8				1	6
		6	4	5		7		9
1	2		9	8		6	4	7
		4	2			1	3	
	3	7	5	4		8	9	2
	5		7	2	8	4		3
	6	8		9	4	5	7	1
	7		6	1		9	2	

EASY - 201

		8		2	4		6	
	6	9	8	3		7	1	
			1	7	6	9	2	8
3		4			2	8		
5	7	1	3	6			4	9
	2		4			1	3	7
6	3	2	5	8		4	9	1
				9	3	5		2
9	8	5	2					

EASY - 202

	9		5		2	8	4	
5	1	4	9	8	3	6	2	7
		3	7	6	4	9	5	
7	6						1	5
		1						
3	4	8		9	5			
		6	3	7	1	5	9	
	7		2	4	6	1		
1	3	2		5		7	6	4

EASY - 203

7	3	4	5	9	6	2		8
		5	1		3	4	9	7
	8	9		7	2			3
9	2			4		1	5	6
		1			9	7	8	
4	7				1			
3	9	6		1		8		
	4		9	6		3	2	
5		2	8	3	7			9

EASY - 204

		1	4	5	2	9	7	8
9	4			8		5	6	3
	8		3	6	9	4	2	
	2		6	1	4			5
8			9			7		
1			8			2	9	6
3	9	8		4				
2			7	3	8	6	4	9
		6	5	9		3	8	

Solutions on Page: 96

INTERMEDIATE - 1

3		9		5	1			7
1		8		9		4	5	
4			7	8		2	9	1
	6	7	1	3				4
2	4	5				1		8
8		1		4	2			6
6	8	2		1				
7	1			2	5	3	8	9
5		3		7		6	1	2

INTERMEDIATE - 2

1		2	6	3		7		
				1	8	6		4
	4		9	5				2
5			8			2	9	7
8	9	6		7		3	4	1
2		7	4		3	5	6	8
4	8		3		6	9		5
7	6	9				8		3
		5			9			6

INTERMEDIATE - 3

6			2		4			9
7		5	3	9		6	8	
		4	6	8	5	2		
5		6	7	3	8			2
3	4	8	9	1	2		5	6
		1	4			3		8
8	5		1	6			3	7
	3	9	5			8	6	1
		7			3		2	

INTERMEDIATE - 4

					9	2	3	6
	9	6	2		3		7	4
8		3		6	7	9	1	
	1			5		6		8
		5		7		1	2	
6	3	8	1	9		4		
	5			3	6		4	1
		1	7		5			9
3	6		9			5		2

INTERMEDIATE - 5

6	5	4			2	9		3
		2		4	9	5	6	
		3	5	1				
	8	1	6		7		3	4
		7	9	8		1		
4	6			2			9	7
5	7	6		9	8		4	
	4	8	7		5		2	
3		9	4			7	8	

INTERMEDIATE - 6

	3	1	4	5		6		7
		4			6	1		8
6	9	7	8					4
	8			2	4			9
4		2	9				6	5
7		9		8	5		4	3
9	2			4	7			1
3	4		5	6		9	7	2
	7						3	6

INTERMEDIATE - 7

5	1	4			9	6	2	
			1		5		8	3
9	3	8	2		7		5	
4					3		7	
3	2	5	4		8	1	6	9
1		7	5	2	6			
7	6			5			9	2
		9		8				
8	4	3	7		2			

INTERMEDIATE - 8

	5	8	9	7		2	4	3
4	2	7	6	8		5	9	
			2	4	5		7	
8	1	3		6				5
			1	3				7
7	4		5	9		1	3	
5		4		1			8	2
	8	1	3			7	5	4
3		2		5		6		

INTERMEDIATE - 9

9	3				8	5		
8	7	1				3	6	2
6	2		3	1	7			
	6	2	4		1	8	5	
4	5	8	9	6		1		3
1			8	3	5	2	4	
7		3			6	9		5
			2	9	4		3	
	1					6	8	4

INTERMEDIATE - 10

4	7		6	8		1	2	
	1	3	7		2		6	
2	8	6		5	4		9	
7		8					1	
6				1	9	3		
	9		8	2	7	5	4	6
9	6					4	5	
			4	6	1		7	9
1		2		7	5	6	3	

INTERMEDIATE - 11

9		2					4	8
5			4		7			
3	7			2			6	5
7		3					9	2
		9	7	1	2		3	4
2	4	1		8	9		5	
8		6	9		5	3	7	1
	9	5	2	7			8	
4	3	7		6	1			9

INTERMEDIATE - 12

	1			3	6	4	2	5
5								9
3		8	9	2		7	1	
6	5	9		7				
8	3	2	6	4	9	5		1
4	7	1		8	3	6	9	2
	6	5	1	9	8		3	
1		3	4			9		7
	9						6	

Solutions on Page: 97

INTERMEDIATE - 13

			5	7	8			2
	2	3	9			8	5	1
	5		1			7	4	
	3	9	8		7		6	4
	6	8	2		9	5	7	3
5		7	6			9	2	
4	9		7		2		8	6
	8	2	3		5	4	1	7
	7				6	2	9	5

INTERMEDIATE - 14

	1	4		9	5		7	3
2	7							1
8			6	7	1		9	2
7	8	5		4	9		2	
3		1					8	4
		6	1	8	2	5	3	7
1	4			5		2		9
			7		4		1	5
5	3	2	9		6		4	8

INTERMEDIATE - 15

8	1	5	3		4		7	2
		2		1				3
3			2	6	8	1		
					6		1	9
	6	9		8	2	7	3	
4	8	7		3				
9	3	1	6		7		2	
2				5	1	3	6	7
7	5	6	8	2		9	4	

INTERMEDIATE - 16

						9		7
	1	7	2		5	6	8	
5	8		6	7				4
	2		7	3	6	5		1
			5	8	1	4		
	6	5	4				3	8
9	5	8	1	4	2	3		6
6	7		3		8			
	4			6		8	1	

INTERMEDIATE - 17

5			1	7	8	2	4	
4			6		3		9	
8		3	9	2	4	5	6	7
	7	8	3	9	2			
	4				1	7	3	
				4				1
2		4		1			7	9
7	6	1	4	3	9			
	5	9		8		4		6

INTERMEDIATE - 18

3	8	4	1	5		6	2	
				7	6	1		4
6	1				2			3
		3	8		4		6	
	9			3	5		4	
		6	2	1	7	3		
2	3	1		6			7	5
7	4	9	5			8	1	6
	6	8		4	1	2	3	9

INTERMEDIATE - 19

6		2	4		8	1	3	5
9		8		1		7	2	6
1	3	5			2	4		9
5			1		7			2
3	2	4		5	6		7	1
8	1					6	5	
	9	6		7	4	3		
		1	6		3			
7				2			6	

INTERMEDIATE - 20

			5	6	4	2		7
		7		2			1	8
9				7	8	5		
	9	5	8		2	4	6	3
6	3	8	9	4	7	1		
	2	4				7	8	9
2	1	9	4	3	5		7	
8		6		9		3	2	
5	7	3	2	8				

INTERMEDIATE - 21

6	9	5	2	4			1	
3	2	8		5		4		
		4	8	9	6	3		
	3	2		1	9		4	5
5	1	6	7	8		9	2	3
8	4	9	3	2		1		
	5				1	6		9
9							7	4
	8	7	9			5	3	

INTERMEDIATE - 22

		9	2		8			7
2	8		9		7	4	3	6
	3	7	5			2	9	8
	4	3	6		1			
8	7			5	2			9
	2		8	7		1	5	
7	1	8	3			9		5
3			7	8				1
	5	2	1	4			8	3

INTERMEDIATE - 23

8	7		3		9	6	4	
9	1	5	4				3	2
3		6		2		1		
		9	1	5	4		6	
1		4			8			7
6	5	8	2	9	7		1	3
5	9		7				8	4
				8	3	9		1
		3				7	5	

INTERMEDIATE - 24

6		5	9	3	8	2	4	7
	4		1		6	9		3
8	9	3	7	2	4			1
		1			2	5	3	6
		4		6		7	9	
				9				
4	6	8		1				5
	2	7	5					
3	5	9		8	7	1		

Solutions on Page: 97

INTERMEDIATE - 25

4		1	6		7			
8		7	1	9	5		3	
5	6			8	4		7	
9	1		8		3			5
7	4	8		5	1		2	3
				4	6		9	8
		2		7		5	6	
	7		5	1	2		8	9
		9		6	8	2		7

INTERMEDIATE - 26

			8		5	9	1	7
	5			3	7	2		
8				4			5	
4			3		6			9
2		3	5		8	7		
			2	1			6	8
5	3	4	7		1		9	2
	9	2		8	3	1	7	
	1		9	5		4	3	6

INTERMEDIATE - 27

4	8	1	3		7	5		9
3		5	6		9		8	7
	6	9			1		2	3
					8	6		
9	3	7		6		8	4	
		6		3		9	7	1
5	1		8		4	2		6
6		8	2		3			
	7		9			3	5	8

INTERMEDIATE - 28

2			1	5		8	3	
5	3				4	1		9
1		9		3	8	2		
	2	1				9		8
6	5			8	2			1
	8		4	1	5	3		6
	9	2		4			8	7
7	1		8			4	9	
8		4	7	9	3	5		

INTERMEDIATE - 29

9						5	4	
5		8	6	3	9	2		
1	7	2		4				3
3	8	5	7	6	1			4
	9		2	8	5	3	6	
	2	1	3		4			5
	1	9	8	5		4	3	6
	5						8	
					6	1		9

INTERMEDIATE - 30

	5			3		2	9	1
2	3		5			8		6
1			2	6			5	7
		1				5		
	6	8	9	4	1		3	2
4	2	3	6	7	5	9	1	8
9				2			8	3
	1		4	9	3	6		
3	4		8		7	1	2	

INTERMEDIATE - 31

	6							
		4	3			5	8	1
5	1	8	9	4		6		2
	9	6	1	3	5		4	
	4			6	8	9	5	3
		5	4	9	2			
3	8	7	6	5	9	1	2	4
	2			7		3		
					3	8	6	7

INTERMEDIATE - 32

	2			1	8	4		6
	3		9				5	1
1		4		5	2	8		7
3		1	4	6	7	5	8	2
7	8		1			6		3
4		6	2	8				9
	1	5			9	3		4
6		3	5	2				
9		8	6		1	7		5

INTERMEDIATE - 33

3		5	6		1	4	8	9
	4	1		9		5		6
					3	1	2	
	3						1	
		4	9			8	6	5
5	6	8		1		3		2
	9	3	5	8	7		4	
4	1			2	6	9	5	8
		6	1	4		2		3

INTERMEDIATE - 34

	3	6				8		1
8					5	3	4	
		9			8		7	
	1	7	5		4		2	3
			1		9		6	8
3		8			7	4		
4	6	2	8	5	3	1	9	7
9		5		4	1	2	3	
1	7		9	2		5	8	4

INTERMEDIATE - 35

	7		8				3	2
6		2	3	7		4	1	
			2		5	8		9
	3	7			4	5		6
9	6	5	7	8		3		
				3	6		8	
7	2	8		5			9	
3	5	1	4	9		2	6	8
4			1	2		7		3

INTERMEDIATE - 36

	8					2	4	7
		1	5	2	8		9	
9		2		7	4	8	5	
8		9	4	3		7	6	
		6	8		7		1	
3	5	7			1	4		
			7			6	3	
5	9	3		4	6	1	7	
7	6		1		3	5	2	4

Solutions on Page: 97 & 98

INTERMEDIATE - 37

5		8	9			1		
9	6	4	2	1	5	7	8	
3	1		8	6				9
6				8	2	5		4
			4	9	1			
4	2				3	9		
7	4	5				3	9	6
1	8	9	3			2	7	
2		6		7	9		4	1

INTERMEDIATE - 38

	9	6		1	3		5	
1			7	5		9		6
		3	2	9			1	7
4	7	1		8	9	6		3
9		5			2		4	1
3				7			8	9
5	4	7		2	1		6	
2	3						7	4
6		8	3	4	7		9	5

INTERMEDIATE - 39

	8	1	4		6	7	2	9
4	2		8		9			
					2	6	4	
		5				4		7
7	4	2	6		5		1	3
9	3	8					5	6
	7			5	8	3		
			3	6	7		9	
	6	3	1	9		8	7	5

INTERMEDIATE - 40

1			3			4	5	2
5		4						9
		2	4	6		8		7
		1	7		4		3	
	3	5	8	1	6	7		4
7	4		5	2			9	8
6	1		9		7		4	5
		3			2	9		1
2	5	9		4			7	

INTERMEDIATE - 41

	8			3		1		
	1		9	2	6	8	5	
6			1		5			
1	4		3		9	7		2
	2		5		8	6		1
9		8	2	7			3	4
8	5			1				9
	9	4			3		1	
2	3			9	7		6	

INTERMEDIATE - 42

				4		7	5	
5	2	4	1		8			
	3		2				1	8
6	5	3	7		4	2		9
4	9	8			2	1	7	
		2	8		6	3	4	5
	4		5			8		7
		5					3	1
		1		8	9	5		4

INTERMEDIATE - 43

		1	8	3		9		7
5			1	7		2		6
9	2	7		6				8
	4	3	9	1				5
		9	5			6		4
	8		6	4		1		9
	5	4	2			7	9	
			3	9	4	5	6	1
	9	6	7	5	1			2

INTERMEDIATE - 44

	3	8	2	6		9	5	1
	2	4		1			3	7
			8	3		6	4	
4	7		1	8	3	2		
	9	1	7	4		3	8	5
3		6	5	2		1		
9	5		6	7				8
8	6	2		5				3
			3	9		5		6

INTERMEDIATE - 45

			8	9		6		3
	5	4	3	2		1	7	
	6	3		4	1	2		9
7	2		1	5	9		3	4
3			4	6			2	
				8	3	7		6
			5			4		
4	7		9			3	6	2
	9	2				5	8	7

INTERMEDIATE - 46

2		7	4		1		8	
	6	8		7	3		4	
9	1		5		8		3	7
1	7	6	3	5	9			8
8				4	6		5	3
		5	8	2	7	1		
	2	3	6	1			7	9
				8	2	3		4
7	8	1			4			

INTERMEDIATE - 47

3	2		5		6	1	9	
9				8			6	5
5		8	2		9	3		4
4	5		6			8		
2			1		8			7
	8			2	5	9		6
1	7	9			3	6		
6				5	1			9
8	4		9	6	2		3	1

INTERMEDIATE - 48

	1				5			
6				2	7	8		9
			3	8		6	1	5
3	4		6	5	1			
8		2		4	3	1		
	6	5	2			4	7	
2	8	6	5	7	4	3	9	
5		9				7		4
4	7		9			5		2

Solutions on Page: 98

INTERMEDIATE - 49

4	3			2		8	5	6
	5		4			3		1
7			5		6		2	
			6			9		
3	1	4	9	7	5		6	
9	2	6	8	1	3	5	4	7
1		7		8			3	
	4		7	6	9	1		
2					1		9	

INTERMEDIATE - 50

9	3		2			1		
2		7		9	4	8	6	3
	8	6	3			2	9	
8	4	1		3	5	9		7
7			4	1				6
5				2	9	4		
							5	9
	5		9		7	3	1	
3	7	9		5				8

INTERMEDIATE - 51

	9	3	5	6		1	8	7
	1	5	7	2		4		
7	4			3				5
5	7	1	4	8		6	9	2
		2	1			5		
9	6		2	5		8		1
	3		6	7		2	1	8
		7	9		8	3		6
6	5		3		2	7		9

INTERMEDIATE - 52

	9	4		2	5	3	1	
8	7	3	9				5	
5							8	9
	5		3			1	9	
4	6	7		1			3	5
1			5	8		2	7	4
9	1		4		8	7	2	
	8		7			5	4	1
			1		3		6	8

INTERMEDIATE - 53

	4		6		1		9	5
	5	1	2	3	9			
9				4	5		1	
					3	1		9
4					7	5	8	
		3	9	5	8		4	7
5	8	6	3	9	4		2	
	9		5			4		8
3	1	4	7	8	2	9		6

INTERMEDIATE - 54

9	7		1			4		
6	3		7	8	2	9	1	5
	5			3	4	2		6
2	1		3	7	9			
		5	6			7	3	
7		3			5		2	
8	4			6				
	6		4			3	9	
3	2	1	5	9	7	8	6	4

INTERMEDIATE - 55

	7	4		9	3		2	5
2		1		4		9		
5					2		7	
4				6	9	8	3	
9		8				6	1	
3		6		1		5	4	
1	4	3	2		8			6
7	9	5	4	3				
6			9	7	1	3	5	4

INTERMEDIATE - 56

8	3		7					1
	2	6				5		
	7	4	1	3	5	6		
		8		2		9		7
	4		5			8		
3					1	2	5	4
4			2		7	3	6	9
7	5	3	4	9	6	1	8	2
	9	2		1	8		4	5

INTERMEDIATE - 57

2	4		7	3	9	1		6
1	9		8	5		3	2	7
	5			2			8	
5	6	1	9	4	2			
7		4	1	6		2	9	
9	3	2						
8	1			7	5	4		2
4		5			6	8		
6	2	3		1				9

INTERMEDIATE - 58

		9	3				6	4
6	8				4	9	7	
4			6		7	8		1
		7	9		1	6	2	8
2	9	4	7	8	6	1	3	
		6			5	4		
5	4	2		6		7		9
9	3	1		7	2	5		
7					9	3		2

INTERMEDIATE - 59

			5		6	1		8
9	1	7		8	4		5	
	6	5	1	2	9	3		
				4		7		
	9			1			2	
4	5	1	2			8	6	
2		8	4	9		5		6
	4					9	8	3
5		9	8		1	2	4	

INTERMEDIATE - 60

		4	1		9	2	7	
1	9	2	6	5	7			
7	3		8		4		1	
5			7		2	9		3
	1						8	2
2	7	9	3				5	
	8	5			3			
3		7	9		1	8	4	
9	6	1			8	3	2	7

Solutions on Page: 98

INTERMEDIATE - 61

4	1		8				3	2
	9	8		2	3		1	
			6	4	1	5		8
8	4	2	1	5	7	9	6	
6		1	3	8	9	4	2	7
	3			6	4			
	8			1	2	3	7	6
			9				5	1
1	6	3	5				4	9

INTERMEDIATE - 62

3	5	1			6			8
2	6	4	3	8	9	5	1	7
8	9		1			6	2	3
	8	2	6	7		9		
			4				7	
				9	8			2
5			9		1	7	8	
1	4			3		2	6	5
	7	8	2		5	1		9

INTERMEDIATE - 63

2	7		9	4	6	5	8	
4		6		5		9	2	
5	8	9			1		6	
	3		8			4		
7		5				6	9	
9		8		7	5	1		2
3			6	8			4	
8	9				3	7	5	
	2	7	5	9			1	

INTERMEDIATE - 64

1	5			4	6	8		9
	2	4	5		7	1	6	3
		8	1		3	5		
2	9			5		6	3	1
		1		6	2	7		5
	4			7	1			2
4	6	2		1		3		8
	3	5		2	9	4		
	1		8	3	4			

INTERMEDIATE - 65

	5		4		9	6		2
				7		1	5	4
		4	5	8	1	7		3
9	7			5		3	4	1
5		8	6			9		7
			9		7	8	6	5
8	1				4			6
		3	7		5			8
4	6		1	2	8			

INTERMEDIATE - 66

	8				4	7		1
	4				9			6
5		6	7	1	8	4		2
7	2		5		6	8	4	3
	6		1			2	5	
8	5	9	3	4				7
4		5	9		3	6		8
6	3	8		7				
9	7				5	3		4

INTERMEDIATE - 67

2	9	4	3	1	5	8		
	6	3	9	7	8		5	
8			4		6		9	3
	1				2	5	3	7
7	2		1	5		9	4	8
5			7	8	4	6	2	1
6	5		2	3	1	4		
3	8						6	2
9					7			

INTERMEDIATE - 68

		7	2	8	5	6	4	
	6		4	7	3	5		
	3	5	6	1	9	8	7	
3			5	6		2	1	4
	5		3	2	4		8	
	2			9	8	7		5
2			9					
	1	3				4	6	8
5		6				3		9

INTERMEDIATE - 69

5		1	3	4	8	9	2	
		3		1	9	8	7	4
9	4		2	6		3	5	1
4		2				6	8	
			1		6	5	4	3
	3				4			2
		6	8		2			
3	2		6	7				8
	9	7		3	1	2		

INTERMEDIATE - 70

6			8	1		4	7	3
7	4	3	9	2	6	8		5
5	1		3		4		9	6
		6	7			3	4	1
		4			1	7		8
	5	7		3	8		6	
		1		8	2	5	3	9
	6				3	1		
3	8				7			

INTERMEDIATE - 71

	3	7	4				2	1
5		8	3		7	9		
	6		9			3	7	5
3	5	1	8		2	4	6	7
			1		3		9	2
7	2	9			4		8	
2	4	5	7			6		8
		3	5	4	6			
1		6		3				

INTERMEDIATE - 72

5			8	7	9	4	3	
				2		1		
2		3	1	4	5	6	9	8
7	3	4		5				1
9				8	4	5		
	5		7		2			3
	4					3		
6	8	9	2	3		7	5	4
3	2		4	9		8	1	

Solutions on Page: 99

INTERMEDIATE - 73

		1	8	9				2
	8	3	6	5	7			1
9	7	4	1	2	3	5	8	6
8	5		9	4				
	9				1	6		8
1							9	4
	2	5	4		6	8	3	
	1	9	2	8				7
6	4	8	7	3		1		

INTERMEDIATE - 74

7				3		9		8
8		3	2	6	7		1	5
	1	4		8	5	3	2	7
9	4			1	2	5		
		6						1
1			6		4			
		2	1		9		7	3
4	6	9	3	7				
3		1		2	6		9	

INTERMEDIATE - 75

	1	2		4	5		3	8
				8	3			2
	3		7	2	1	4		9
	6		3		7	8		
8	4			9				1
	7			5		9	2	6
9		6	5			1		3
7	8	1	4	3	2			5
3		4		6	9		8	

INTERMEDIATE - 76

3		8	4	1	5	7		9
4	7	1	3	9	6	5		2
	5	9		8	2			3
2					8	6	1	
9		4						7
		5	2			8	9	4
7	9		8		4	3		
	1		5	2	9		7	
5	4		1		7	9	2	8

INTERMEDIATE - 77

5		2	6	9		1		
4	7	9		3			6	5
		3				8	9	7
7		1	3	6		4		2
	5					9	3	
		6	2		9	5	7	
9			1	7	3			8
3	6	8		5		7	1	
	2	7	4	8		3		9

INTERMEDIATE - 78

1	3				4	7		
				9	5			6
6			8	7		9	2	1
5	7	2	9	1	8	3	6	4
3	6	4				8		
			3		6	2	5	7
2		6	4		9		7	
4	8	1	6	3		5		2
		3	5		1			8

INTERMEDIATE - 79

8								9
	7			4	9			
1		9	8	5	6	7		3
	8		4		2	6	9	
4		3	9		5	8	7	2
		6			8	3		5
9		4		1	3	2		7
7				8		9	3	
	6	8	2	9			1	

INTERMEDIATE - 80

4	3	5		2	7	6	9	8
	1	2	6	8	9			
		6				2	7	
6	5	9		1	3			
2	8		4	6			1	
1		4				5	6	
9	4	7		3		1	8	
3	2		8		6	7		
5	6	8	9	7	1		2	4

INTERMEDIATE - 81

4		3	8	1		6	2	7
6	1		2		7	3		8
2	8	7	3		6		4	1
		1	7			4	3	6
8						7	1	
	7	2	4			8		
		5			8		6	4
9	4				5	1		
	2	6		3	4			

INTERMEDIATE - 82

	5		2	7	6	4		
6	7	1	4			5	8	
	4	2						
5		4		9	2	3	7	
9		8	7				6	
1	2	7			3	9		8
4	8		9	2	7	1	3	
	9		1	5	8		4	7
	1	5		6			2	9

INTERMEDIATE - 83

2		9		5	7		6	3
	4					7		
7		3	4	6		9		
				9	8			
3	9		2		5		7	6
	7				6	5	9	4
1	2			7			3	9
9	5	6	3	8	2	1		7
4	3		6	1		2	5	

INTERMEDIATE - 84

4	7			3			2	6
2	3	5		6	9	1	7	8
1	9	6		7	8	3	4	5
3		7			5	2	8	
	4	9		2				
			8	9			1	4
		1		5	4	8	9	7
9						6	3	1
7	8	3			6		5	

Solutions on Page: 99

INTERMEDIATE - 85

5			9	3	7	6	2	
	2			1	8		5	9
9	3		6					4
	8	3	5	7	6	1	9	2
6	7	2	3	9	1	4	8	5
			8	4	2	7		6
2	6		7					
		4						
	5		1		9	2	4	

INTERMEDIATE - 86

4	9	1	3			6		8
2	8	5	6	4			3	1
		7		8	1	2		
		3	4		9			2
	2	4	8	3	5		6	9
8	7		2	1	6	5		3
7	1			2			9	6
	5				8	4		7
		2					8	

INTERMEDIATE - 87

4	3	7	9	8	5		6	
9	2	6		4		5	7	
	8				6	4		9
		9		7		3		4
2	5	4		1	9		8	
1				6	4	9	5	
		2	6	3			9	
3	9	8	2		1	6	4	
6		5			7	8		

INTERMEDIATE - 88

9	1		8					
		6						
2	8	5		7	4	6	9	
		2	7	8	3	9	1	4
	9			6		5		2
1	4	7				8		6
7		8	5		9	1	6	
6	5		2	3	8	7	4	9
		9		1	7		5	

INTERMEDIATE - 89

3	2	6	8	5	4	7	1	
1	4	5			3		2	6
	9		2	1	6	3		
6		4	5			9		
5	1	7		6				
9			7		1	5	6	
8	5	1	6			2	9	
2	6	3						
4	7		1		5	6	3	

INTERMEDIATE - 90

4		7	2	8	9		5	
9		5						
			3		7	8	9	
			9		1		4	5
	4	1	8	3		9	2	7
7	5	9	6	2	4	3		8
	1		7	4	2		8	
			5		3	4	6	1
5		4	1	6		2		

INTERMEDIATE - 91

2		4	5			9		
3	1	9	4		8	5	6	2
	8	5	3	2	9	7	1	4
1	3					8	5	6
7			1	8			2	
	5	8		3		1		7
	6		8				7	
				5		6		3
				6		4	8	1

INTERMEDIATE - 92

	7			8	3	6	1	5
8				5	6	4		
1	6	5				8		
3	9	7		6	5			4
5	8	4	3	1		7		6
2	1	6	7	4	8	5		3
	2	1			4		5	
		3	8	9		2		7
	4	8		7				

INTERMEDIATE - 93

	4	2		3	8		9	7
	9	1	4	7	6	3		2
6			2			4		1
4	1	5			7	9		
	7	8			2		1	6
2					1	7	3	4
	8	4		9	5	2		3
	2	6				8	4	
7				2	4	1		

INTERMEDIATE - 94

	1	6		3	4	2	5	
		2			6			3
9		4			7	1	8	
6		8	5		1	4		
		3					9	5
	9	5	7	2		6		8
	4	7	8		2		6	
	6	1	3	7	9	8	2	4
2	8	9	4	6			7	1

INTERMEDIATE - 95

7		1	5		8	3		
	8		4	6	3		5	7
4	3	5	2					6
5	1			2	4	7	9	
			1	8		5	6	4
8	7		9			2	1	3
3	2			4	5	8		
				3	9		2	
6	5		7	1				9

INTERMEDIATE - 96

3	8			4		9	1	5
	9			8		6		4
4	2	6	9	5		3		
			5			2	6	
6	1	4		2	9		8	3
9	5	2		6		7		1
7		9				1	5	
5	4	1			6	8		7
2	3			7		4		

INTERMEDIATE - 97

8					9	3	1	7
9	7	3	8	1	6		5	
1	5						8	6
4	8	5	7	3	1	6	2	9
3		9			2	5	7	4
6	2	7	9			8		
				9		7		
		4	2	5			9	
5		1			7			8

INTERMEDIATE - 98

4	9	1					5	3
7				6	3		9	
5		3	9					7
	5	6		3				
3			6	7	8			
8		9		2			3	6
	8		7	9		3	4	1
9		4		5		2	7	8
2		7	8	4	1	5		9

INTERMEDIATE - 99

5	8	9	7				1	
	3			1		4		
1	4	2	8		6		9	7
6				8	3	9		2
	2		4		5	1	7	8
	1		2	7	9	5	6	3
		4	5			8		1
		1	3	4	8	6	5	9
8	9		6				3	

INTERMEDIATE - 100

	4	3	2	8	6	9		
2		8			9		3	6
		1					2	8
7				4	1	6	5	
1	3		8				9	4
			3				1	2
8	9		6	1				7
6	1	7	5		4		8	
3			7	9	8	1	6	5

INTERMEDIATE - 101

		3	8		2	9	1	7
	7	1		3	6	8		5
		2			7			6
8	3	7	2	1		6		4
1	9					7		
2		5				1	9	8
		4	3			2		1
5	1				8	4	7	
3		8	4		1	5	6	

INTERMEDIATE - 102

7	3	4			5		8	2
9	8	5	4	6	2	3	1	7
	2	6				9	5	
	7		1	2			4	
			3		6	1		
4				5	8	2		9
6		7				5	9	3
	4		5				6	
	5	9	6		7			

INTERMEDIATE - 103

	3			5	9		7	4
6	5		7		4		2	
7		4	2	3	6			1
5			4				8	7
		8		7	1	4	6	
		7					3	9
9				2	5	8	4	3
8	2	5	9	4	3	7		
3		6	1	8	7	9	5	

INTERMEDIATE - 104

4					9	5	8	
5		9			8		2	
8	1	3					4	9
	8		3	9		4	1	5
1	3	4	2	5	7	9	6	8
9					4	7		
		7	4			3		
			9		3			
	9	1	5	6	2	8	7	4

INTERMEDIATE - 105

6			5		9	7	8	4
9	4			7			3	5
7		5		3	4	9		
2	7	8					6	9
	9	4	8	5		1		2
5		6			2	3		
		7	1	2	5	8	9	
8	3	2			7	4		
1	5			8			2	7

INTERMEDIATE - 106

7	2		5			6		8
	9	6	7	4			3	5
	5	1		2		9		
9	4	2	8		6		5	7
	8	3	9					
1		7			5	8	2	
			3	5			9	6
4	3				7	5		2
6	1			8		7	4	3

INTERMEDIATE - 107

	8	7	5	2	6	1	3	
5	6	2	4	1		9		8
	3	1	7		8	2	5	6
3			2	4	7			
2			9	3	1	8	4	
1		4		8		7		
	4	5						1
8		3				4	6	7
	1		3	7		5		2

INTERMEDIATE - 108

2		8	1	4	5	9	7	6
7			3	2		4	5	
		5	9	6			3	2
		9			2	6		4
		3	5			7		9
4		1						3
9	6			5	3	1		8
8	1	2	4	7		3	9	5
3		4		9			6	7

Solutions on Page: 100

INTERMEDIATE - 109

1	7	8	4	9	3		6	5
	6			7	5	3		1
3	5		8					
				1				
		3	7	4			5	2
	4			2	8	9	7	3
5		4		8		7		6
2	1	7	6					9
9			1	3		5	2	4

INTERMEDIATE - 110

2	4	8						
1	9	3	2		7		8	5
5						2	1	
	2	1	5	7	3		4	
			8		4	1		
7				9	1			2
4	3	6	7			9	2	
8	7		3		9		5	1
9	1		4	8	2	3	7	6

INTERMEDIATE - 111

5		2	4		3	7	8	
		4	7			1		9
	9				2	3	4	5
4		7	3		9	5	6	
6	5			7	4	9	3	
	3	9		6	5		7	
2	7		5			6	9	3
	4		6	3				2
8	6		9			4	5	

INTERMEDIATE - 112

6	2	5	4	1			7	
7	4				3		5	
					5			4
	1	6				9	4	
5	7	9	8	3	4		6	2
2	3	4	1				8	
4		1	9	5	2	7		6
					6		1	
		2	7	8	1	4	9	5

INTERMEDIATE - 113

1	6		5		8	3	9	
9	5	3		6	7	2	8	4
8		7	3		9		1	
4						1	3	5
				1				
2	1	9		5	3		4	
6	9	4	2	3		8		
7		5	4			6		3
3		1			6	4	5	

INTERMEDIATE - 114

5	4		7		1			6
		9	8		4	7		
		8	5	9	6		2	
6	3	4			2	1	5	8
	9			1	5			2
1		5	6	4		3	7	
	7	6	2		9	5		
2		1				9	6	7
	5		1				8	

INTERMEDIATE - 115

5	8		9	1	3	6	2	
7	3	9			2		4	1
1		2	8	4			3	
4				8		7	9	3
		7		2		4		6
6			3	7	4		1	
2	4	1	7	9	5		6	
			2	3		9		4
9		3			8	1	5	2

INTERMEDIATE - 116

	8			9			4	
	3	6	5		4		7	
			3	2	6	1		5
4	6	9	2				5	7
	2	7			5	3	9	1
3	1		9			4	2	6
	7	1	6	4	2	5		8
2	5	8		1	3		6	4
6	4		8		9	7		2

INTERMEDIATE - 117

			8		4		7	
7	4	2	3	6			5	8
1		8						4
	2		9	5	7			
8		9	6	4	1	5	2	7
	7		2	3	8	9	4	
9			4		5	7	6	2
					3	8		5
2	8	5		9	6	4	1	3

INTERMEDIATE - 118

3				1				5
1			3	5	9	6	4	7
5	9	4		7	6			
8						4	3	6
	1		9	4			5	
4		2	6		5			1
9	7	1	4	3	8			2
		5	7	6				
6	8				2	7	1	4

INTERMEDIATE - 119

8	2				1		6	3
		6	3		8		7	
3	4	5		7	2		9	
		8		4	3	6		9
1				8	9		4	
4			5			7	1	
	8	4		3	7	1	5	
6		3	2	1	5	9	8	4
			8	6		2	3	

INTERMEDIATE - 120

	5		4	7	8			
6	1		3		5	8		7
8	7		6	1			4	3
4	3	7	5			2		
	2	8	1	6	3		9	
	6			2				
7	9	5	8		6		3	
3	8		9				7	
			2	3	7		8	

Solutions on Page: 100

INTERMEDIATE - 121

4	2	9		3		1	6	8
	8				6		7	
3		6	1		9	5	4	
7		4			8	2	1	
		2	6	7	3	8		
		8		4			5	7
	1	5	3			7		
	4	7	8		2		3	
2		3	9	1			8	5

INTERMEDIATE - 122

9		7		8		4	2	5
2				9	5	3	1	8
8	5	1	4			7		6
	6	5	3	2				1
1		3		4	7		5	9
	8		5	6	1	2	4	3
6					3	5	8	
	9	4	2		8			
				7				2

INTERMEDIATE - 123

		6	9		7	2		
7			8	2	4	1		5
2			3			7	9	8
3	7	4		9		8		
	6	8	5			9	7	4
5	2	9			8			1
4	8	5	2		9		1	7
9		7				5	2	
	3	2	1		5		8	

INTERMEDIATE - 124

3			6				2	
	8		1	2	5	7	3	6
6	7			8	3	9		5
8			7		4			9
	9		3			1	4	
1				5			8	7
5	6	4	2	9				3
	1	7				2	5	
2	3	8		4	7	6	9	1

INTERMEDIATE - 125

6	2		4	8		5		7
4						2	1	8
5		7		1	2	4		9
				5	7	3	9	6
7	5	2			6			1
			8		1			
	4	6	7	9		1		
	9		5	2	3	6	7	4
		5		6	4	9		2

INTERMEDIATE - 126

	2	3		1	6		8	5
5		6			2	1		
8				3		6		2
6		1	5	4		2		
3	5	7				4		
			3		1	7	5	8
					3	8	9	
9	3	8	7			5	1	6
		5	1	9		3	2	7

INTERMEDIATE - 127

		7	4			5	9	
				1	5	4	2	8
			8	9				3
		1	5		8	9	3	
	3		1			6	8	5
			6		9		1	4
		6		8	4	7	5	1
1	8		2			3		9
4	7		3	5	1			2

INTERMEDIATE - 128

7	4				5	3		2
2			4	9	3	1	8	
1	9	3		7		5	6	4
	2	9			7			1
4						6	2	9
8	3		9					
	8	4	2	3	9	7	1	
9	1	7	6	5		2		
		2	7	1	4	9	5	

INTERMEDIATE - 129

1	6		3	7	5	4	2	8
2	5	3	8		4		9	
					9	1	5	3
6		7		4				1
3	2		7	5			6	4
5	9			3	6			7
	1	5	4	6		8		
4						6	1	9
			2	9	1	3		

INTERMEDIATE - 130

3				8	1	6	4	7
	4	1	5			9	2	
9	6		2	4	7	1		5
	9	2	3	1	8	7		6
	8			7				
1			6	9	5	4		
	1			2	3			4
	5	4	1		9			3
6	3		8	5				9

INTERMEDIATE - 131

6		2	8	7	1	9	4	3
	9	3		6	5		1	7
1		8	9	4		5		6
8				1			5	
	1			9	6			8
2		9				1	3	4
9		1	7	5	4	3	6	
5	2	4		3	9			1
7	3	6	1	2				

INTERMEDIATE - 132

6	8	4			3		5	9
		9	4		1	6		2
		2	9	6	8			
	2		8	4			6	5
	3		2		5	9		
		7		9			4	1
1	9		5				2	
7	6		1		2	4	9	
2	4	8		3	9	5		7

Solutions on Page: 101

INTERMEDIATE - 133

		2		4	6	9	3	5
				3	5	2		
4	3		1					
6	4	3		1			2	8
	5	8	3		4	1		6
	2	1	6	7	8	4	5	3
	1	4				3	8	
3	9	6	4	8		5		
	8		9					4

INTERMEDIATE - 134

4			6			9	3	2
6	2	8				1		7
	3	1	2			8	5	
	6	4			5			3
5	9	3	7		2		1	4
8	1	7	4			2	9	5
			8	6	3	5	7	9
		6	9		1			
3						4		1

INTERMEDIATE - 135

9	7	3	8	4	2	5	6	1
4		6					2	9
		5	6	7				3
2					3	6		5
	3			5			8	
6	5	8			4	1	3	7
			4				1	6
3	4		9	6	7			
8	6	2		1	5			4

INTERMEDIATE - 136

		9		8	3		7	
8		5	9		2		1	4
			1	5	6			
5	7	3			1			
1	9	6	2	3	8		5	7
4				9	5		6	
3	5	7	8	2	9		4	6
2		4	3		7			8
9					4		3	2

INTERMEDIATE - 137

		1	7	9	5		2	
8		9	3	4	2	6	5	1
2		4	8			9		
	8			2	9	7	4	
1	2		4		8		9	5
4		6	5			2		
5	4		2	3				
	6	3					8	
9				8	4	5	3	7

INTERMEDIATE - 138

		3				5		
	9	4	3		7		6	
7			5		1		4	9
3	7	6			9	1		
1	5	8	7	3	6	2	9	
2			8	1	5			7
6		5	4	7		9	1	
9	2	7	1		8	4		3
4						7	2	6

INTERMEDIATE - 139

7	6	5	9	8			1	
1	4			5	6			8
	3		1	4			6	7
3			6	9		4		2
	9	4				1	5	3
	2		4	3			7	
					9	8	4	1
	1		5	7	8	9		
		6	2	1		7	3	

INTERMEDIATE - 140

2	1		3	4				8
5	4	3	6	8	9	7	1	2
	8				1	6		3
		8	1	6		2		
9				3	2			6
	6			7	8	1	3	5
1		5			6	8	2	
6		4	8	2		3		
8	2		4		3			9

INTERMEDIATE - 141

7	5			6			9	
		4	9	2				3
8	3	9	7					
	2		5	7		1	4	
	7		4		9	3	2	5
5		8		1		9	7	6
4	6			9	7	5		2
2			6	3		4		9
	9		2			6	8	7

INTERMEDIATE - 142

8					7	2	3	
		5	4	9		8	6	7
	9	6	8		3			5
4	6		1	8		9	5	3
5	8	9		4	6			2
	3		9			6	8	4
		3	7	1				
6				3	9	4		
9				6	4	3	7	8

INTERMEDIATE - 143

9	6	8			4	5		3
1		7	2	9			4	8
		2			5		7	1
6				2	8	1		7
3			4	5	9		8	
2		5		7			9	4
7		6	5	3	2	8		9
5			6	8		4	3	
8		3	9					5

INTERMEDIATE - 144

8	9	7					1	5
4	5	6		2		8		3
		1				7		
			1			6		
		4		3	6	1		7
6	1		4	7	9			8
7		2					8	1
1			2		7	5	3	6
5			6	1	8	9		2

Solutions on Page: 101

INTERMEDIATE - 145

		1	6					5
9			3	5		8	2	1
		5	9				7	
		8	4		6	3	5	2
3	4		2	7	5		1	
	2	9		3	8	7		6
6	5		8			1		
		4		2			6	9
7	9		5	6		2	8	

INTERMEDIATE - 146

					8	6	7	
	4	7	1		9	3	8	5
		1	7	5	3		4	
7		6	9		4		5	8
3	9		5		1	7		6
	1	8		2	7		3	
	8	2	4		6		9	
	5		3		2	8		
6	7	3	8					4

INTERMEDIATE - 147

8	9	4		7	1	2		3
		3	9	2		8	1	7
2		1					4	5
				8	9	1		
1			2	6	3	7	8	4
4		2		5			3	9
7			3		5	4	9	8
9			7	4		3		6
3		8	6					1

INTERMEDIATE - 148

	3	6		5	4	2		9
	9	8		2	1		7	6
					6	8	3	
2	7	5		3	9	4		
		9		4		3	6	7
3	6		1	8	7	9		2
6	2		4		8			5
			9			1	2	3
		1	2		3	6		8

INTERMEDIATE - 149

	6		5	7			8	9
	5		2	6	3	7	4	
					9	2	5	6
	1		4	3	7	5		
5			8				6	
8	4		6		2			3
4	9		7		8			
1		5		4			7	2
7	2		9	1		8	3	4

INTERMEDIATE - 150

9	1	5	6			8		3
	6	3	5	1	2	9		
	7	2					5	6
			1	2		6		8
						5	3	
5	8			9		4	1	
6	3					7	4	1
7		8	9	4			6	
1	5	4	7	6				

INTERMEDIATE - 151

8						4		2
6	5		3				8	
7	2	9		4			6	
	7	1	9	6	3			
			4		7	6	2	
4			8		2	7		
3		7	2	8	5	1	9	
5	9	8		3	6	2	7	4
	6	2			4	5		

INTERMEDIATE - 152

		1	4		7	6		
	9	2	5		3		8	4
	4		6	2			1	
	2	6	8			1		
	3	5		9	1		2	6
1	7	8	3		2		5	
2	6		1		8			3
		3		4	6	2	7	1
	1	4		3	5		6	8

INTERMEDIATE - 153

3	1	6	8	7	5	2		9
7		5	3	9	4			8
	9	8	6			5		
							6	5
9	5				6			
	7		5	4	9			3
	6	9	4	1	3			
2		7			8	1	5	
	8			5	2		9	6

INTERMEDIATE - 154

1	6	5			9	4		
	3	4	8	1				
2			4	6				1
7		8	6		3		1	
6		1	5		7		3	2
3	5		9		1	7	6	
	2	6	3			1	4	9
	1	3	2		4		7	
4		9	1	5	6			3

INTERMEDIATE - 155

6	3	5	8	1		7		
				2	9			
2	1		6	7	5		3	4
1			5		3			
	9			8	1	5	2	3
	8	3	2		7	6	4	
3		1		4	6	2		7
	4		7	5	8	3		6
	6	7		3		4		

INTERMEDIATE - 156

	3	6			5		7	
			9	7	8		6	
	1		3		2			
			2	9		1		5
5	9				1	8	4	7
3			5	8	4	6	2	
1	2		7		9	4	8	6
		8	1	2	6		9	
7	6		8		3		1	2

Solutions on Page: 101 & 102

INTERMEDIATE - 157

9		5				4		6
3	1			6		9	7	5
	4		5			1	3	
	9		4	5		6	2	7
			2	1	6	8		3
	2	8		7		5	4	
5	6		7	8		3	1	
	8	4				2	5	9
	3		9	4		7		8

INTERMEDIATE - 158

4		6			3		9	
9		8	4			5		7
3			8	9		6	1	
6		7		5	4	1	8	2
5		4		1	8			3
1		3	6	2		7	4	
			2		7	4		1
2		5	9	4				6
7	4	1					2	

INTERMEDIATE - 159

	9	7	5	4	8		1	6
2	8	4	6	9	1	5		7
6	5	1	3	7				
				8	9	3		2
9							5	8
	3	6				1	7	9
4	7	8		1				3
5	2	9		3		7		
		3	8	2	7	9		

INTERMEDIATE - 160

7			6				2	8
1	4						3	
		2	3	8		7		1
2		9	8			5	4	7
						3	1	6
		6	1		5			
6	9			2	1		7	3
8	2	7	5		3	1	6	
4	3		7			2		5

INTERMEDIATE - 161

4			8	9	7			
6	9	2					3	7
	1	7	3					
9	7	1			4	2		
		6	2	5	8	1	7	
	8		1		9		6	4
1	3	8					4	2
			4	8	3	5		6
5		4		1		7	8	3

INTERMEDIATE - 162

9	1	8		5		2		
		6	7	8	1	4		
	5			9	6	3		8
	4	9	3			6	2	7
1					4	9	8	
7	6	3	9		8	5	4	1
		4	1	6	9		3	
3	9		8				6	
	8	2	5		7	1		

INTERMEDIATE - 163

2	4		6	9	1		8	
6				8	7		2	9
	8		4	3				6
7				4		2		1
3		8	7	1		6		5
1	5	4	3		6	8	9	
	3	1	9	7	8	5	6	
			2		3			4
5	9	2		6			7	8

INTERMEDIATE - 164

7						3	2	
		9	2	8	3		7	
3	4	2		9	5	1	8	
6		3	8			4	9	
	5	7		4	9			
4	9		5	6		7	3	
5	7	1				9		3
8				5	7			1
9			1		4	8	5	

INTERMEDIATE - 165

			7	9			6	
			2		8	7		4
4		7	5			8	2	9
3				5	9	1	8	6
5	1	6	3		7		4	
	4	8		2			5	
7	3		8	4	2		9	5
6		4	1	3			7	
2	8					4	1	

INTERMEDIATE - 166

		5	2	8		6	3	
3	4	9		6	5			
2	8		3	9		1	4	
5					9	7	1	
4	7	2		1		5		
1	9		7		3	4		
8	3	1		7		9	6	4
		7		4	6			1
6	5	4	9	3		8		

INTERMEDIATE - 167

9	7		8				5	6
	1	4	7		6		3	
8	6	2	1					
2	4	6			3		1	5
7	8	1	5	2		9	6	3
3	9	5				4	7	
	3		2	6			9	4
	2					3		
4	5			1	8	6		7

INTERMEDIATE - 168

		5	7	3		4		9
2	3			4	1	6		8
	6		2	5	8	1	7	
4	5	6	1	2	9		8	7
			4	8	7	5		1
	7					9	4	2
5	4	3						
7	8	1		9				5
	9	2	3	7	5	8		

Solutions on Page: 102

INTERMEDIATE - 169

3				5	6			7
2	7	1		9		4		
			4	7	1			
9	1			3	2			4
	3		8		7	9	1	2
7		2	1	4	9			8
	9			2		8	7	1
	2	7			3	6	4	5
		4		1	5		2	9

INTERMEDIATE - 170

	1	2	9	3	4	8	5	
				6	5	1		2
6		5	8	1			4	9
		3			9	5	6	4
8	2	6	5		3	9	7	1
5		9	6		1		2	
3	5	4	1				8	7
	6			2		4		
	8	1	4		7	6		

INTERMEDIATE - 171

1	3	2		7			8	4
7			8				6	
	6		4	2	5			
2		8	1	3	7	6	4	9
	9	6			2	7	1	
4		7	6	9		2	5	3
							3	
9	7	4		5	1	8	2	6
			7			1	9	5

INTERMEDIATE - 172

3			9	5	8		2	
9		5		6		8	3	4
	7		2					9
7		3	4			1	5	8
4		6		8	5	9		2
	8	2						6
2	5	1		7			9	
	4	9	5				8	7
		7		9			1	5

INTERMEDIATE - 173

6	8				4	1		7
			6	2	1	5	8	3
		5	7	3		4		9
7					3	9		8
	9	1	4	7		6		
	3		8	9	2	7		4
	6	8			7	3		5
3	2		5	1		8		
9	5	4			6			

INTERMEDIATE - 174

6		1			7	8		
	5	2	8	6		7	4	
	9		5	4	1			
9	6	7				3	1	4
8	1		9		4	5	6	
5	2	4	6		3	9	8	7
		5	4	9	2		7	
2	7						5	3
	8	6	7	3		1		9

INTERMEDIATE - 175

7	3	9	1	6		8	5	
			8		4		9	2
2	8				9	7	6	
9						1		
		3	5	8	1	2	7	
1		8		4	7			6
		2	4			9	1	
4		5	7		8	6		3
3		1	2	9			8	7

INTERMEDIATE - 176

4	9	8						
6	7	5		4	8		3	9
		3	7	5	9	4		
			4	8		9		1
	1			7	5	2	4	3
	4		3	9	1		5	
	3	4	8	1	6	7	9	2
		2	9			6		
	6		5				8	4

INTERMEDIATE - 177

	1		8	7		2	4	9
7	2	5	3	9		1	6	
	4		2		6		5	7
				6	8	9		
6	8		1		2	4		
4			9	5	7			
5	9		7	8		6		4
2		1	6		9			
8		4	5		3		9	1

INTERMEDIATE - 178

	3	9	8		6	1		
	8		3		4		7	2
					5	8	3	
	6		5	8	7			
4		3	2	6		5	1	8
9	5		4		1	7		6
3		6	1				4	7
		5				3	6	1
2						9		5

INTERMEDIATE - 179

	9		6		1	4	3	5
7	4			9		2		1
6			2		5			7
			5			3	4	9
4	5	9		3	6		2	
3			4	8				
8	3	1	9			5	7	
5	6	4	7	1	2	9	8	3
9	2		3			6		

INTERMEDIATE - 180

		3	8					6
8	5	6	4		7	2		3
	1		2			5	8	4
	9	1	3	4	2	8		7
7	2			6	5			1
	3	5	1	7		6	2	
	7	2	6	8		4		5
5		9	7	2				
3	8	4		1			6	2

Solutions on Page: 102

INTERMEDIATE - 181

	5	9	4		3			
8		3	9	6		5		1
2	7					9	4	
	1	4		3	9		6	8
6		8			2	3	5	4
3	2	5			6	1	7	9
	8	1		7		6		
	3	2			8	4		7
	6				1		9	

INTERMEDIATE - 182

	5	1	7	8		4	3	9
	3	2				6		8
			3		1		5	7
			9					
8	6	3	4	2			9	
	2	9	6		8			
2	7	4			6	9	8	3
	8			7	9	1		5
	1	5			3	7	2	6

INTERMEDIATE - 183

	5	7	6	9	3	2	1	4
6	1	2		4		3		9
					1	5	6	7
		8					7	
1	7	3	8	6		4		2
9	6		4	7	2			
		9			6	1	2	
			9	5	8			6
7		6		2	4	9	5	3

INTERMEDIATE - 184

	1		5		8			
2		8	7	1	3	6	5	4
3		5	4	2			1	8
	8		1				6	3
		7			4			
6		1						
1	4		2		7	5	8	6
8		9			6	4	7	1
		6	8	4		2		9

INTERMEDIATE - 185

	4				1			
6	2	5		7		9		
			2	5	6		7	
1		4	6	8			2	7
	6				7	1	3	4
2	7	3	4	1		8	9	6
	8		1	4	9	3	5	2
3	1							9
			7		2		8	1

INTERMEDIATE - 186

5	2	1	3			4	7	
9				6		3	1	
	7	6	1	4				5
			7		5		2	9
2			4	9	3	7	6	1
	9			2	6	5	3	4
	3	2	9			6		7
7				3				8
4	6	5		7	8		9	3

INTERMEDIATE - 187

1	2			6		5		
	4		1	5	7	2		6
6	5			2	3			8
			2	7	6			1
7		5	3				4	
2	8	1		4			3	
8	7		4	1	5			
5	1	3				7	8	4
4	9	6	7		8	1		5

INTERMEDIATE - 188

1	4	3						8
		7				3		
8			3	4		2		
		1		5	3	8		6
5	2		4	6	7			3
3		6		1			7	5
9	1		6		2		8	4
7	3	4	1	8	9	6	5	
6		2		7	4			

INTERMEDIATE - 189

		3	4		8	7		
4		7		1	6			3
9			3	7	2	1		
		1	2		9		3	6
		9		8		5	2	7
	5		7					9
	3	5	8				9	
1		2	9	6	5	3	4	8
8	9	4	1	2		6	7	

INTERMEDIATE - 190

		4	6	3	9		1	
5	1	3		2	8	6		7
				7	5			2
1	6	5	7	9	4	2	8	3
7	9	2	5	8	3	1		
4					1			9
				1		4	3	
		1	8					6
9	5	7	3					1

INTERMEDIATE - 191

		3		9	4	5	1	
2	4		8	5	1		3	
	5	1				4	8	9
4	6	2	1	8		7		5
7			4	6			2	
	3							6
		7	9	1		3	6	
3		4	7	2		9		1
	9		3			2	7	8

INTERMEDIATE - 192

				8	3	2		4
1	4	9	5	6	2			7
8	3	2	4		7	6	9	5
		1	7		4	8		
		3	8				2	6
5	8			6				
9	7	6	1	4	8			2
	5	8	2		9	4		
2		4			5		7	

Solutions on Page: 103

INTERMEDIATE - 193

		9		3	2	4	8	
3			1			6	9	
5	6	2	8		9	7		1
4		6	7		5	9		8
	2					5	1	4
1		5	9			3	7	
			2		6		5	
	9	1	5		3		4	
2	5	3			1	8		9

INTERMEDIATE - 194

1						7		
	4	3	6		2	8		
		2					4	3
5	1		7	2	9	6	3	
3	2	9	1		6	4		
7			4			1	9	2
8		1	2		7	5	6	
	6	7			8		1	
4		5	9		1	2	8	7

INTERMEDIATE - 195

3			8	4	6	7	1	
	7	6			1		5	
	9						4	2
9	3		6	5		1	2	4
2	6	4			7			8
5			3	2		9		7
	2	9		6		3		
			7	8	2	4	9	
		5			9	2	7	6

INTERMEDIATE - 196

			4	8		5	2	
9	2				7		3	
8	5	4	2				7	9
	9	7	8					3
6	4	3	9	2	5	1	8	7
	8					9	5	
4	1	8	3	6		7		5
			5				1	2
		5		9	8	3		4

INTERMEDIATE - 197

		1	4	6	2	5	7	3
5		2		7	3		9	4
	3	4	9	5				
	1	5	2			7	8	9
			3	8				1
	4	7	5				6	
	7	3		2	8	9	4	5
2			1		4		3	
4		6	7		5			

INTERMEDIATE - 198

6		8	4		1			
		2	3	6	8		9	
9			5	2	7		1	
2	9	4			5			8
7	3	6	8	4	2			1
8			9	7	6		2	
3		7		8			4	
				1			8	3
4				5	3	7	6	9

INTERMEDIATE - 199

	2	1	4			5	7	8
7	6		9	8	5	2		1
		5		1	2			9
6		7	3					
3	9		8	5	1	7		
	1			2	7	3		5
5	3	8	1		4	9		6
1					6			
2				9		4		3

INTERMEDIATE - 200

	3	4	7			2	8	
		1	8	3	9	4	5	
6	8			2			1	3
	6	7		8		5	2	4
3	1	8		4		6	9	
2	4			6		8	3	1
			3				6	8
		3	6	7		1		5
1	5		4	9		3		2

INTERMEDIATE - 201

	4	6	1		8	7	9	
	8		7				4	
	1	9			4		6	
	7		8	4	6		5	2
4	5	8	3				7	9
6			9	7	5	1	8	4
	6		2		1		3	7
	2		6	5	7	9	1	8
					3	5		

INTERMEDIATE - 202

		4		1		8	3	2
3	2	9	6	4	8	7	1	5
			5	3			6	4
8	3			5				
9		1	2					
6	5		1		9	2		3
		3	8	6		5		9
4			7	2	5	3	8	
7	8	5	3				2	6

INTERMEDIATE - 203

	1					6	3	
6	8	3	1	5		9	2	4
2		5	6	4	3	8	7	1
		4		1	5	2	6	9
9		1	7				5	
				6		4	1	
	3		8	7		5	4	2
	4	6				1		3
5			4	3		7	9	6

INTERMEDIATE - 204

	1	7	4		2			6
				7			4	8
	3	4	1		8	2		
	7	3		2	4		8	1
	6	8	9		5		2	7
4	5		7		1	9	6	3
	4	6		1	9	7		5
7					3		9	2
	9	5	2		7	8		

Solutions on Page: 103

HARD - 1

	8	6			2			
	4		8			2	1	5
	5		4					
	6	8		4		3	7	
		3	2				6	
5		1			6		2	
8	1	5		2	4	9		
	2		9	3	5		8	6
	3	9	7			4		

HARD - 2

4							7	
		1			9			
	2	6			1	4	5	8
8			4					
		2	8	9	7	6	4	3
		3	1		5			
	6	8	7	3		1	9	4
2	1		6				3	7
				1	8	5		2

HARD - 3

3			9	7			5	
	8	7		6		1		
9	6		2		4		8	
		6				5	7	3
		2		9	3		6	
5		4	6			9	1	
4	1		7	2		8	3	
	5	8	4		1		2	
7	2	3				6	4	

HARD - 4

1					3	7	6	5
			7	6			8	
		3		5	8	4		
5			3			8		
2	1	7		4		3		
4	3	8		1	6	9		
	2	5						4
6	7		4	3				
8		9	6			5		7

HARD - 5

7		8				5	1	
				4				
			7		1	9	6	
4	5					8	9	
		9	4	5	7		3	
3		1			9		5	
9			3		8	6	2	
	8	7		1	6		4	
6	2		5	9	4	7		

HARD - 6

		4						
	1	5	8	6	9	3		
		6	5	4		9		8
5	2					6		
4		9						1
6	7		9		8	4	5	3
9	5			8	2	1	3	4
	6	8			4			7
1	4		7					

HARD - 7

	7	2	6	4			9	1
	5	4	3				6	2
6	9	1	5		2	3	4	
		6			7		1	
		8		9	3			
	1	7			5		8	
			7	5		1		
	6	9	2	8				5
	2				4			

HARD - 8

		2	3	5	4			8
8			7	1			5	
3				6	9			1
	2	8	1				9	3
						4		5
4	6	3				2		
1		9	6	7	8			2
		7		2	1		3	
2	8		5		3			

HARD - 9

	9					7	4	2
	3	5	1		7	8	6	
		4						
8	2		9	5			3	7
9	4	7				5	8	
		1	7	6	8			
		9	8			4		
			4	1	3			8
4			5		6		7	1

HARD - 10

	4	1	2		7	8	9	5
7	5		6			3		
					5	2		6
4							6	3
	9	6				4	8	
		3					2	9
9		5		7	2	1		
		8	1		6	9		
2		4	8	5	9			

HARD - 11

	7	9		5				1
1			4	3	8		9	2
3	8			1			4	
		5	1		3	2		7
		1		2	5		3	9
2					4			
9		7		4				5
6	2			8				
5		8	2		9		1	

HARD - 12

8				2			3	
1			8	5	7	6	2	4
	6	2		3				8
	3					8		
		8			6		1	
2		7	5					3
7					3	5	4	
9		1				3	8	7
3		4		1	8		9	6

Solutions on Page: 104

HARD - 13

4	1		8	3			2	7
6		2						
9		8			7	4		
5	8		9	1		2		
			4					5
	6				2	1		9
	4				8	3		2
3		6		9	4			8
8		9		2	1		6	4

HARD - 14

9		5	7			2	3	6
			5	2	3			
	3	8	6		9			
3			1		5			
	4	7	9			8		
	2		8	3	7	6	1	4
		3			1	9		7
8			3	7				1
7						3	6	

HARD - 15

	8	2	3		4			1
	7			6	1	2	9	
			8		9		3	
7				8		3		
2	3			9			1	
		9			3			4
4	5	7			8	9	6	
6	9		7					3
		3	9	4	6		8	

HARD - 16

	6		9		1	7		
		9		2		5	3	
3	7		8				6	
7					9			
	9	3		5	4	8		7
	4	8			3		5	1
		5	4		8	3		
	3	6	5		2		1	
	8		3				9	5

HARD - 17

9			1				4	8
5	1	4			2			
			7		4	5	6	1
	5	6	3	4		8		
		8	6	2	7	4	5	9
				1	8	3		
	7			8		6	9	5
			9				3	4
3	9	5	4					

HARD - 18

	6				5	3	1	2
		8	6			4	7	
9	2	1	4	7		5		8
	4			8		1	2	3
3	5	7				8		
8			3	4		7		6
	8					6	4	
4	9			5		2	3	
			1			9		5

HARD - 19

		4						
6	5	9			3	1		
				4	9			8
8	4			6	2	9	3	
2	7	3		9		8		5
		1			8	7	4	
5		8	6	1	7	4	2	9
			9	3				
4	9		2					

HARD - 20

				9			1	
		1		8				2
8	2	7			3			9
7		6		2			9	
	9			5	1	3		6
				3	6			
4	5	9	3		2	8	6	1
6	7	8	5		4	9		
3	1						7	

HARD - 21

	9	1		4				
5	3				2	8		
7		8	3		6	1	4	
4	1	2				5		6
9				6		7		4
	7	6	5			9		
	8	5		2				1
2			6		1	3		
	6		9			4	5	

HARD - 22

	7	2				6		
	8	1	5	4	9		3	
3	5	4	2					9
7				2		5		1
				5	1	4		
4					7		9	
	6				5	9	1	7
1	4		7					6
5			1	6	3		2	

HARD - 23

8		9				7	3	4
7				4	3		2	
3			2		7	8	1	5
		8		2			4	1
4			1	8	6		9	7
		3	4			2		6
	6		5					8
	8			6	4	1		3
		4	7		8		6	2

HARD - 24

		1	9		2			8
7			8					
	8				5	1		2
		8			7		4	
5		6		9	4		1	
	1			6	8	3		9
	7	4		8	9	5	3	1
1			4	2	3	7	8	
					1			4

Solutions on Page: 104

HARD - 25

5	6	7		8			9	1
		2	3	7	5	4	6	
	8			1		2		7
7	3	4				6	1	5
					7	9		
	9		4	3	1	7	8	2
			1		3			
3			6	9		8		4
	4	9				1		6

HARD - 26

2	6		4			9	1	
	3	1	2	8			4	
	4	5				6		
6		4			3		7	
8			5	4			6	
	1					4		
	8				9	1	3	
		3	8				9	5
	9	7		3	5	2	8	6

HARD - 27

8			2		5	1	9	3
				8	1		4	
	1	5	4					
4								
	7	6		2	3	4		
	8	3	9	5	4		2	
		8		9	2	3	7	1
7	5		3			2	8	4
3		1						

HARD - 28

	8			3	6		1	2
1				4	2		7	8
		7	9			3		
	9		2	6	4			
		4		1	9		6	3
6			3					4
4	7	2	1	9	8		3	5
	6				3		2	
5			6					

HARD - 29

5						8		6
				5				
6	1	8	7	9	2		5	
		2	1	4		7		
1	6		8			5	3	
		3				9	1	
7	9		3		4	1	2	5
		1	2	7	5		4	9
2		5		1			8	

HARD - 30

	3	8		1			9	
	2		8		6	1	5	3
6		7		4		5		1
2	1	5	7	8	3	9		4
						8		7
3	5	1		2		6	7	
					9	2		5
4		2	5			3		

HARD - 31

	4		8				3	
	3		7	4	2	1	9	8
8					6			
				7	8	5		3
				6	5	7		
	5					6	8	2
2	7	6			4			1
	8	9	5			3		7
	1		6	9	7		2	

HARD - 32

		2		6	3	8		4
6			1	2	8	3	9	
5					7		2	1
	1	4		7	2	9		6
9	5	6	8		4	7		2
		3	6					
3		5						9
	8							
2	6			8	1	5		3

HARD - 33

1			5			8	2	7
7	5		1		8		6	
				9	2			4
6	8		2		4			
		3	8					
	7				9	5	4	
		5			7	2	3	6
8	3			2		4	1	9
4	2		9			7	8	

HARD - 34

	2	1				8		
5	9			1		4	7	
				5	8	1	9	3
	3	5		7	1		8	9
						5	3	
8	1		9		5	7	2	
6				4	9		1	
	4	9	1					6
1	5			6				

HARD - 35

7								
	3				2	8		
1	2	9	3		8		5	6
3	7		1			6		2
				2	5	3	7	9
	9	2	7	3				
	6		8		3	1	2	4
5	4		2		1			
			6		7	9		

HARD - 36

						9	1	
	9	3	7		1			
4			9		2			8
		4			8	2	3	
7	1	2	3		6			
9			4	2			6	1
2			5	8		1	4	
3		6	2		4		5	
		5			9	3		2

Solutions on Page: 104 & 105

HARD - 37

2	6	8	3	7	4	5		9
4		1	9				7	6
							2	
9	2				1	3	6	
7	1	3	6				9	5
		2		9			3	
6		7		5		1		
3	8	4			2	9	5	

HARD - 38

6		4		9	5			
9	2		1		3		6	
8		3	4		7		9	
			3	5				7
7				4			1	
		1	7	2		6	8	
1	7		8	3	4		5	
				7	6		4	
	4	6	5		2			

HARD - 39

	7				6	2		
	4	6	9		8			
8			5			6	4	3
9	1	8	2		5	7		4
		2	7		4		1	8
	3					9		
4		3				5		6
	9	5	6		3			
		7		5	9			2

HARD - 40

	6			4	9	7	1	8
		1				2		3
8		2	3	1	5		4	
	9	4	8					
7		6	1	5	4			
2		5	9	3				1
		9		6	3		2	5
				8		3		
3					2		6	

HARD - 41

	9		2					
6	1		7	8	4		3	2
	3		6		9	4		1
7			3		2		1	
5	4				1		2	7
				7	5	8		3
9		1	5		6	2	4	
	6			4				
4		8						9

HARD - 42

	8	1			7			
				1	5			8
5	4		9		3			
		3	1	4				
2	1				8	7	3	6
		6		7	2	8		4
6				3	4		8	1
			6			5	7	9
1	2					6	4	3

HARD - 43

					8	9		
		9	5	7				
	5			1	9			6
	8	6		2	4	1		9
	4		9					8
1	9			8	5		4	3
	6	4			2		3	
3	7		8	5				4
	2			3	7	8		1

HARD - 44

6	2	3	5		9		4	7
9	4					3	1	
			4	6	3	9	5	
	8		1	4				9
	3				2	7		
2	5	9		8		4	6	1
	6							4
	9	8	2				7	
	1			3				

HARD - 45

5			1	9	4			6
	4	1	2			5		8
	2				3	1	4	9
			6	5				1
		2	7	1		3		4
	1		4			8	5	
	3			4	1			5
	6	8	3			9	1	2
1	9	5				4	7	

HARD - 46

3					8	5	6	
7	8	2		5	6		3	1
		6		7	1		2	
6			5					2
		5						
8	2	4						
4		8	7	1	9	2		
	3			2	5	1		6
2	5	1	8			9		4

HARD - 47

4	5	6	9	8	3			2
			2	6		4		5
7	2	8	1	5		3	9	6
	4	3		9				7
			3			5	1	
5	6				1	8		
	1		6			7	2	
		5	7			9		
			8	4				1

HARD - 48

		1	9		4			6
	6	9		7	5	4	3	8
	3		8	2		1	7	9
		6		9			2	7
		8			7			
	2		3			6	4	1
6	8			4				
	1	2	7					
	9				2	3	8	4

Solutions on Page: 105

HARD - 49

6	8	2		5		7		9
		5	8		7		6	2
	1	9	3		2		5	
8	7	4		3	9			5
5								
3			6	8				4
2	4			7	6		9	
9				2			4	7
			9	4		2	8	6

HARD - 50

		2			7		8	3
	5				3			
7	3		8			9	6	2
	8		9	5	1	4		6
6			3		4			
5	4				6	3		9
		6	5		9			1
3	1			4		2	9	
	9	7		3				

HARD - 51

2	6		5		7			
3	1			2			4	
	7			3			9	
1	3	7		9		2		
9		6	4			3	5	
5	2			6	1			
	4		9	1		5	2	
6		2				9	1	
		1		5		4		6

HARD - 52

	1	6			4		5	
		2	3	5				4
				1				
	8					6	9	2
					5	7		3
		3					4	
5	3	8	1	9	7			
6	4	1				9	7	8
9	2	7	4	6	8	1		5

HARD - 53

7	9		3					
	5		7		2	6	9	3
6				5	8			
5			6	8				
		6	1				8	
8	1	9		2	3			
3	6		4				7	
9	8	4	2		7		5	
1	2	7		6	5	4	3	

HARD - 54

7		5		8				9
			3	7				
	6	8	9		1		3	7
	5	4	7	6				
8				2		7	6	
2		6	1		8	3	9	
6		7			5	9		1
			8	1				
		3	6	9		4	5	8

HARD - 55

		6	1	3	8	7		5
5	9			6				1
7	8			9	5	4		
4		2	8				7	3
		5	4			9		
9	3						8	4
3	5							
6				8	1	2	5	9
		9		5		3		

HARD - 56

9		2	1			6		7
	8							4
6		7		8				
	1			7				8
		6	2	5		4		
	2	8		9		7	6	5
8				1		2	4	
	9		8	6	2	1	5	3
		1	9	4	3	8		6

HARD - 57

	3	4	6		7			
	5							4
2				9				
	8	6	7	2	1		3	
3	7		8	5		6		
		5		6	9	7	4	
8				7			9	
			9	8	2	4	5	
5		3			6	8	2	7

HARD - 58

	2	8						
3		1					4	6
7				9				1
	6	2	4	8	5		9	
5	3			6	1			
	8	4	2		3	6	1	5
2			7			3		
			8	3	9	1	2	
8						9	6	4

HARD - 59

6	4	9					2	
2	5					8	6	
		1	4	2		9		7
					5	1	4	6
		8	2			7	9	
		6	3				8	2
	8	2	6	7	3	4	1	5
1		4						
			1	4		2		

HARD - 60

6	2	8		1				
4		9			3	5		1
		1	6					
	1	5	3	7	6	4		9
7	4		5	9				
9	8				4	3	5	7
					7			8
	9	4					3	5
8	3		9	5		6		4

Solutions on Page: 105

HARD - 61

				3			4	1
			5	1	2			3
6	1	3			9			
1		9	4	7	3			2
3				6	8		7	
2	7						6	4
		1	9	5	6			7
			7	8	1		5	
	5	6		2	4			

HARD - 62

9	2							4
						5	2	9
		4		5				
	4	8	6	3			1	
	7	9	5		1	2		3
5	3	1	9	2	8	6	4	
1			8				9	
	9			6	2	7		8
	8	2				4		

HARD - 63

	1			5		9		6
	6	3		4	2	1	8	7
9	7			1	3			
4		2		7		5		1
	8		2	9		7		
			4					
1	2			3			5	
		6			9	3	1	2
		5		2	4			9

HARD - 64

	5			8	4	2		
6		3			5	8		
	8	2						5
1			5	9	6	3		
		9			2	4		7
			8	4		1		9
5		4		7		6	9	
2				5	3	7	8	
		8	1		9		4	

HARD - 65

1		2	4	9		5	3	7
3		7		5	8	6		
		6	3	7			8	
			7			4	9	8
	4					3		
9			8	2		1		
8		5	9	4				3
7		9	5		3	8		1
		4		8	7			

HARD - 66

		6	9		2	4		7
				4		3	6	9
	7	4		6			2	
8		7		3				5
6				8	4		9	
	4					8		
	6		3			1		8
7				1	5		3	
1	3	5	4	2	8	9		6

HARD - 67

				7				6
	6	3	2	9	5	1		
1	4	2						
3	8	1	9			4		5
			3			7	1	9
		7	5	6				
2				5		6	3	
9			6	1	3	8		2
6	3				2		7	

HARD - 68

			2	3	7		1	4
			6	4				7
	4				9			
4		8	7	9	3			5
3	1			6	2		8	
7		9			5		3	
		4		5	1			6
5							4	1
		1	9	2	4	3	5	

HARD - 69

	6	5		8	3		1	4
1							7	
3		8		7	1		5	
9					7		8	
2				9			6	1
	8	4	1	2	6		3	7
8		1			9			
4		7			5		2	8
6				4				5

HARD - 70

		3	5				6	4
		4	3	9	6		8	
5	6	9			1			
		2		3	5			
3			8	2			1	
8	9			4	7		5	2
			2	5				8
2			9			1	4	
6	5			1		2	3	

HARD - 71

1		6					2	7
	2							
4				2		8		
		7		4	3	5		1
9			2	6	5	7		3
8	3	5			1	2	4	
5					9	4		
	4		7	8	2	9	3	5
		9	5	1				

HARD - 72

		6						8
4	7			3	2		1	
	1	9	8	5	7			
8		7			4	1		3
6		2		8	5	7	9	
	4	1		7	6		8	2
7						2		
	8	4			3			9
		3		2		8		

Solutions on Page: 106

HARD - 73

9				5				
			2	6	7	5		
							4	
4			6		5	9	7	
1	9	5		7	8		2	4
7		2	4		9		3	
		1		9	6	4		
3	5	9	7				6	
	4	8				1	9	7

HARD - 74

7	3	9		6				8
	4		7	3			1	
6		5		8				
	6	7	2					4
1	5			4	7	9		2
			3	1	9		6	
8	2		5		6		9	1
5	9		4		3			
	7				1		2	5

HARD - 75

5	8		4					9
	3		9	7	2			1
	2	9	5	8				
		8	1			6	4	7
3	4	7	8		9			
1			7		4	9		3
				4	5		3	
	5							
6	1	4			8		7	5

HARD - 76

		3		8		6		1
7	1		2			4	8	
	2			1	5		9	
	4	5			8			2
	8		1	4			6	7
3			5	2		8	1	
2		6					7	8
	7		8	9		3		6
				7		2		9

HARD - 77

			5	7			6	
6	3			2		8		1
			8				4	7
		5		4		3	1	
9		3				6		
	4	2	3		9		8	
		8	1	9			3	
3	2				7			8
4		9	6	8	3		7	2

HARD - 78

		2	7					3
					5	4		8
	5		9	4	6		2	1
4	6				3	2		9
				7			4	
2	1	9	4	6	8			
1	9							2
8	2	6	1	3	4	5		7
					2	8		

HARD - 79

	5		9	7				
7			3	2	1	8	6	
		2	5	8				
8			1	5	7	6	4	9
5	7	9		6				2
4					2			3
			2					
	3					2	9	4
2	4	1	6	3		7	5	8

HARD - 80

9		1	3			5		
	2		9	1				6
8	3							
	5				1	6	9	
1	9			8			7	5
6			5		7	2		3
	1		6	5	3			2
	6			7	9		5	4
3	7		8					9

HARD - 81

	5	2		4	3	7	8	6
1		3					2	5
			5	9				
					6	1		
4	1			2				8
		9				6	4	2
	3	1	8	6	4		9	
	9	4	2	1				
8	6			3	5		1	

HARD - 82

		3	6	1	2			
8		6		7		9	1	2
1				5	9	7		
		7		8			4	
3			5		7	1	2	
2	1		3	9	4	6	8	
4			9	3		5		
			7	2			9	3
6			1					8

HARD - 83

	9	7	1			4		3
1	5		7	9		6		
		8	5	2		1		
				1	9	3		5
9			4	3				
			2	6		9		1
4				5				6
8		2		7		5		4
		3	6		8		9	2

HARD - 84

2	5	6	4	1			7	3
			2			6		9
	1		3		6	2		
4		5				9		
			6	2		5	3	
6	3	8		5				
	9	3	1	6		4		2
5		7	8	4		3		1
		2		9			8	6

Solutions on Page: 106

HARD - 85

6		4				7	3	2
5		8		3				4
	3	9	2	4	6		8	
1		3	6	7	2	5		
	8	5	3		4	9	7	
4				8		3		
			1			4		9
8		1				2	5	
9	6	2						3

HARD - 86

6		5		8	3		4	
	1					2	8	
			6	7			1	
					5	9		
9		4	8			1		2
3				1		6	7	8
2	3	1	4		8		9	
	9			5	1		6	
5	8	6			7			1

HARD - 87

	4			7			8	
3	5	8						4
	7	6				2		
			4	1	8	5		
		1	6			8		
			7	3	2			9
9		7			6	4	1	5
5	8			9	7			6
6	1	2	5		3	7		

HARD - 88

8				5				2
7			3			9		
9	4	5	7	1		8	6	3
3			6		1		2	9
4	7	9			3		8	
2	6		9				5	
5					6		7	
	9			4		6	3	
6		4						1

HARD - 89

6		1		7	2		8	5
		8		6	5	1	2	4
	2	4		1	3	7		9
7						6		1
		9		3		4		
				9	7			2
8		5		2				3
	4			5		2		
			3	4	9	8		

HARD - 90

			7	4				
3					6	7		
	4			9		1	5	3
1	7	8					3	
9		2	4	6		8	7	
4			3			9		
7	9	4	6	3	2	5	1	
5	6	3	8					
				7	9		6	

HARD - 91

7	9	8				1	4	5
	5		7		4	2	3	9
3	2			1	9			6
		3					5	7
8			4	9				2
9				3	5	8		4
	3	2	1	7	6	5		
	1			5				
			3			6	7	

HARD - 92

		2		7			9	1
		1	9		6	5	3	
5		3	8		4			
	3		2				5	9
		8				3		7
	5	7			8		6	
		5	7			1	2	6
8	2	9			1	7	4	
		6		4	2			

HARD - 93

			4			6		8
6		5		2				4
	4			7	6			3
3	2		1		5		6	7
5		1		8				9
8	7			6	3		1	
4	8		6	5	9	2	3	1
		3		1			7	6
					7			5

HARD - 94

3	7	1	6	4	2		8	9
2		4				3	7	
	9	8	1	3				
				9				2
					3	8	1	7
				7	6			
			9				3	
1	2	5		6				8
9		6	7		8	1	4	5

HARD - 95

	2			3	4			
3	6	8	5	9	7	1		4
	4		6		2			3
				6	3		7	8
	1	3					4	
6	8		4	2				5
			9		5	4		2
	9	2	3			7		
4	3	5	2	7	6	8		

HARD - 96

		5	7	3	9			4
4	9	3			6	1		
6								3
				2	4		7	1
2	3	4			7		8	
5	7						4	6
		2	3	5			9	8
9	1	8	4			7		
3			9			4		

Solutions on Page: 106 & 107

HARD - 97

	9	2				8	3	
7	3		8					
1		5					6	4
5		4		7		1		3
		9		2		6	7	
3	6	7	1					
			2	5	7	3	9	1
9	7		3				5	6
			6		9			8

HARD - 98

	6		5		1		7	8
	8			3			4	6
			6	4	8		9	
	5	3			7	8		
			9	5		6		7
		6			4	9	5	3
	9			2		4	8	1
3			1		5			9
6		8						2

HARD - 99

6	1		5		4			9
2	4		9			6		
			1	3				4
8		1	4					2
	6				3	7	8	
	3		6	1		9	4	5
	7		3			1	9	8
	9			6	5	2		
3				9			5	

HARD - 100

		1		9				
			5	2				1
		2		7		8	3	5
	9						2	7
6	3	7			2			4
		8		3				6
7	6	3	9	1	5		4	8
	5			4		1	6	3
		4	8	6	3	7	5	

HARD - 101

6	8			1				
4		3			5		6	9
5	7	9	3	2			8	1
7				5	3		2	
8	2	6	9		1			
			2	4	8			7
		8			7		4	6
			8	3	4	5		
					2	8		3

HARD - 102

							7	
	6	4			7	9	2	
2		7			1			8
4	2	8		1		3		
9		3		8		7	4	5
	7	5	9		4			
5	4	6	1	7	9	2	8	3
7	8		2					9
				4	8	1		7

HARD - 103

	6	1			8		4	
3	8			7	4	6		5
	5							
	9		3			4	7	1
6	2	4		1	7	3		9
1	7	3		4				
		6				2	9	
9	1	8			2	7		6
	4			9		5		

HARD - 104

8				6	4		2	5
9		6	7		2			
7		4	3		8		9	
5			6			9	3	7
	7		2	3				
	9					1	6	2
4	3				7		8	
1			5		6	4		3
2	6				3		1	

HARD - 105

		7	5		6	8		4
	5	8		4	9			
4	9					1	6	5
5						2		
7		6	8				1	3
3	2	1	6	9	4		7	
	7		9		5			
2	1				7	6	8	
				1		3		

HARD - 106

2		4	7		5			
5							6	7
								4
		8	6	7	3		9	2
3	6	7	1	9	2			
		2	4		8			
7				2		5	1	
8			3	4			2	6
		1	5		7	4	3	9

HARD - 107

8	4						9	1
2	5	9	6	3	1	8		4
		1				5	6	2
	7	2		5				
	9		3		6	7		5
		4		9			1	3
9		3		6	7	4		8
							2	7
6	2	7		4	5			

HARD - 108

2	7		8	5	3	6		1
4	8		7					
		3				2		
		1		3				5
	6	7	5	8			9	
9	5	2		7				
	2			9	7	8		3
				1	8		2	
	3	8	2	4		9		6

Solutions on Page: 107

HARD - 109

3		4	6					
	6	1		9	5	2		4
9								
4		7		8	6			3
6	3			2		1		5
2	9		4			7	6	8
5		3		6			9	2
		9				3		
8			9	4		5		7

HARD - 110

	7		6			1	2	5
2	6	9		5				
				2	3	9	8	6
7		8	3	4		2		
		5		8		4	3	7
6	3			1				8
	4			6			5	
	1	2			8			9
9			5			3		

HARD - 111

1		9	5		6			4
4				1			6	
		3		8	9		5	
5							1	
2				3	5	8	4	7
9		7	1	4	2	5		
3		5	7	2	1	4	9	
8	9			5	4		7	
7		1	3		8			

HARD - 112

6	4			3				
		5	7			2		4
7				2	6	3	5	1
	5		9	1	2	7		
3	9		8	7	4	1	2	5
		1	5		3			
9							7	
	2	4						3
5	1		3	4				2

HARD - 113

2		3	4					8
8		6			3	1		
4	7			8				
5			3			6		7
1	3	7			6	5		2
6	2	8		5	7	4	1	3
3					8			
7		4	6		1	8	2	
9			7				6	

HARD - 114

					6	1		
5		7					6	
2	6	1	4		7	9		
6	9	2	1	7			3	
8		4	3	5	9	2		6
		3	2		4			
		6	7	9	3	8		5
	3		8					
9	2							1

HARD - 115

9	3	6		5			2	4
			3	9			1	6
		2		7	4		5	
7				6		2	3	
						1	7	9
	2	9	1	3			6	8
1					3			
		4		1	9		8	7
2	5		7	4				

HARD - 116

			7			9	1	5
				3		2		8
				1		4		7
1		8	4			7	2	
3	6	2		9	7	1	5	
	7	9		5	2			3
	3						4	1
8			6	4			9	
		4		2		8	7	

HARD - 117

			8	5			3	
1		8	7			6		
6		3		2				
			5			7	4	9
9			4		2		1	5
4	3		9					6
8	7			9	4	2		
5		4	3		1			8
		9	2		5	4		1

HARD - 118

3		7	5			8		4
	8	2			9			
1					6	7	2	
		1	6	4		2		7
2	3	6	7	8	5	4		9
	7	4		2		6		3
					7	9	3	6
	2							
		3	1			5		

HARD - 119

1				8		7	6	3
		6		2		8	5	1
	8			6	7			
9			8	7	2		1	
	6			9				
			5		6	9		4
2	1		6	4	8		3	
	3		2		9		4	8
		5			1			2

HARD - 120

	7	1			2	5	8	9
		6	7				4	3
8						6		
	1	8	2			9		
		9			1		6	5
					6	2	1	
4	3	2		6	8	7		
		7	3			4	9	8
9			1		4	3		

Solutions on Page: 107

HARD - 121

				8	3	5		1
1				2	9			
5	7			1	4	3		
2	1		4	6	5	9		
8				3			4	
	4	5					3	
4		2		5	8		1	3
					6		2	9
9	8	1		7		6		

HARD - 122

		6	4		1			
3				5	9			
8						5	2	1
				9	5	3	4	
		9			4		5	7
6	4	5	3				1	
	6	1	5	4		2		3
5	7		8	6	2			4
	2		9				7	

HARD - 123

8		3	5		1			6
	1	4		6				
5			2		8	1	9	3
		2	1	3				
1		5	8	7	4	2	6	
4	7			5				
					7	4		
3		8		2				7
		1		8		9	3	2

HARD - 124

6	3		8	9		4		2
						1		
	5	2	7	4				
	6		2	7	8			
			5				2	8
		8			4	9	6	5
1	4				3		8	7
8	9	7		5		2	4	
	2			8			9	1

HARD - 125

		3	2			8		7
2	7					3		5
4		9				1		2
3	6				4		8	1
7	4			8	2	9	5	6
	9	2		6				4
1	3		4					
		7				4		
		4	5	7	3	6		

HARD - 126

3			7		6			1
	1				4	3		
5				1			6	9
9			3		8			7
		1	6		5	9	2	
6			9	2	1			4
1	3		4				7	6
	8				9	1		
2	6		1	3		4		

HARD - 127

	3				4		5	
		8	7			6	1	9
6	1	5			2	7	3	
	7	4		1				
5		1		9			4	7
8		3			7	1	9	
4		7			6			
	8		5				7	2
	9				1	4		5

HARD - 128

1	2			3		9	5	
			9		4			
8		3		6	5			
5		2			8		3	
	6			5		7	1	
7	3		6	1	9	2	4	
3						8	9	1
		7			1		6	4
9		1	4		6	3	2	

HARD - 129

9	1	4			7		5	3
8		7	1				6	
	5		9		4	7	8	
4	7	5		1	6			8
				5	8		4	
	8	6						
			4					2
		9			3	8	1	6
	3		8	6		9	7	

HARD - 130

3		9		7				
	6	2	3	5		8		7
5		4	9	8		3		6
	4			6				5
	9	1	7	3			4	2
			8		9			
	5		6		3	2		
2	3	7	4			5		1
				2		4		

HARD - 131

7	8	2		3		4		
9	1							
6		3		5		2	1	9
5	6	4		8		7		1
	2				1	5		
3		1	7					
2		8						
	7			9	2	8	4	3
4			6		8	1	2	

HARD - 132

			1	3		2		9
7	9				2	1		
1		2		9	4			3
				2			3	
9	4	5			8			2
3				4				5
6				5	3		8	
4	5	8			9		2	
		9	4	8		7	5	1

Solutions on Page: 108

HARD - 133

4			7			1		
7	1			8	5	3		
			4	2				
8	9				2	5	7	1
5				1	4	6	8	
6	2			5	7	4	9	3
			5	7		9		
9				4			3	8
	8	6		9				

HARD - 134

2			7	1	4			
9	7				8		6	
		4			6		3	
	2	8		3	9			
4				8	1		5	3
		5	6	4		2	8	
1			8	9	3		7	
				6	5	3		1
5	6		1		2			

HARD - 135

7		5				9	8	
6		8	4	3	9		2	
2		9	7					3
5					2	4		
					3			8
	2	6	9			3	5	
	5	7		4		6		2
4		1	3				7	
3			5	9	7			1

HARD - 136

	8			7	3			6
7	3	5		6		8		
	9					3	4	
	5	7	8		6		9	
3	1	8		9		6		
6	4	9	1	5	2	7	3	
	2	1	6			9	8	3
9	6		5	2	8	1		
	7		3	1			6	5

HARD - 137

8	6		7	3		1	4	5
	5	9		2			3	
7		4		6	5			9
2		1				5		
	7			8	6			
		3		7			2	8
	9	7	4			6		2
	2	6	3		7		9	
	1	8	6	9				3

HARD - 138

		3	6					
	6	4		2		3	9	
2		9	1		3	7	6	
5		2	3	6	9		4	
						5	3	
9						6		
3			5	4		2		6
6	8	5			2	4	7	
4					6	9	5	

HARD - 139

	3	2		1		9		4
	1	5	8	4		3		
9				3				
	8		5	2			4	7
		4	1			2		8
	2		9	8		5		1
4	9	8	3			7		2
		3	4		7		1	
5				9	8			

HARD - 140

1					4	6	8	
	4							1
		8	1			3	2	4
5	1	4	8	3		2		6
	2		6	1	9			
			5	4		1	3	
		1	9	2		8		
7		2	4		3		1	9
	5	9		8				

HARD - 141

9			3			6	1	
3	6				9		7	4
		8		2	1	5		3
1		7	9		2	4	5	
	9							
5		4					3	9
7	4		2		6			
2				3	8	7		5
		3		1	5			2

HARD - 142

7		2		8				
		6	4	2				3
	9		5			6	4	2
1	6			5	2	3		7
		8	7	3		9		
					8	5	2	4
4			3	9	5	7		8
			8	4				
9			2	1			3	

HARD - 143

		9				3	8	7
5	3	8		7	1		4	6
6				8				
	8	6				1		
9	5				2	6		8
2			7			9		
	7		5	3		8	2	9
3	6			4	9	7		5
8		5						3

HARD - 144

		3			6	9		1
		9		5			6	
6	8			3			4	2
	2	5			1	8	9	
8	3				7			
1	9	4				7		
	6		9	7		1	5	
			4	1	3			
3	1		8	6	5			9

Solutions on Page: 108

HARD - 145

	7	8	9	1		5		
		5			2		9	7
			3					4
					8	7	6	
		6		7		4	8	
7							2	5
9			8	4	3	2		6
6		2	7	9		3	5	8
	3	7		2				9

HARD - 146

		3		6		7	1	
	7	1						
	4	2	7				6	
	2	9	1			8	4	6
			6	7				
1		8			9	5		
		5		8	7	6	3	4
4		6	9	1	3			7
		7	5				8	1

HARD - 147

	5				9	4	3	
		9	5				2	
7							8	
9			7	1	2	3		
	7			3	8		9	
1	8		9			2		
6	3	8	4		5			
	1			7		8		9
	9	7	8	6	1		4	3

HARD - 148

		2	3		6			5
8		7			5	9		
		3	9	4			6	2
7	8					3	5	
	2			8				9
3			5	6		2	7	
			6		2			3
6	9		8	3	7	4		
2			1	5				7

HARD - 149

		4			9	6		
6	3		8		2	9		1
	9		1			8	5	
	8			9	7			
	7		4	2			3	
	2	1	3	5	8			
3	6	8	7				9	2
1	4			8		3		5
2	5			4			8	

HARD - 150

9	5			3			1	
6				4	1	3	5	7
1		4			7			
2		5	3		9		7	
			1					4
7	6	1	5	2			9	
	9					7	3	1
	7		8		2		4	9
4			7			2		

HARD - 151

8	1	4	6	7	9	5	2	
	9			4	2		8	1
		7			1			
1	5		9		4			8
2					5	3		
		3			8		1	
7				1	6	4		
	6				7	8		2
	8	9			3		6	

HARD - 152

	4	1	9	7			5	
2	5			8	6		3	1
			3	5			7	4
8		4	6				2	
1				4		3		9
			2			4	8	5
		9	1	6			4	
4		2			8			
	1	3			4		9	

HARD - 153

5		8					1	
9			8		1		4	
1		2	3				9	8
3		6			9		8	4
		5	6	4				
		1	2			9		6
	1			2	4			9
		4	9		6	7		
6		9		8	3		2	1

HARD - 154

							6	7
6	7	1			5	8		
5	4	2					9	1
3			1	7	6	2		9
		8		9	4			
9		7	5	2				6
7		3	4	6	9	1	5	2
4	9				2			3
				5	3		4	

HARD - 155

6	4	9			7			
		1						9
			1	9	4		6	3
4	8	5			3	6		
1		7	4			5		8
		3	5	6	8	7		4
5	1	6		4	2			
3								6
8				7	6		4	5

HARD - 156

7	1	4	2	8	3	9	5	
3		8	9					1
			1	7	4		3	
8						5	1	
	5							
9	4	3					8	7
4	3			2				5
	8	7	3	5	9	1	2	
2		5	8		1			

Solutions on Page: 108 & 109

HARD - 157

						7		6
	6	1		5				
			8		2			
	9	3	5			6	2	
	5			8	6	4	9	
6					9		8	
9	1	4	6				7	
	7		2			9	1	8
5		8	7	9	1	3	6	4

HARD - 158

2		3	5	1	8			
	5				6			8
	1				7		4	
		1		5	3			9
4					2	1	6	7
		9		6	4	8	3	
	9		6				8	
	7	6		8	1			4
3	8		2		9	6		

HARD - 159

	7	3	8		4		2	6
2			3		7			4
				6	2		7	9
	4	8						
1		5	7	8		4		
			2	4	5			
	9	6		2	3	7		
3	1	2	4		9	6		
	5			1	8			

HARD - 160

		8	9			6		5
			4			9		
2					3	8	4	
		6	8		5		7	
7					6	5	9	
		5			4	1		6
	1				2	7		
8	2	7	1	6	9	4	5	
	5			3	8	2		9

HARD - 161

					8			
5		8	4		2			9
9	3		7	5				
					7	4		1
4				1	6	9		3
	1	9		3	4	5	7	8
1			8	4		6	3	7
		6				8		
	5		6			1	4	2

HARD - 162

						7		8
2	7	3	4	8		1		
8		6		7	1	2	3	4
	6			2	5		8	3
3		1					2	
5				9		6	7	
	2					8	1	7
	5	9				3		
			7	4		5	9	6

HARD - 163

4			8	9	6	2	7	1
		8	1	5		3		
		6	3	2	7	8	5	4
5	2							6
	8				9	4	1	
	1							7
8			9					
9	3					6	2	5
7	4	5		3				

HARD - 164

	8		3	7				
9	3	4			1	8		
		6		9		3		
				2	3	9		5
3	6	1					4	7
5				7	6	8		
8	1	3	7		6	4	2	9
6				4			3	
4	2	7						1

HARD - 165

	6	9		1		2	8	
7	5	1	4	2		3		
8				5				7
	4		9				1	
		8	2			6		3
2	1		7	3		4	9	8
		4	1			9	2	
					2	5		
3					4		7	1

HARD - 166

			2	8	4	1		
6			5	9				3
8	5	2		6	3	4	7	9
					9	7		
		8	6			3		
				4	5		1	
2		1	3				4	
	8					5	3	1
3		5	4			2	8	6

HARD - 167

	5	2	3			1	4	6
	7			6			3	
3		6				9		
6	3	8	5			2	9	
4	2	5	1					
				3	2	4	5	
5			9			7	6	
2	6				1	3		9
	9			8	6	5	1	

HARD - 168

9		5	6	4	8		3	
8				9	7		6	
7		2		1	3	4		
3		1		2				6
		8			4	2		
		6			1		5	9
	8							
			7	3		6	8	4
	3	7	4	8	9	5		

Solutions on Page: 109

HARD - 169

	2			1	6	8		9
7	3		8			1	6	4
1					4	5	7	2
6	9	3	1		5	2	8	7
	1			8	7		9	3
4	8							5
		6	4					
	4			3				
			5	6	9	3	4	

HARD - 170

		8	2					9
		6					2	4
2	3			5		1		7
8		1			5	2	9	3
		3					1	5
9				6	1			8
6		9	5	8		7	3	
	5						8	
	8	7		2	9	5		6

HARD - 171

			9	6	1			
		9			2	4		
			4	7	8		6	5
8	2		7		6	5	9	3
5	6					1		7
	1	7	2		3	6		4
7		6						1
		8	6		4		7	9
3							5	

HARD - 172

4	8						9	
	9	5		8	3	1	6	2
1	6	2			5	3	8	4
	1		6				5	
	3	7	2			6		
							7	
				2	6	8	4	1
5			8				3	
6	4	8			9	7		

HARD - 173

			3				7	6
3	2				4	5		
1		7		2		4	8	
		5				3	9	1
7			2	1	9			5
		1				7	2	8
	4			6			5	7
	7		5	4		1		2
		3		9	2		6	4

HARD - 174

6	2							
		5	8		9		6	
	9						2	
			1		8	7	5	6
		6		7	5			
5	7	8	6				1	2
	6					3	4	5
4	8	2	7	5		6	9	1
				1	6	2		7

HARD - 175

9				8				3
	7	3	4	9		5		2
8			3			9	4	6
4	1	2	7	5	9		3	8
3					4			5
		8	2					
				3				
7		1	9	2			6	4
	3				5	8		7

HARD - 176

	9	2		7		4	8	
	7			4				
5	4	6			3	9		
7		8	1	2		6	5	
9	1	4				7		
2	6			3	7		9	
4		7				8		
1		9		6	4			3
6	8			1	2	5		

HARD - 177

8			2	6			4	
	2	6	8					
	9				1	8	6	2
		3	5	8	2			1
	5			1	7		3	8
			4			2	5	6
		2	3	5			1	7
	7						8	
6	4		9			3	2	

HARD - 178

		8	1					
			9	3				6
		2				7		5
				1	8	4		
	8		7				5	
5	1	9		6				
9	5	1			7	6	3	8
8	2	4	3	9	6		7	1
3	6	7				9	4	

HARD - 179

9								7
1		7	2				3	6
	4	2				8	1	5
6	7	4	3	1	2	5	9	8
			6		8			3
5		3		7		6	2	
					3			9
		9	5	2		3		4
				9			6	

HARD - 180

		1	9			2		6
					6	1		3
	5	6		2	3		7	
2		9		4	5	3	8	
	3			1		5		7
	7	5		8		9		4
5		3		6		7		
6					7		4	9
4			2					5

Solutions on Page: 109

HARD - 181

			8	5				4
7		2	6				3	5
	5	4		7	2			
		6	1				4	
9		1	2			3	5	7
8			4	9	7			6
			7	8		5		
		8	3	4			6	1
6				2	1			3

HARD - 182

				9	4	7	8	
								2
		4			3	1	6	
2	8	9			7			6
	7	6	9	8				4
4	5			2	1	8	7	
	1	5	7	4			2	
3	9	2	5	1	8	6		7
8	4							

HARD - 183

		7				9	1	
9	1			6		8		
8					4	3		6
4	9					7	8	3
		8		4	2		6	
6		3		7	9		4	
5			4		1	2	9	
	3	1		2	6		5	
2	4	9			7	6	3	

HARD - 184

6						8	9	2
		5		7			6	4
8	1	9	4	2	6		3	7
1		8					7	5
		6					2	
9	5				4		8	
		2					5	8
		1	5		3		4	9
5				4	2		1	

HARD - 185

	8		4	9	1	7		
9		4	2	7		8		5
3	2		6	5	8		1	
6		1						8
	7	8	1					6
4				6		2		1
	4			8		1		
1	9		5	2	4		8	
			7			3	5	4

HARD - 186

	8			2				
7	5		3		8			
9	4			1	5		2	3
5				3	7	2		1
		4		5		3	8	9
1						7	4	5
		5			2	1	9	
	6			8	3			
2		7		9		6		8

HARD - 187

			1				2	
		3		7	4		1	6
	8	6	3		9			7
5	9	7	8	3	6		4	2
2				5	1		7	
6				4				9
	5		7	9				4
3				8		7	9	
4		9				3		

HARD - 188

7	5			9			1	6
4	9		6			2		3
	6	2	4	5			7	9
	7			3		9		
	4		7		6		8	
3	8		2					
6					5	1	3	
			3	2	7		9	4
9	3							8

HARD - 189

	4		3		1	6	7	5
5	3			8	6	9		4
			5	9	4		3	
1	8	5						
	2		8			7		
6		9	2	4	3	1		
			1	6		4		9
7					8			3
8		4					6	7

HARD - 190

7						4	5	2
	8					9		
		5	2				6	7
	9	6	4	5	3	2		
	3	8	7		2			
4			9	6	8	1	3	5
			8	4		3		
8		3	6	2		7		1
	2	7						

HARD - 191

		2					9	8
	6		2	7			1	4
4			6	8	1		3	2
5		6	7		8	9	4	
	4		3			8	2	
				9	6			
	1	9	8		7			
6	5						8	
7	8			6	4		5	

HARD - 192

	8	2		5				1
7		6				3		4
3			8		4			7
5	6	9	1	4		8		3
1	2			9	8	7	4	6
8		7		3				5
6					9	5	7	
			4	8				
9		8	3					

Solutions on Page: 110

HARD - 193

8	9		7	1			2	
	5		2	9	4	1		
4		1				6		
1	8	5		2				
	7	4		8				2
6		2				8	1	
3	1			4				6
	4	9		6		2		
	6			5	7	9		1

HARD - 194

			5	9		1	3	
2	6		3			5		
	5		2	1			4	
			7	2		4		3
	7			5	3			
3	1	2	6	8	4		5	7
			8	6				9
7	2	8		3		6		
	3						2	4

HARD - 195

		6	9		7		3	
9							1	
								7
1		9	5				7	6
5	2		8	7	6			
		7			9	5	8	2
3				9	5	7	2	
	4			3	1	8		9
7	9			6	8	3	4	

HARD - 196

	5			6				8
8		2	7		5	1		6
1	3				8	4		5
	8		1	3	7		5	
5		1	6	8		7		
6			5		4	2	8	
4	1			7				9
	2							7
	6				9	5		

HARD - 197

				7	4			
4		9	8				2	3
					6	5		4
	4	6		8	9	3	7	2
		3	1	6		4	5	
9	5							8
		5		1				7
2	9				8			
3		4	6	2	5	9	8	

HARD - 198

1		8				3	5	
7		5		1	9	8		6
				8		9		
	5	2			4			9
	1	6			8		2	
	4	7	5	2	1	6		
	8		1					
	7			9		2	8	
	2	3		6	7		9	1

HARD - 199

		5	8	2				
						1		
				7				2
6		2		9		5	3	8
	9			8			4	6
	8	4		6	3	7		
3	1	8					6	
2	6	7	4	1	8	3	9	
4		9				8	2	1

HARD - 200

			9					5
4	9	3			5	2		
	5						4	6
		6	8	3				
					6		3	
8			2	1	4		6	7
9	8			5		6	1	3
7	6	5					9	2
3	1	4	6	9	2	7		

HARD - 201

2	1		9		7			
			2		8		6	7
7			6	5			9	
9				6	2		1	5
	2	5		8				6
3			1	7		9		2
4	9			2		5		
8		2	5				3	
		1	8		3		2	

HARD - 202

				5	8		2	1
2	6	5		4			8	7
		8				4	6	5
5	4		3			2	9	6
	8			1				3
3		9			6			4
9			8		4		7	2
	2			6				
	1	4			5			8

HARD - 203

6		1			4		8	
	8						5	
			9	8	7		4	1
1						8		
	4	8	7	3				6
3	9					1	7	4
7	6	3	2	4	9		1	
9	1				5	4		
	5	4		6			2	

HARD - 204

				6		1	7	8
7			4		8	3	9	
						4	5	2
	7		8		6	9		5
1		5						3
8	9	6	2	3	5		4	1
9						6	8	
		2	3	8	9			
		7			1		3	

Solutions on Page: 110

VERY HARD - 1

		5		1			2	
6		9		3	2			
	2	3	4	8				6
		7				8	1	9
	8		3	2				
			8					
					9	6		4
7		6	1	4	8			5
		8	2			7		

VERY HARD - 2

1					7		5	4
		7		3		9		6
2	6		8	9			1	
				1	4			5
4			3	2			8	
			7		8	4		
			5				4	1
8	5					7		
	4	2		7				

VERY HARD - 3

1	4							
		8	2					
	6				8			3
	5	1	6		4			
2	9			7			8	
6			9			5	1	7
		5			7	9	6	4
3	7	6		4		8		
			8					1

VERY HARD - 4

	9	3				2		5
5	8				7			3
7		4	5	9	3			8
4	3	6			1	5		9
8	7		6			3		2
2		5		3				
		8				7		1
3	6		1					4
1			4	7				6

VERY HARD - 5

	2		8	1	6			9
1					3	8	4	
					7			1
		3		9	5	6	1	
		4		8				
5	1	9		3				
		1			4			8
	4			7	8		5	
			1	5				3

VERY HARD - 6

		3	4				8	7
4		7	8		5	9	6	
				7	6			
				4			3	5
					8	4	1	
5	6	4			3			
3			5				4	
	5		2			7		1
	4	1			9			

VERY HARD - 7

1	7	4		5			2	
5				6	2	9		
6	9	2	3		7			
			1	7	5		4	2
		5				8		6
				4		5		
						1		4
3			2					5
8		7					6	

VERY HARD - 8

8	9		7				4	2
	2	4		1	5	6	7	9
7	6	5				3		1
		2			7			8
		6		9				
			3				9	
				3	6	8	5	
				7	4			
6		3					1	

VERY HARD - 9

7	3		1	4			8	
	8	4			3		7	
5		9		7	8			3
	4	5		3		6	9	
9			8	6	7		4	
				2	4			
1					6			
4		8	7	1	5			2

VERY HARD - 10

		4		9	5		3	6
	9			6		2		
	6					4		9
9			7		3		6	
	5	3	6					7
7					1			
	8			5	2	6		3
				7		9	5	
5						8		1

VERY HARD - 11

5		2				7		3
	9	1		4	7		2	
8		7			6	4	9	
	2	8			5		7	9
	5			8				2
3	7				8		5	
	6	4	7					
2		5				1		

VERY HARD - 12

			8	1		6		
		9				1	3	4
				3	4		5	
1			2			9	7	
	8		3					
3	9		1				6	8
	1						9	
5	3		9					2
		2	4	8	5	3		

Solutions on Page: 111

VERY HARD - 13

3				4	7		1	
8						7	2	
	2	7	9				3	
	7	2	3		4	1	6	
5		3				4		
4				9		3	5	
	3		8			5		
	9							
		1		2	3		4	

VERY HARD - 14

3			5			7	4	
9	8			7				
	5			3	4		2	
4		7				1	8	
1	6	9						2
	3		7			9		
	7	8		4			1	
	9			6				
2	4	3				6		

VERY HARD - 15

5			8					2
	1	3	2	6	7		4	5
	9							
		9	5	8		2	6	
4		2				8		
1				3	2	5	9	
				5			3	
7	4					1		
9		1		2				6

VERY HARD - 16

			7	8				4
	9	2				8		
8			2				5	
		6	9	2			3	
7	4		1		6			9
	2	9		3	7			
			3	7		5	9	
				1			6	2
			4				7	1

VERY HARD - 17

	7	6	8	5	3	1		
			7				8	4
	1				6		5	7
9				4	7			
2		4	3		8		1	
	6							
			6				3	5
6	3		9					
7			5				4	6

VERY HARD - 18

	1	8				2	4	
	2		5	7			1	
						5		
1		6	2				8	4
8				3	7			
2	9		4		6		3	
	6				5			
7	4							6
		5	7	6	1		9	

VERY HARD - 19

	9						4	
		7	9	4			2	6
	4	5			3			
		2	8		7			
			1	6			5	
6				3	4			9
5		1	4	7				2
7				2			8	
8		4		9		1	3	

VERY HARD - 20

		4				6		
	7	9	8		4			
1		6	2	9			3	8
2	3	1	5			9		7
		7		4	9		5	
		5					1	6
5		2			8	7	6	
		8		5	2			
4		3		7			8	5

VERY HARD - 21

4				9	7			
		2	5	4	1		8	
8			2		6		9	5
					2			6
7	1	6	9		8	2	3	4
2	3			7	4		5	9
3			7		9		1	
		5						8
	8							

VERY HARD - 22

	6	1	5			9		
4		9		6	2			
7								2
6			8		9	3		
							2	5
	4			7		6		
3	1	8						4
5	2		7		1		9	
9			4			5	3	

VERY HARD - 23

4	6			2				
2	3	7	6		1			8
	9					2	4	
			9	8	2	6	5	
	8		3					
			5	1	7			9
		4		9		3	8	
	1			5				2
	5	3			8	4	6	

VERY HARD - 24

5						3	2	
	9				1		4	
1		3	4		7			9
9	5				6	1	7	
								3
		8	9		4		5	
3		5			8		1	
	1	9			5	6		7
		4		7		8		

Solutions on Page: 111

VERY HARD - 25

		5	9		6			
8			7					9
			4			5	7	1
	2	4	1			7		
1		8						3
5	6	9				2	1	
		1				8		
2			3	1				6
	3	6	5			1	4	2

VERY HARD - 26

	3				7	9	1	
		9	3	6	4	5	8	2
	5	2		9	1	3		
6		4						9
		3				2		6
7					6	8		
	4	7						
			2		9			8
						4	2	

VERY HARD - 27

		8	7					9
4		3					7	8
	7			3	8		4	
		5	9				8	3
	9	1		8	7		6	4
8				4	3		1	
9		6				8		
	3			5				1
		2		7	6		9	5

VERY HARD - 28

8		2	7					6
			2	1				5
				8	3	4		
9			3				5	8
3		5			8			4
6	2				4			
2			8				4	
		4		3	1			2
		1	4		6	3		

VERY HARD - 29

	7		5		9	6		
						1	9	
			6					4
	9				3			1
5	6			4		8	3	2
3								6
	4	6	7	1			5	
	1	9		5				
7	5		2		6		1	

VERY HARD - 30

	7		4				8	1
	6		8		9			
				7		2		
9		5		1	4		2	
	2	3			8	4		7
		7	6		2			
			3	4			7	9
	4	9	2			1		3
	3	6		5	7			

VERY HARD - 31

		1	4	7	5	3	8	
	8		6					9
	5				2	6		1
		5		3			7	
3		6					9	
8					9	4		3
	7					2		
	6	9						7
		4	7			9		5

VERY HARD - 32

		9			8		5	
							3	6
3			2		9	8	4	
4		7		9	1	3		
5		3					9	
		8	6			7		4
		4			6		7	
	1		3	4				9
	5			8		4		

VERY HARD - 33

					4	9		1
	4		8	9				7
1		9						3
7	9		5	1			6	8
				3	6	1	9	
6	1	2		8	9			
		4	1		3		2	9
						7	1	
			9			8		

VERY HARD - 34

	6				3			
			2				1	
1	9		6	4		5	7	3
3					7		6	
	7	6	4	5	1	9		
9	1							
5	8	1	7	3	4	6		9
							5	1
		7	1					

VERY HARD - 35

		4	9				2	
		3		2	8	9	7	
8		2					4	
		8			5		3	
	7		3	4	6	8	1	
3					2	6	9	
			1			3		
	3				9	4		
					4	7	6	

VERY HARD - 36

	9			1			8	
1					8			
	2		3	7	4			1
		2	1			5	6	9
3	7				9			
				4		3	1	
		5		2				
	1	7			3		4	6
8		4			6		5	

Solutions on Page: 111 & 112

VERY HARD - 37

8		4		9		3		
	6	3	4	1		7		
7	2	9		5				
3						1	5	
9			2					7
6	5	8	1				2	
2	8							6
					7			
			5	8	1	2		

VERY HARD - 38

3	6				8	7	2	
		4	1	6	2	5	9	3
2	9			7		8	6	4
5			4			6	3	
8						4	5	7
4	7				5			
		8		2	4			
	2		7		9			
			6				8	

VERY HARD - 39

5				6	1	9	4	
		2	3		7		1	
	1			9	4	2	7	
					3	5		
7			9			1		
1	5	8						
					9	4		6
		5	6					
9	7		4	3	5			

VERY HARD - 40

6	5				1			9
8	1					3	6	
			5		8	4		
								3
2	6					7	5	
		8	7		5			
5		9	6				4	7
	4		9			6	3	8
	8	6	3		4		9	

VERY HARD - 41

	6		7	4	3			5
	1	5	6		2		3	
7	3	9		8		2	4	
			4					
9		7			8			
				7	1	5	9	
	9	3		1	7		6	2
		4						
	8	6	9			1	7	

VERY HARD - 42

6		7				2	1	
4		5				3	7	
	1		7		4	6		
	6			9				
	5			4	7		2	
8	7			1				
9	8		4			1	6	
5	2	1			3			
7	4	6		5			3	

VERY HARD - 43

		2	5		1			
	6						4	5
7	8			3				
	1	9	4	5				6
5	2	3						
				2		8		3
				4	2		1	
	3			9	8		2	4
	4		3			6	8	

VERY HARD - 44

5	6					7		
2	1			4				6
7			2					4
	3	2	4		1	9	7	
	5			3				
1	7	6	5	8			2	
						2		
			9	2	3	8	5	
				7				1

VERY HARD - 45

	3				6		8	
	5	7		3		6	9	
8	6	9				2	3	
4		6	1			3		7
7	2	5						
3				8	2	4		
			5			9		
5	7		3	6	9			
9		8					6	

VERY HARD - 46

					6	3		
	6	8					5	
7	1	3						
	2		7		4			9
8	4	6	1			2		
		9			5	4	3	
	5		3	8				
	3				7			8
		7		1		5	6	3

VERY HARD - 47

	4	9		5	8			2
7	8	2				1		9
5		3	9	2				8
9			6	3		8		
	2		1					
		6			4	9		7
								6
			8	7	6	2		5
						7		1

VERY HARD - 48

			9	3	2			
4			8			9	7	2
	9			7	6	5	3	
				8			5	
7								6
6	4			2	9	7		
2			7	4				
8			2					5
		4		6	8	2		

Solutions on Page: 112

VERY HARD - 49

		9		5	3			
1		4		9	2			
		7						8
				4		3		
	2			3				7
	7	1	5	8				
	5				4	8	2	9
9		8	3	2			1	
		2	9	6		5		

VERY HARD - 50

	8		7		6		2	
1		9			3	6		
6			1		2		9	
	6			1	8		3	
9					7	1	6	
								2
	1	2			9		5	
4						7	8	
		5	3	8			1	

VERY HARD - 51

	4		3		9			7
9	2	6	4	7	5			8
3								
		4	1					
	9		2	5				3
				3		1	8	
		8				4		
	5		7		3		6	
		3		9	1		7	2

VERY HARD - 52

	2	7		6		9		
8				9	5	7	1	
1						3		2
6				8	1		9	
						8		
5	8	1		3	4	6		
2	1							9
		4	3		9			5
	9				8	2		

VERY HARD - 53

	1		7			2		
7	2		6		1		9	5
9				5				8
			8		3		6	
					7			
	5			1		7		
8	3							1
	9	2	4				5	
4		5	1	3		8		7

VERY HARD - 54

5					3			
1		7	5			2	6	
2		4			9	1		
		2		5			8	
	4				7			
		1	3				9	
		3	4		6	7		
9	7	5	2				4	8
	1	6		7				

VERY HARD - 55

			4					
	5	2			1	8		9
9		8						6
2			9					5
8	9			3				
4			1			2	9	7
6		3		1	2			
5		4			3	6		
7			8	4				3

VERY HARD - 56

6				7			4	5
9	4			1				
2	5	7			4	8		
	8			9	6	4		3
3			7		1			
	9		3			1	7	6
				5	3	6		8
1	6			8				7
	3					2	9	

VERY HARD - 57

8			5	3			2	
	2			1	6			8
				8			9	
			3			2	1	7
3	4			5		9		
1							3	4
	7	9	1	2				
	3	6		9				2
		1					7	9

VERY HARD - 58

	1	5		4		8	9	
2	8		6			4	3	
4		3			5		7	
1	5	4		8				
7							4	
		6				2		
	4		1		7			
	6			3			8	7
		2		9	8			

VERY HARD - 59

8	7			3	4		6	
		4		6	9		5	8
	1						3	2
2				4			9	
7		3		1				6
	8		6			5	1	3
	4			9				
			7					4
5	9	2						

VERY HARD - 60

			9	8	3		5	
5					2	3	9	
3		9						8
8		3	2	5	7			9
			1				2	
	2		8	6			3	1
	6			7	5			3
		5					1	
		8				7	6	5

Solutions on Page: 112

VERY HARD - 61

1								
8	7	4	2	1		6		3
6	2	5					4	9
			4			7		
			8			5		
5							3	
	9			4		3	8	5
2	6		9		3		7	
4	5	3	1	8		9		

VERY HARD - 62

	6	8	4	7			1	
4				1	3	6	5	2
	1	3			9			
						1	3	
3		5						
							9	
9		7	2	6			4	
		4			8	2		
6				9		5	8	7

VERY HARD - 63

					8	7	9	
7		5					3	
	8						1	6
6	4			1		8	7	3
5				4	7	1		2
	7		2			9		
	9	3		8				
			5			3	8	
		8		3	6			9

VERY HARD - 64

			9		1			
	3				8			2
2		8		3		5		
			7	9		1	5	
1		9		4			6	8
4	6	5	1	8		7	3	9
				1		3		5
	5		2			8		
7		1						

VERY HARD - 65

	3		2			7	5	
1	9			5			4	
	5	7				1		
	6					8		
			7	9				4
	4				6	5		9
	2					6	3	1
		8	4					5
		6		3	2		8	7

VERY HARD - 66

2	8	3		4	5		6	7
4				6	7		1	
		6					4	2
	4			1				
				9		7		1
	1	5	7		2			
				7	3		9	
				5	8		7	
7		4	9		1			5

VERY HARD - 67

	3	4		5				1
5				1	9		4	
7				4			8	5
	4		1	7		9	5	
9				2				3
3		1						
8	7		4	6			2	
	9			3	2		7	
		2	5					4

VERY HARD - 68

	6			9	1	4	2	
		5	6	4			3	9
	9					8	6	
8	2		1			5		
		6	5	7			9	
5		7				3		
9		4		2	7	6		
			9					
	3			8				

VERY HARD - 69

	3				1			
	7			6		9		2
6		2	7		8			4
8		6	1	7				3
7	4			2	3			
		1	4	5				
		3			7		4	
5				3				6
4	8							5

VERY HARD - 70

		2				5	8	
	5		4		6		3	
8		6	5				4	
		8			5	6		
4	7	5		9		3		
9				2		7	5	
5			2			8		
		4			7			3
6			3				9	5

VERY HARD - 71

	9		5			7	6	
4	5	3	9	6			2	1
			2		1		5	3
	2	8		1	4		7	
9			6				8	
	6						1	
			1			2	3	
1								7
6				5		1		8

VERY HARD - 72

	5	3						
4			1	7	6		3	
6				5		9	8	
	4	7	8			3	9	
	2				1	7		
					7	6		5
3		5	6			4		
7	9							
				1	9	2		3

Solutions on Page: 113

VERY HARD - 73

				3				1
		8				5		2
					2		9	3
6		7	1					5
4				2	7			
9			5	6		2	3	7
2		9		1	8		7	
1		3				6		
8			3			1		

VERY HARD - 74

		6		3	7			1
8	3				1		2	
		7			4	5		
	7			2	6	8		
6				8	3	7	5	4
	8	3	7	1		9	6	
	2		1	5			4	
		1			8			5
3			6		2			

VERY HARD - 75

4		8		9				
	7						8	2
			8			3	4	
5	8		9			2		
	3		7	6			1	
			1	2		4		3
1	4		2		9			6
8	9	7	6	1				
			5	4		8		1

VERY HARD - 76

								6
	1	9	6	2			4	
6			5	1		2		
	4		1		2	7	6	
8				4			1	
	7			8	6		9	5
1		2	4		9	8		
	9	5	8				2	4
							7	

VERY HARD - 77

		2			5		3	
4	9	7		6	3			
	3	6	2					8
9		1	7		8	5		
							2	9
3	8	5	6			4	7	
	6				4		8	
2			9		7			
		8	5			3		

VERY HARD - 78

	8	6	5					
2		1	7				5	6
5	7		4	6	3	1		
		2		7			4	
			3	1	5	2		
	5				6	3	8	
8	9			3			1	5
							3	
7								

VERY HARD - 79

	6	7	9				1	2
		1				7		6
	9	5					3	
		4				6		1
		3			6	4		
8	1	6		4	5			
		8		2			5	3
					4	2		8
1	5	2						

VERY HARD - 80

2	4			3			8	5
7	8	1		5			3	
		9	6	7		4		1
8				6			5	4
	7		8				1	
		5	3	4		2		
4						1		
		8					6	
	9	7	2			8		

VERY HARD - 81

	4		8		9	6		3
9		6	1	7				
			2	6	3	1	4	9
3	1		6	9				4
	8	5	3		2			
						8		2
	6		7	2		3		
	7		4	8			9	
					1	4		7

VERY HARD - 82

9				2				5
			3	8	9	4		
					4		2	3
	8			5		1	3	
	6							7
1	7		6					4
			9		5	3		
5	2	3	4					
7		1	2	3	6			

VERY HARD - 83

5	4		9	3	1		7	
					8			
9	7		6					
		4		6				
8	2	7			3		9	
6					4	5	2	
	9			8	6	1	5	7
1						8	4	
7			4				3	

VERY HARD - 84

5			1			4		2
4		7		9			5	6
1		3		4	6			
8			7					4
		1				6	7	
					4			3
		4	9	6		5		
9					7		6	
6			3	1	5			

Solutions on Page: 113

VERY HARD - 85

6			4		9	2		8
4		9		1				5
	2	3				6		
					3		8	
		8	9					1
			8		7	5	3	
2		5	7	4			9	
8					1			7
			3			1	6	

VERY HARD - 86

5		3	7		4			6
				5	2	3		7
	7	2		3	6	1		
2			1	7		8		
	5			6	3			
	1	6			5			
9			5		7			3
					8		7	2
	4	7	6		1		8	

VERY HARD - 87

8	4		2		6			5
			3		7			4
		5	4		8	2		
	1	7			4	3		9
	9			3		8	6	
		3					5	1
	5				1	6		
4	7	2		8				
	6					9		2

VERY HARD - 88

7	4	5				3	2	1
				1	7	5		4
6	1			2	4	7	9	8
	8		4	5		2		
2							1	5
		7		9	1			
			7	8				6
8	6	4						
			6		5			

VERY HARD - 89

4							3	8
	1	5				2		
7	8		2	6		5		
			5	3			6	
5			4			9	8	
		9	8		6	3		7
			6				2	3
		7						
	2	4	7		8		9	

VERY HARD - 90

6	4	5	8		7		3	
			1		5		6	7
	9		4		3			8
9	2	4	3			6		5
		6	9					
3	5	8						
	8	3				7		4
		9	2	7		3		
	6		5				1	9

VERY HARD - 91

5		3						
	7		9				3	4
	2		8	3		1		
2			1		3		4	9
7	8	9					2	
	3			7	9		5	
		7	4		5		6	
	5	2						7
8			3	1				

VERY HARD - 92

4			3		7		8	9
9				8	6			
	6	8	4		9			3
				3				
1		4		6			3	8
						4	9	
3				4	8	9	5	7
			2		5		6	
		7						1

VERY HARD - 93

	4			6				
	6					4	5	
		9		4			8	3
	5	1				2	4	
4	8		5			9		
6		3	8	2			1	
9	3	5				1		
	1			5	2		9	
		4	3					

VERY HARD - 94

	6	1		2				
							1	8
			9	3	2		4	5
	4	8	6	5				
5		3		1		6		
3				4		9		1
		4		8	1		6	
	8	2		9	6		5	7

VERY HARD - 95

				1				5
	3	7					1	
1			7					
			8	9	1			
		1		4			3	7
			3	7		6	8	1
4	7			8		5		
	8	5					9	4
3	1		4	5			6	

VERY HARD - 96

9		8	4	3		7	6	5
		1				4		8
	4					3		9
5	3	7		2				4
6	9	2						
				4		9		
1				6	2	8		
	5			1	9	2	7	

Solutions on Page: 113 & 114

VERY HARD - 97

1							4	
8				5		9		2
2				4				7
	4	8			5	6		
5	1			3	6			8
	3		9		8	1		
	8				3		2	
7				8		5	3	9
	6		7					1

VERY HARD - 98

2	8				1	6		
	9				2	5		
	5	3	4	8				2
7	3			6	9		2	
9			8			3	6	
	6	4	3	2			1	
	2			5		4		
3		9			8	1		
5				1		2	9	

VERY HARD - 99

8	1	4		9		3	6	
			6	4				5
6	7						1	
7	2	8			4	9		6
				7	3		4	
5		7		1		2	9	
4						6		
		1		6		8	7	

VERY HARD - 100

1	5							7
					5			
	2		3		8			4
	7	4						
8	9		4	3			7	
			2			4	8	9
4		3		2		1	9	6
5			9	6	4	8		3
			1			7		

VERY HARD - 101

	6		2	5				
	8				6		4	
2							6	3
6			8				9	7
1	4			6		3	8	
9			3			6		5
	9	6		8				4
	5	3				9		8
		2			3			

VERY HARD - 102

					8	4	3	
			3	4				1
		6			9			5
	8		1			6	4	
			5				1	8
	1				4		2	7
9	6		4				5	
			8	9	5		6	
	3	5	6		2	8		

VERY HARD - 103

	1	5		8		9		
				2	1			8
4				3	9		1	
	2			6		5		
	6		1	7				3
8			3	4	5		7	
	3	7		5	4		6	
6				1	3	2		4
				9				

VERY HARD - 104

9		5			7	8		
			1		6		4	
	4				5		2	
5						4		6
3	1					7		2
2	9		6	7			1	
8						1	5	
		1	4			2		
	7		5		9	3		

VERY HARD - 105

9			8		4		1	
1		4				6		
6							4	2
8		3		7		4	6	
5	4	6				3	7	9
	1				6			
	9	8	6					
							9	
		5		4	1	7	3	

VERY HARD - 106

6		9				2	7	
8			5		7		6	1
			2					8
	4					1	9	6
7					1		4	
3			6					2
1				3	9			
9				6			2	7
5		8		4	2			

VERY HARD - 107

	3					2	1	6
	5		2		8		7	3
4			3					9
6	2			7	1			
5					6	9	4	7
8		7	4		3			
		5	1		2		6	
						5		4
				4				

VERY HARD - 108

8			6				4	
		6	7	5		9		8
						2	7	
2		3			1			
		7	2				5	4
		8	4		6	1	3	2
	9		1	2			8	
		1	9		8	3		
					5			

Solutions on Page: 114

VERY HARD - 109

	6		4			5		3
		4			9		8	
9		5		1		7		2
8		3	2		5	1	7	
		6			7	3		
		2					5	6
	1	7						
		8	1		2			
	3		6	8				

VERY HARD - 110

	4				7			8
		8		4	9	2	1	7
7			5	2	8	3	6	4
3			7				2	
6		7						3
	5				4			
9		5	4		2			
	7	3	8					
4					3		7	

VERY HARD - 111

	1					5		
	5							2
7	9		3					
1			5		6		2	9
		8		9	4	3		
		5		3	8		4	1
				1	3			7
6				2	5			
3		1	6				8	5

VERY HARD - 112

	3		5			8	7	
		9	1	3		6		5
	6		9			2	1	
6		8	7		5	4	3	
			3		2			
	1						9	
3		2	6			1	8	
8				7		3		
	4	7						

VERY HARD - 113

			9	7		4		1
				8		9		
	4	9		2	6			5
	2			6	7			
	6		3		9	7		
7	9	1	8	5				
		6				5		
	8	7		3	5			
2	1					3	6	9

VERY HARD - 114

			3	7	8	9	6	
9					2			
			9		6		4	2
	3		4		5			
				3			5	
5	4						8	6
		1			4	5		
	5	7		9	3		1	
8	2			5		7	9	

VERY HARD - 115

6	5	7	1	4		3	8	9
8				3	5			
	2	9						
						8		2
9		6		2	1	7		
		2	8	7		5		
	9				3			8
							7	
			4	5	8	9	1	6

VERY HARD - 116

	6	8	7	9				3
	7	2		3				
		3		2		4		7
				4	3	8	7	
		7	6		2		9	
5	4			7	8		2	1
		5	3					8
	8						1	2
6				8				

VERY HARD - 117

1	5		8				3	
9		8			4	6	5	
		2	6					9
	9					5	2	
7		4			3	1		
			1		6		4	
8		3					1	
	1	9				2		7
4				6		3		

VERY HARD - 118

	1					8	5	4
		7			4	9		6
	2	6	8			1	3	7
	5		7	4		2	1	3
	4				8			
7	9	1		5		6	4	
8		4	5				9	
	6				3			
	3					7	6	2

VERY HARD - 119

4				1	6			
		1	4			6		
	8		7	5				4
				4	2			
1	4					8		
7	6	2	8	3				1
	7						4	6
	3		6		5		2	8
		6		2				7

VERY HARD - 120

	4	5	9	8				
		7				4		
9				4	5	6	7	8
		3	2	5			8	
		8	1	9		2		3
1	6							
7		9				8	1	
3		4						7
		6		1		3	4	

Solutions on Page: 114

VERY HARD - 121

		7		5		4	3	
2	1		9				5	
		6		7	1		9	8
7	3				5			
		5	4	9	8			2
4	9				7		6	
3			7		6			
		9			3			6
6				4				

VERY HARD - 122

4			7	2				5
					8			3
5	3						7	2
9			3	7	1			4
7	4	5			9			
3			8			2		7
	5	6		3	7			8
8	9		5		2	7		
1			4					9

VERY HARD - 123

7				1		3		
	2		4			9	8	
	8	9				7		4
			5			4		7
					6		2	
	5	4	8			6		
4				5		8		3
	1			9		2	4	5
5		7				1		

VERY HARD - 124

	3	7				2		5
		2						
9				6			7	3
6	2		1		7	5		
		1			5			
						7	2	1
3	7	6				8	9	2
		4	9		6		5	
1		5			8			

VERY HARD - 125

					5			2
		5	8		2		7	
	9	2	6					
9	6	8	5	1	4	3	2	
		3		8				5
			2	3	6			
					3			8
6	3			5	8		4	
	1	7					3	6

VERY HARD - 126

9	1	3						
		5			8	4	1	
7	8			6	1			
8			7		9	3		
	4		6			1	8	
1				4				
2	6							4
					7	6	9	
5		8		9			3	1

VERY HARD - 127

			7		9	4		2
	4		8		1	3		
1		7			6	9	8	
7			3		8			
	9	1					2	
4	8					7		3
8				1		6		
			5	6			4	
6		4					7	

VERY HARD - 128

							1	
			7	8	5		6	
	2					8		
2		4		9	1	3		
5	3			7				4
	1	6	4		3			5
	4		1	3	2	6		
	6	2		4				
1		3			7		4	

VERY HARD - 129

			6	9		8	5	2
	9	4	3		8			
	2		5	7	1	3	4	
4			1		5		3	8
				3	2	7		
3		2				4		6
	3	5			7			4
6		7					2	1
2	8							

VERY HARD - 130

7	4			8		1		
			9	4			8	
9		8		2	5		6	
6					4	5		
		4			7		2	
	7	9		6	2	3		
		6						2
4				1				9
2		3	4		8			

VERY HARD - 131

3	4	2		8				
1			3		4	8		6
	9					3		
			1		2			
7		1		4	3	6		
9					8		1	
	5			3		1	6	
2	8			1	5	7		3
		3	4			9		8

VERY HARD - 132

	2		7		5			
9		4		1			2	
3			9		4	1	7	
5			2	9				
4	6		3		8			2
	8	7			1			9
	9		1	3				
8		2	4			7		
					2	6		1

Solutions on Page: 115

VERY HARD - 133

1			4	9				5
	6	3			1		2	4
7	4		2	3			6	
2	9		3				8	7
						2	9	6
		8					1	
8				5				1
						6	7	2
		6	9	1		3	5	

VERY HARD - 134

		9				1		2
		8		5	2		7	
	7		8				3	5
	9						4	
	1			4		5		
6		4					1	
		1		6	4	3		
	8		3		7	4		
	4			1	8		6	7

VERY HARD - 135

4	1	9	8			3	5	
		3	9	4				6
		5		1			4	
	6	8	2	9		1		
9					1			
						2	7	9
					9		8	
2						5	9	
3	9	7		5	8			

VERY HARD - 136

	4				5			1
				7	8	3	9	
			2	9		4		
7					2			8
9			8				7	
5		2			3		4	9
6	3	7					5	
8			6		4			
4	2		5	8		9		

VERY HARD - 137

	2	6		5			9	
			2		1		6	3
7	3	8					1	
4	1			3	5		8	
		7			2		4	
			6				2	1
2							7	
		4						6
	9		3			4	5	8

VERY HARD - 138

1							3	9
			8	1	7			6
7	6						8	
2								
		3		6	5			7
6	7	4	1		3	8		5
3		7						8
	5		3	4				2
4	8				2		1	3

VERY HARD - 139

			2	7	4		1	5
		5				6	4	8
			6			3		
					3	4		6
	9		5	4	2			3
3		4					5	2
	6		3				8	
1					8	5	6	
2	5		4		1			

VERY HARD - 140

5				6				
6		3	4			5		9
2				7				
8			9			1	3	7
9			1		6		4	5
	4		7					6
				5			8	
4		1		9				3
7		8			1	6		

VERY HARD - 141

			5		4			
	9		7				1	
1	4					6		
				8				
6		3					9	
9	7				1	4		
7	5	1	6	4		2		
2	3	6			5		4	8
4	8		1				5	

VERY HARD - 142

1						6	5	
			9	1		4	7	3
	6		5	4	8	9		2
7		6				8		
9			6	3		7		
		1		8		3	6	
6	9					1		
		8	7	6		2		4
					4	5	8	

VERY HARD - 143

			5		4	3		
			6	1		9		
	6			9				1
2	1					7	9	4
9			4				8	
3	7	4	8		9	6	1	5
				5	8			
5	2		9					
		1	2	3				

VERY HARD - 144

8			2			5	1	
	5			8	9			
9	6		5	1		3		
4		6				2		7
1				2				3
3			7		8	1		
						8		1
	4		8					
5		8	6		2		9	

Solutions on Page: 115

VERY HARD - 145

					2			
		2	5			1	4	
5		1			6			8
	3	9		7		5	1	
1		6		3				4
7		4			8		9	6
3	9	8						
	4						3	7
2	1				3			

VERY HARD - 146

				2	3			
3		7				9	1	
5		4				6	3	8
8				3	1			4
1			4			2	5	9
				6			8	1
		3		8	7			6
		8	3	4		1	9	7
	4					8	2	3

VERY HARD - 147

1				8		6	2	
6	3	8			2			
					1	4		8
		6			7		1	2
			9	2		8		6
						7		
5				6	8			1
7		2					6	4
3	6	1		9			8	7

VERY HARD - 148

1								4
9	2	5		4		1	8	
		3	2	5		9		
7			8	3			5	1
			9			6	3	
		1			5	8		9
6			1		8			
							6	5
	7	9	5			4		

VERY HARD - 149

		6	9	2		1	4	
	9	1		6	7	2	3	
5						8	6	
		5	6		3		9	
	6	8		1	4	7	5	
1						6		
2			7	5				
6	5	4				9		
3	8							

VERY HARD - 150

			5	9	1		7	
5	8	3		7			4	
9	7	1		8		5		2
2			4	1		6		
	5	9					3	
	4		7	3			2	
			9			7		6
			8		3	9		
8		4						

VERY HARD - 151

	2	8		4		5		
			6	2	9	8	7	3
		3	1	5	8			
	6	1		3	7	2	9	5
4	5			6		1		
					5	4		7
	4						8	
							5	
9	8	6						1

VERY HARD - 152

6		1		8			2	5
8		3					6	4
9				3		8	7	
			2	4		1		8
7		8		9				2
			1			5		7
		4				7	1	3
			3					6
					1	2	5	

VERY HARD - 153

	3	2	5			6		8
6	4		8	9				3
	8						7	
			7		9			
				8				
9			1			3	6	7
5					8	2	4	
		3		1	2			9
	7	4	9	5			3	

VERY HARD - 154

				4	1	2		5
		6		8			4	
		3				7		
2						6	9	
		9						8
	3	8	7	5		4	1	
	6	4	8	9				1
1			6	7	3	9	2	
			5	1				

VERY HARD - 155

		1	9		7		6	
		2					8	
5			6	2	8		4	
	2	4	3		9		1	
					4		2	6
			2	5	1		7	9
		6		9				4
	4		7			1		
			1	4			9	7

VERY HARD - 156

					9			1
5				8		3		
	7	6		1			5	
					2			6
	2			3	7		8	5
	3	8		9		2		
			5	2		8		9
7	9		3		8			
	1			7	4			3

Solutions on Page: 115 & 116

VERY HARD - 157

7	6		8					
			4	3				
	4	9	6					8
2	3	7	9		1		5	
	5		7	2	8		9	
				6			4	7
							1	2
1				8		7		4
		4		7		5		

VERY HARD - 158

		6	1	4			7	
		4			9	8		5
8	9			7				4
9				1	6		8	
	4				5			
			4					
	1			6		2	5	3
3				9				
6			3	5		9	4	8

VERY HARD - 159

	9			6	3		4	
		8		9				3
	1			2		6	9	7
5				3		4	6	
3		6				7		
			4	7			8	
				1			7	2
	5	9	2		7			6
7			6		5		1	

VERY HARD - 160

	2	9			8	4		
		8			6		1	2
5	1						8	
9			4	2		7		8
	3	7	8		5			
8					9		3	
					2		5	
1			3	8	4			
7		2				9	4	3

VERY HARD - 161

8				7	9		5	
	2				3			
		6		4	5	1		
	4		8		6	5	7	
					2		6	8
		9			7			4
5	3		9		8	2		
7					1			5
		1			4		3	

VERY HARD - 162

	1	5			4		2	3
8					2	9	7	4
	4		3	8			5	1
	5					7	8	9
2			1			3	4	6
					7	2		
1	9	4		6				
					1			
	7		4			1	9	

VERY HARD - 163

3				4	1		8	7
		7	9					1
						4		9
5	4	3	8		9			
9				1	7			8
	7	8			3			
			7	8				
	3					8	4	
		5		9	4	2	7	

VERY HARD - 164

		6	1				4	5
	8	9		7	4	2	6	
1							9	
			2		8		3	
	2		7		6	5		
	9				5	7	2	
			3	4		1		2
					2	6		
	5					9	7	

VERY HARD - 165

5				8	6		3	
		4		7	1		6	
	8		4	3			7	2
4		7	8				5	
3						7	8	
								4
	1	3		6			4	
2		6			7			8
		9		2				6

VERY HARD - 166

		6	4	9	3	7	5	1
		3				9		6
7			1	5		2		3
4	2							
					9	6		
		8	2			4		5
	7		9		5	3	6	
9					8	5	2	
		2	3		4	1		8

VERY HARD - 167

6				1		8	9	
	4		8		3	5	7	
	7				9	6	1	
		7	4					6
	3	6			1	9	4	
9			3					
			9				8	
	9			3		4	6	
7		1					3	

VERY HARD - 168

		3			5		8	
2	7		1	9		4		
9	6		4	3		1	7	
				5				
4			2		9		3	8
	3				1	2		
3				2	7		1	
	9		8	1		3		
7								

Solutions on Page: 116

VERY HARD - 169

5				7				
7	9		4					1
4	8	6			3	7		
	5	7		6		8		2
8		4	7		5			6
	6				1	5		
	4			1	6		7	
9		1	3	8				
6						2	1	

VERY HARD - 170

5		9				1	7	
	7			9				5
8	4							
	8			4	5	9		
2					8	7	5	
9	3		2		1	4		
	9		1			3	2	
7				3				
3	5			8				9

VERY HARD - 171

9	3	4				6	7	
8		6		3			4	
7			8	6		5		
2	6	1	3				8	
4		3	1	7	6	2		
		9		4			1	6
						8		
					9		6	
6	4			1			5	

VERY HARD - 172

4		9				1	5	
		7					9	8
1			4	8				3
5		3	6		8	9	4	2
	4	6		2				
	1			9	4			7
		1	9				2	
						7	3	9
7				5	3			

VERY HARD - 173

2		7	9		8			1
				1	6		5	
		5		3				
5	9		1		7			6
7				8			1	
		6			2		4	
4		2	5		9			8
			8	7	3			4
	7					6		

VERY HARD - 174

5				2	6		7	
	9	4						6
	2		4			5		
3				7	9	6		
		7	3	4		1		
				8	1			4
8		2					6	1
			9		7	2	8	
9	1	5			2	3	4	7

VERY HARD - 175

	7					8		
		2	7			5		9
		8		2	9	3	7	4
						2	8	
8	4	3		6	2			5
		7					9	6
	8		2		6	1		7
				9		6		8
		6	5		7			

VERY HARD - 176

	8			5		6		2
							4	
9		7					3	1
5			1		7	3		
1		3		9	8	7	2	
8		9					5	4
			4			2		
		8	2				7	
	2		9	8				3

VERY HARD - 177

			7					
			6			4	8	
	5				8	7		
8	6		5	4				9
							2	
9				6	2		1	4
4	8			9				
3	2	9			7	6	4	
		5		8	4	9		1

VERY HARD - 178

1					9	4	2	
3			1	6	2		5	8
	2		5				1	9
	6				8			7
9	1			5				6
8			7		6			4
5	7				1	9		
6			3	4			7	
4		1						

VERY HARD - 179

8					1		7	3
4	2	9	3	6	7		1	
						9		
		7			5		2	
				4	2		9	
9		2	1	7				
					8		4	9
			7			2		1
1	7		4					5

VERY HARD - 180

	5					4		7
9			1				2	
	7	6	9	4	2		1	5
	9		7					4
5	1				4		6	
			6	5		9		
4	2	9		3	1	6		
		1					4	
						1	3	9

Solutions on Page: 116

VERY HARD - 181

	3	6		2		9		
							8	7
9	8		1		3	6		
5	2	3			9			
	9		4				2	
4	6	7	5			1		3
	7	5			8		6	
		2	3		1			9
8	1				6			

VERY HARD - 182

9	4		3	2				8
1								
				7	8		1	4
3	2					8		
	6			8	9			
	9			4	3			6
6	1		8	3	7			9
	8		6			1	2	
				1	2			7

VERY HARD - 183

	3		9					
					1			5
		2		7		3		
6			8			1	2	7
1					7	4	5	3
		7		2		6		9
7		3		1		5	6	2
			7	5		9		
		6						4

VERY HARD - 184

2	1	5	6				9	
3	4		5		2	6		8
		7	9		4			2
	2		8	9	1	3		
8	9				6	7	2	4
								7
			4					
	3	6	7			5		

VERY HARD - 185

		2	6	3				
			5			9	2	
5				4	2			
7	5					4		
4							5	1
8		6	4	1	5	2		
3	7	5	8					
6	1			9	4	5		
	9	4		5			7	

VERY HARD - 186

	3			1	4	8	6	5
				2			3	1
						2		
2	5		1			4		
		1	4	6			5	
	6	4				1		
7					1	6	4	
		6	2	4				
	1			8	3	5		

VERY HARD - 187

					3	9		
			1					8
	4			5		6		
8	5	6		1		2		7
	7		5					
			8	6		4		5
9	8	7	2	4				
		2			5	7	4	9
	3			7				2

VERY HARD - 188

	1	3	2			8		7
	7	8						5
4	5			3				
		7				9		
9		6		2	4	7		8
	2		9			5		
	8		7			6		3
7	6			1			8	
3					6			

VERY HARD - 189

7		8	5	4		3	2	
5	1					6		
3	2			7	9			1
		3				9		
4	9	1		8				6
2	7		1		4	8		5
	4		7			1		
1	3			5			9	
6		7			2	4		

VERY HARD - 190

			6	2			5	9
2				9		1	3	
			3		1		2	
8		1	2		6			
			1	3				5
3			8			9	6	
	7				2			3
			4		7	8	9	
4	6					5		

VERY HARD - 191

			6		1		7	9
9		7		8			6	3
3			4					
4	3				6	5	2	7
	2			5			8	1
	5	9						6
			2			7		8
	7	3						4
	1			9	8			

VERY HARD - 192

7	2			5				9
3			7	8			1	
		6	4					2
5		9	2	1				8
	6	1		3			7	5
			9					
6	8				1	2		
	1					6	4	3
		7	6	2				

Solutions on Page: 117

VERY HARD - 193

5			7				4	1
	7	2	1			8	6	5
					2		3	
8		4	9		6			3
6			8			4	1	
	3	7						
				9			8	4
4	8		2		1	3	7	
		5			8			

VERY HARD - 194

5		6	2				1	
			5		9	8		3
	3		6			5		4
7		2				1		
			7		5		3	
3	5						4	
	2	5				3		9
4			1	5		6		
	8		3				5	

VERY HARD - 195

							8	
	9							
	8	7	1		6			5
	5			7			3	1
3	2	4	9	1			7	
						5	4	9
8		1	3				5	7
			6	9	7	1		
9			8	5				3

VERY HARD - 196

9		6				5	1	
			3		2	9		4
	1			9				7
	9	1		3	5		7	
3	8			6				
6			9			1	8	
	4	5						
	2		1	5		6		
		9				7	2	

VERY HARD - 197

9	3				2	7	1	4
	4		9	8	1	5		
					4	9		
	7	6	4		3			5
	5		7	1		4		
					9			6
		5			7			
4	9		1		5			
7		3			8	2		9

VERY HARD - 198

			5	7			6	
6								
1			8				3	7
3	9			8	5	7	1	6
4	1	5	6			3	2	
7	8				3			4
	6	9	3					
			9				5	
				2	6			

VERY HARD - 199

	8		7	3	5			2
7	1						6	
	4				8		7	3
6						2		4
1			6	2			5	
	2	7	3					
3	6	4	5		7			
8	7	2	4	1				
5				8	3		4	

VERY HARD - 200

5		2				7		
6		1		4				8
3					5	9		
2	1			7		3		
			9	3	2		1	
7		9		1	6			
4					9	2	7	
	2	5		8		6	3	
9	6	7			3	1		4

VERY HARD - 201

5	1		4		3			
					7	5		
9					8	2		
		9						
6		1			9			7
7	3						8	
2	7			4	6	8	9	3
			3	8	2			
	8		9		5	6		1

VERY HARD - 202

8			7	9	4			
			3			4		
	9	6	1	5	8		7	3
2	6	8	9	1	3			4
7	1			4		3		
			8		7	6		
	3		5					
				8		5		7
	8							1

VERY HARD - 203

1				9	5		3	4
7	6		1	4		2	9	
	4							8
	3	4		1	2	8		
			9			1		
		1		7			2	6
2	1			5				
3			6		1		4	
4		9		3			6	1

VERY HARD - 204

9	1							4
			1	3	8	9		2
7	2	8				5		
3		9		6	7	1		
	5			2	3	6		7
		7			9	3		
6					2			1
				5	1		8	
	3							

Solutions on Page: 117

INSANE - 1

9	8		7		4			
		5			2	1		6
3			9	6				
			2					7
		9						
8					9			3
			1			4		
		3	4			5		
				8			1	

INSANE - 2

8						2	9	
	1		5				8	
3					6		5	
4								
2			3		9		4	
		9				3		
	5		4	1				
					2	8		
			8	6				9

INSANE - 3

	6		9		7			1
	8				1	6		
2								
3	1							7
		5	8			4		
				7				
				8	2		7	
9	7			4		5	6	2

INSANE - 4

	4	5			8	6		
		9		6	7			
				9			5	
					4			6
		2				4		
7	8							
1	9						6	
	7				1			
		3	6			2	7	

INSANE - 5

1			6		2	9	3	
3	4				5		1	
2	7							
						5		
	6		2	7	1	4		
		7	5	6				
	8							6
						1		5

INSANE - 6

	7			8				
5		4						
		9		6	4			
		7	2				9	
3	6				9	4		
					3			6
		6						7
1	8						2	
	2		3			1		

INSANE - 7

			1				4	
4				8				
					7	5	6	
9							7	
8		7			9	3		
5								
2	6	3		4				9
					3			6
		9			1	8		

INSANE - 8

	4	8				1	7	
		7						
			9	1				
	1				8	6		5
							8	
		9	5	7	2			
							5	6
	3		7				1	
	2			3			4	

INSANE - 9

2	6					9		
3							6	5
	7		4		6			
				7	9			
		4		8				
6				5		7	3	
	9				5			
				9	3		2	
						1		4

INSANE - 10

8			7					
		9		4		6		
	1	6			5			
		5				3		
	8		2		7		9	
	2							
						7		
	4	2		8	9		3	6
							8	1

INSANE - 11

	5	6	7	1				
	3				2			9
	7		5				8	
				5				
9					1			
		7				4		
	4		1	9				7
				2	3			5
					6	8		

INSANE - 12

1	4		6				9	
			2					
6	7				5		1	3
								7
				5				
8	1						3	9
					9	8	5	
5		6	8		4			
	2							

Solutions on Page: 118

INSANE - 13

	7					4		
							2	7
			6	1		8		
1	2	4		8			3	
			5					
		3				9		
						2		
		8		9	2	6	1	
		5	3				9	

INSANE - 14

9	4		5	3				
						8		7
			6			3		
2	9	4						
					1			
			9		7	6		8
	7		2		5			
8				1				
	3	9	8					

INSANE - 15

	1	4						
	7			6	8	4	3	
					2	6		
8	6	1	5					
						5	1	
4					3	2		
					4			
7	3	2						
			8		6			

INSANE - 16

	7			3		2		
	9	6				3		
3								6
			1				9	4
	3				9	5		
4			7					
				1				
	6	5	3				8	
						7	1	2

INSANE - 17

3			2	6				
								2
	5	6		8			1	
	7		1	4	2			3
		5						
	4		8					
	9		5					
		8	9				2	
		3			7	5		

INSANE - 18

					9	8		
		2			4			6
	9	6						
9	4					6		5
	7						8	
			1		5			2
		4				1		
	6			2	8	4		
		7					3	

INSANE - 19

		3		5				2
		9			6			
	7		4	6				5
	1				2		6	3
3								
	5		9				1	4
1					8		5	
	8			1				6

INSANE - 20

	7					3		
			5				2	7
			1					
2	1		9		6			
			2				3	
						2		6
1			8					
		8		2			5	
6		4		9		7		8

INSANE - 21

						2		8
3	8	2						
						5	4	
1	5		3		6			
					7			
	9		8					
7			9	6				
	4	1				7		
				3	4		9	1

INSANE - 22

1		8	6			3		
	9							1
				5	2	6		
5				9		1		
	4				6			2
9	5			2				4
					4		5	9
7	3							

INSANE - 23

		3			8	6		
	7		2					
				3			7	
	2	8			9	4		6
	4	6		7			8	
	1							
	5						3	8
		4						
6					7		2	

INSANE - 24

8		2					9	
					7			5
7	3						8	
1		7			6			2
			1		5		7	
		3			4			
						1		
9		6	2	5		3		
						2		

Solutions on Page: 118

INSANE - 25

			8					1
9	5							
			9				6	
					2			7
	2				3			
6	1						9	
4			5			3	1	
				2			4	8
7		8			4			2

INSANE - 26

		2			4	3		
			3			8	9	
4								
		6	1		9			
	1		8					
5		7				9		
		9			1		5	
7							2	
	8			9		7		6

INSANE - 27

6								
		5		4		8	9	
					3	5		
5	2			3	4			1
		9				2	8	
					6			
		2						
	5	8			7		4	
				2	1	3		

INSANE - 28

	1			5				
							4	
				6	1	2		5
			7				3	
		2			3			
1		3						7
6	2	7						3
	5				4	8		
9			2		7			

INSANE - 29

				4			6	7
		3						
			8	2		4		
	7			3				
			6				4	
					4		1	9
7			5		8			
5		9			6		3	2
			2					

INSANE - 30

	3						4	
7				9		2		
	9	2	1	3				
5					9		2	
6					4		5	
				5	1			
		5					6	9
		7						
					2		3	8

INSANE - 31

			7	2	8			
		4			6			
				9				7
		5	3				4	
	7			5		6	2	
6	4		8		9			
9								5
		8			5			1
	1							

INSANE - 32

							5	7
			4	2				
	6	3			1			2
6			3			1		
7		9					6	
	4	2				5		
5					6	3		
		7				4		
				1	8			

INSANE - 33

			5	7				
9	4				1			
		8	6		2			
		9				2	7	
			3	1	7			
	5					1		4
1	6							
						4		
	9	7			8		6	

INSANE - 34

9	8				2			5
				7				
6			4	9		7		
	2			3				
8							4	
	5		7				8	
7					3			1
	3		1	2		5		
			9					

INSANE - 35

				1				
	6	3		9				2
	2							7
			9					
				6		5		
	5		1	2		8	3	
4		8				7	2	5
					7			
2							8	1

INSANE - 36

		1						6
	5	4			7		1	
9								3
					1			
			7	2	6			
		6	8					9
	2			4		6		
	9				5			1
				1		5	4	

Solutions on Page: 118 & 119

INSANE - 37

					4	1		
	4	8		6			3	
		6			3			
	8				9			7
		7				9		
	9			5		3		1
	7						6	
			5		1			9
		2	8					

INSANE - 38

	3							
				1	9	2	7	
		1	5		8			4
		5			4		1	
	7		1			4		
				8			9	3
	9							
	2					7	8	
7			6					

INSANE - 39

					5	2	4	
	9				2			
4			1				9	8
8	5		3			6		
	6				7			
		3	5			7		
3			9					
	2				4			
				1				6

INSANE - 40

		5						
3	9							7
		7		4	9			8
				8		9		
9		4						5
				2	7			
	2	8	3			6		
			2				5	
6	7							4

INSANE - 41

		4		2		9		
	7	5						2
6							3	
	8	6	9					
			5			4		9
				7			6	
				1	8			6
						7		
	2	1		4				5

INSANE - 42

				9	7			
	7	4						
			3					4
7		6			8		9	
				6		1		
1	8							
		5						3
	2		8		4		7	
	4			2			6	9

INSANE - 43

		9	3					
	4	8			6	5		9
6			5					7
9		2					8	
	8				3			
	6		8		9			
		4						
					8		2	
7			2	5				

INSANE - 44

		9				5		6
				6	4			
	5			3				
8					7	3		
	4					2		1
		7		1		9		
	1			5				
7	2	6					3	
			6		2			

INSANE - 45

			5	7			9	
	1	3						
				6			4	
			9	5		4		
5	8	2	3	1				
9								
							5	
				8		2		
2		9			7	3	6	

INSANE - 46

8	5	6			1			
		2		5				
1			6	3				7
					9		2	
				8	3	4		6
7	9						1	
								4
6							8	
		1			2			

INSANE - 47

3	5		6		9			
	6		5		8	4	2	
						6		
7								
		9		4		2	1	
		3						5
6								
			9					2
4	1			6			7	

INSANE - 48

7			6			9		
					4	2		3
	8							
	6			1	2	8		
	3			4	6			7
				9				2
6			3			1		
				7				
	1	3					8	

Solutions on Page: 119

INSANE - 49

1								9
		7					1	
4				3		6		
	3			4		8		1
				5			4	
6			8				7	
5		6	1					
		4	7				3	
	2						8	

INSANE - 50

	6			3				1
			6			2		
			7					
6		5				1		
	2			1	4			5
		7			9	8		
					7	5		4
				4	2			
	9	3					8	

INSANE - 51

8			6					3
9		6				8		
	5							6
	2		4					
			5	9			1	
	6	3				5		
					1		7	9
	7			3				1
					7	6		

INSANE - 52

						5	8	3
	1		9				4	
	3						1	
6		1						
				8				
	7	5				3		
		8	6		3	9	5	
	4	3	8	2				6

INSANE - 53

						5		8
				1				9
3		6	5					
	7			5				
5				7		6		4
		9		2				1
8				3	7	1		
9			4				8	
								7

INSANE - 54

						7		
5	2							
	3		1			4	5	8
		9	3					
		6		7				
	4			5		1	2	
		2		3	9			
						2	4	
				4	1		7	

INSANE - 55

		7	1					
	2				3			
	4	8		2	5		7	1
8	7		2	4		6		
								8
				3	9			
9		1			4		5	
6							4	

INSANE - 56

7		2						
	9				3	8		
5							2	7
	7							
				6	2	1	7	4
		8	5					
		6	2		1	9	3	
9								
				5		2		

INSANE - 57

1							7	
		4					5	
				8				2
				9	3		2	
6		8		4		9		
5	9				8	7		
	6						3	
	3				4			
		5			9			4

INSANE - 58

		7	5			9		
				6		2		8
							1	
4		6						
2	5		1		4			
7	4	1	2	5				
5	6			1			7	
					8		4	

INSANE - 59

	6							
			9	4		6		5
			7					
	3		1		4		2	
	2				5		7	
6				8	7			1
				2			9	8
		4	6		3	5		

INSANE - 60

					9		2	
		5				4	7	
8						5		
6		7						
4		1			5			3
			1	6				
	6	3		9				8
2			5	1		6		
					4			

Solutions on Page: 119

INSANE - 61

			7					
	5			1		2		
9		1					4	
				6				
	8		1		3	4		
						8	2	9
		4			7		6	8
		7		3	4		9	
							5	

INSANE - 62

1		6		4				
							6	3
3	4				9			5
7		5	8		6			2
		9	7					
					5	8		
	2			1				
		7				3	9	
	1							

INSANE - 63

	4			9				
			6					
		9	7		3	2		
		4		6		9	5	
				8		3		
6					7			2
9		7					8	
		1	9		2	4		
		3						

INSANE - 64

6								
		9	5			8		
2						5	7	
	6					9		4
4						7	5	
8					1			3
				9		1		
		2	4	6				
		3		5			9	

INSANE - 65

	5		9		2	1		
	1			3			7	5
	6							3
	7	1		5		4		
4								
	9		5	8		6		
2		7					4	
		5		1				

INSANE - 66

8	5		6					
						1		
	1	2	3					
		4		2			5	8
9		6		4		2		
		3		6				
						9	8	
							6	
		5			2		7	4

INSANE - 67

					3	4		8
3	6	1			9			
		4			5			
	3				8			
	4					2	7	
				7				3
				4	7			
		8		5				7
	2					9	6	

INSANE - 68

					6		7	
		8	3	5				
2			9					
	7	9			4	3		
	4		8	6				
3						1		
5						2	8	
			1		2	6	9	5

INSANE - 69

2	4		3		5			8
					6	7		4
								2
8								
3			8	5		4		
6	2						3	
			4		3			6
		9				1		
	5			6				

INSANE - 70

4			3					6
				2		5	4	1
			6			9		2
	4	2	5					
6		5				1		
1							8	
			4				2	
		3		8				
				3		7		

INSANE - 71

5						3		4
				7	6	5		
	1	8				9		
					9			
2				5			1	
	3			4		8		
			6			4		2
				3	7			
		4		1	2			

INSANE - 72

	8						5	
	6				5	8		
		2			7	9		
		4	7			1		
	3							6
		9		4				2
	7	8						1
3			1					
			5	3		2		

INSANE - 73

		8		2			
		9						7
						9		5
	8		5				4	
2	4		6					
	5				1		9	3
	9	2	1	6	8		3	
	7			3				

INSANE - 74

	2	3	9				1	
	1			8				4
	6							
7		2	5					
5				6			7	8
	9							
			1	9		2		
			4			5		9
					2	8		

INSANE - 75

		4	3					
					9		1	
	9				8	7		4
2		3			5		7	
	6							1
	1		4			3		
							3	9
			1		4			
		6	2	3				

INSANE - 76

5							3	
		2		9				
			4		8			7
8				5				
		9	8		1			
	6				2			4
7					9		6	
9		8		7				2
	4		3					

INSANE - 77

		3		7				9
								6
4				2				
9		8	5					
3	2				9	7	6	
	6			4				1
2						9		3
			6		3	2		8

INSANE - 78

			4			2	7	8
6						4		
2	3							5
					2			
		9	7	5	4			
	5	7	3					
9					3			
						8	9	
					5	7		2

INSANE - 79

		8						
7						2		6
				6				4
			9					
2	3	1					8	
4			2					7
		2					7	
	1		6		5	3		
6	4				3			2

INSANE - 80

					8		2	
					9			
		8	7	3				4
	9		4			7	6	
	7			9				
		6	5					3
9		5						
7					2		4	
	1					8	5	

INSANE - 81

		1	2	3				7
	6	8						9
		3						
			1					
9		6				2		
					8	7		
4			6					2
		5		1			8	
				9	5		3	1

INSANE - 82

	7						5	
4		2		7	6			
								1
2	6	8	7	4	9			
								8
		3	1		5			
							3	
1			2					
5					4	7	8	

INSANE - 83

						7		
	8			9	2	4		
	7	6						1
	4	3	5					6
	6				1			
		5		6			8	
9				4	8	6		
5							4	
					9			

INSANE - 84

				7	2		4	3
1						2	5	
	6							
		4			1	9		8
5			8			3		
			4	6				
4		7						2
8	1			4	7			

Solutions on Page: 120

INSANE - 85

	4	3						
5	8	9		3				7
				8	1			
7				5				4
3		6				7		2
						6		
			7	6		1	9	
		2		4				
			9					

INSANE - 86

					2			5
			8					
7	4		3					
		2						8
		1					4	
				7	5	6		
9			5		1	8	2	
	8			4	7	3		
		7					6	

INSANE - 87

			2					
	3	6			9		4	5
	8		6	7				1
					4	2		
				8			5	9
5								
			3		7	9		
		1						
				5	8		7	6

INSANE - 88

7		8						
6		3	7					
		5			2		9	
		4		5			1	7
	8					4		3
			6					
					4	8		5
			8	6		3	7	
				2				

INSANE - 89

			9	1	3			
		7				4		6
	3	2			6			
		5			9		8	3
8			2	5				
	1			6	8			
	2		4				1	
6							7	

INSANE - 90

			9	8	3		7	
	9			6				
							6	3
				7				
5		9						
	7	4	8					1
			4		5		1	
2	8				9			6
		5		2				

INSANE - 91

	6						1	2
		9				3	4	
4								9
		1	6			7		8
	7		8		9			
	2		1	4				
			7			6		
		5			3		9	1

INSANE - 92

				4	1		2	
6		4	9			1		
						3		
	2			3		4	9	7
		3		6				8
		2						
	9			5		8		
1				2	9		7	

INSANE - 93

			6				5	
		9	1		7			3
5						2		8
							3	
		4	9			8		
1					5			6
	2						7	
	6	1	2			4		
		7		3				

INSANE - 94

				8		2		
9		4						7
8		2			3	5		
		8			4			
2	6						1	
					5		6	
				5	8	3		
3				7				
		9				1	8	

INSANE - 95

		7					6	
							2	7
3		5					4	
		1				3		
		9		8				2
	5	6						
	6	3			5			
	1		7	2				8
			3			7		5

INSANE - 96

		4	8		1			
	7	8		3	4			
	5		6					
						8		
					9			4
	4	3					5	7
3						6		
		2		9				
		1		2		5		8

Solutions on Page: 120 & 121

INSANE - 97

				7				6
4		7			3			8
1				6		9		
			7	4				
	8	2			6			
		5	3	8				7
		6					2	
						6		5
9						7		

INSANE - 98

	1		8		5			
	6		4				1	
	3		6			9		
							9	3
6			5		4			
							2	
4								9
		2				8	6	
	8	3	7					4

INSANE - 99

			3		7			
8			6					
5		2				7	6	
		5						2
9	4		7				8	
								9
	9		1		6	4		
		1						8
			9		5		2	

INSANE - 100

		7			4			
			7			1	6	
	2			9				3
						8		
	1	8					2	
			2	5			3	
6			3					
	3				7	4	5	
		1	6		9			

INSANE - 101

		3					6	
			1			9		
7			6			5		
						1	2	
	5	4	9					
2							9	8
	3	5		2		6		
8	7			1		3		
				4				

INSANE - 102

			9	3		4		
		5						
1		9		2		6		
			7		2		3	9
		7						1
	6							
			2		7	9		
2		1	8			5		
7			4					

INSANE - 103

			9					7
9	1			5	4			
			1		8			
3	6			9			5	
8		9		1	2			
1								4
							6	
	4						1	
2				3				5

INSANE - 104

	6			3				
	8		7			2		
		4				1	6	
6		5		8				
7	1		6		5			9
	3					4		
	5				1			
4			2					
						8	1	

INSANE - 105

						7		
		5		7		3		2
		6	5			9		1
9					2	8		
4							3	
							2	7
			8	3				6
3			7				4	
	1			5				

INSANE - 106

		4		3				7
	5	1						
			8		2			
	8			9				3
	4	9					1	
	1			5	3			4
		7				3	9	
9			7		8			
						6		

INSANE - 107

5							1	3
		9				2		
			6		1			8
	8	2		7				
					9		2	
		7	1		6			
					7			
9			4	5	8			
	3						7	6

INSANE - 108

		9			3			4
2			7					6
		3		1		8		
	1			4		3		
	6					4		9
3					8			
5			9		7		8	
							7	
1	3							

Solutions on Page: 121

INSANE - 109

		3			4		9	
			5	2		3		
			6	1				
5		4				7		8
		6			8			
1						5		
		9						5
			1					7
	4	1	8	6				

INSANE - 110

			9					
			7	5				
6		4				8		
		3		7			8	
4			1		9	3		
						4		6
		8					3	
	3	1		4				
	9	6	2				5	

INSANE - 111

5			4					6
	3	1						2
			1	2				8
2		4			5		3	
		5		9				
	1	7		8				
					7		4	3
							1	
6			5					

INSANE - 112

	5			9	7	1		
2	4							
	1	9					6	
							9	
	6		1	8			2	
			4					
			8	1				
6	3						4	2
5	8							3

INSANE - 113

6		4		9				1
		8						6
	2	3			7	9		
	5			7				
7		9	6	5		8		4
				1	5		7	
4				8		2		

INSANE - 114

1	4				8			6
					4			
	5		7			1		
8			5	1		7		
		3			7	6		8
3	1		9			4		
	8				2			
		7						3

INSANE - 115

					7	2	8	
		8		9				4
	2		4				7	6
		4	2					
	5						4	1
1			3		9			
	3		7	8				
		2		1				5

INSANE - 116

		4						1
	5				3	9		2
	1		5			7		
		1						
7			4		9		6	
			7	6		5		
5								8
	3	7			6	1		
			3					

INSANE - 117

							4	
	1		8	2				
			3		7			
		7		9	3			8
							7	2
	3					1		
				8	4	7		
2								3
		8	5		9		6	4

INSANE - 118

9				4				3
	4	7				6		
			1	8				
	6	5						
			9		2			
						4	7	1
5					6		9	
4		1			3		5	
	2						4	

INSANE - 119

	8			5				
		3				2	6	7
		9			3	1		
	7			4				
		6						8
	9	5		7	2			
6	2							
			1					
		1			8	4		9

INSANE - 120

			9					
		2	5			9		6
3				2				
		3	6				9	
		1						
4							1	3
	9	6		1			4	
7			2	3		6		
							8	7

Solutions on Page: 121

INSANE - 121

7	3	5	4		8		9	
	4			3				
				9				
				6				
	5					9		
	8				4		3	2
8								3
6				5	7		4	
		7	1					

INSANE - 122

			2	7	4	6		5
	2							
			6				4	
					1			2
	5		4	3	9			
7								1
1							9	
		6		8				
3	9			6			7	

INSANE - 123

7	9		8			4	1	
4	8				1		2	
2								4
		5				8		
				4	6	1		
		8	9	6				
3						2	8	5
								1

INSANE - 124

	9			2				
		3			4			8
					1	9		
				8		6		
					6	2		1
		1	3			8		
5	4		6			1		
1								
		2			7		3	6

INSANE - 125

	4				9	6		
7				2				
2	3							4
8					2	3		
		9		5			7	
		3		4				6
	6					2	5	
9					7			
3						8		

INSANE - 126

		5	7					
1						7	2	5
3					6			
7					8		1	
				3				4
8				7	1			
5				8			6	3
		4					5	9
	2							

INSANE - 127

1		7		2				5
		2		5		3		
					1			
			3					7
					4		8	
				1		2	6	
5							4	1
8								
		9	4	3	2			8

INSANE - 128

			7		2			1
		5				8		6
	8						9	
				3				2
					1		4	
3	4						5	
9			2	7				
	5	6	9					
	7						8	5

INSANE - 129

	9	4						
				9		5		
			7		5			3
	1	7	2					
				3	6	4		
					1			9
		9		1		6		
3	8		9		7			
					2		1	

INSANE - 130

	5			1		4		
						8		6
			8			3		
	7							
		1						3
3	6		7		9			2
	8			9	7		4	
	2	5			6			
6								8

INSANE - 131

		7	5					6
					6	2		
		4						
8			7				5	2
	3			8		1		
				9				
3			2					1
6	2			7		8		
1			6	5				

INSANE - 132

	8		5		7			
			2		3	4		1
		7		1			8	
	7	8		4	9	2	1	
	5	4		2				
9								
	6							
7			3					9

Solutions on Page: 122

INSANE - 133

	2					3	6	1
9	8						5	
			4					
	7							
			6					
		9		5	4	6		
			8		2	5		4
	1		5				7	8
		2					1	

INSANE - 134

5				7		4		
8	3			2		1		
					3			
							3	
	4		9		5	6		
	1		2					4
	9					3		1
	2	6						
4							5	7

INSANE - 135

7				9	5			
	3					2	8	
8		6					7	
			1	3	7	5		
3	1		5					
	4	5	6		2	1		
							2	
		2			9			

INSANE - 136

	4		9					3
	7	2	6	3				
		1					6	5
8		7	2					
				1		8	7	
					6			
				4		9		
			5					2
				8			4	1

INSANE - 137

				5			7	6
8		6						2
			3					
		7			5			
			7		6			4
	3		9	8				
		2					8	5
						2	3	
		8		1			6	7

INSANE - 138

				6	3	1		
3				7		2	5	
	6							
							3	
	2				8		7	
			1					
1	4		7					8
		7						2
6	3	8			4			9

INSANE - 139

	1			3		2		5
				7				
4	8	5		6				
		6	7					4
5	9							
			1		2			
	2		3		5		1	
3					7			
						5		8

INSANE - 140

2		1	5					
	9			8		3		6
			3					
4		6						5
7	2				3			
							8	
		7						1
	6				8	7		9
8			2	1				

INSANE - 141

2			6			4		
	6			5		1		
8			2				3	
	7							
					3			8
		4				5	2	
3	9	2						
				3			1	6
4				2				5

INSANE - 142

		7	4		2	6		
3		4						5
	2			6	1			
		1		2				9
				4	9	8		
					7		4	
8				9			3	
						1		
			3					7

INSANE - 143

	5					4		3
	6		1					
1		8	7		9	2		
5			8	6			7	
		4	2		5			
	8						5	
							2	
				9		3		
	4					9		

INSANE - 144

		6	5					
3		7	6				9	
			8	1			4	
		3			9			1
4		9					2	
1				8				
2				7		3		
			4				8	
				5		6		

Solutions on Page: 122

INSANE - 145

	3							4
8			6				9	
	9	5		1			2	
			4			2		
				9				5
6					8	1		
		3						8
	4						7	
5		1			4			6

INSANE - 146

	1							2
2	5			3		8		
	3			7			9	
			4					3
					9	2	5	
		3		2	8		7	
8				9				
		7			5	4	6	

INSANE - 147

	6	4		2	1			
	2	7					9	
8		9	7					1
	8		3			5		
	7	1	4					
							3	
						9		4
			5		6			
						1		8

INSANE - 148

						9		
1		8				4		3
		5			4			
3		4					1	2
7								
						6	3	4
			6	7				
	5	2	8	9				
9	7			5				

INSANE - 149

				8	7		6	
	9				2			
				4			3	9
			4					
5	7							
3		4	1		6		8	
2	3				5			7
8		7		3	4			

INSANE - 150

6					3			
	3	8	7				9	
	2				8	6		7
					4			
				9			4	1
	9					2		
				1	2			8
			4		6			
	7	2		3				

INSANE - 151

6			4	8		5		
7			6				3	2
	9						1	
		1	8					6
				5				
							7	4
		8	9			2		
		6		2		3	8	9

INSANE - 152

					4	6		
			7		9			
					1		5	
	1							
	5				2	3		
7							8	1
1		4	5			2	6	
2	9	3		4				7
		8						

INSANE - 153

								7
	1	2		8		5		
	5			1	6			
	6					3	2	
				7	4			
						1		
5			8		2			
	3						8	
	2	1	4	9				3

INSANE - 154

6				3			1	
	4		6			8		
			9		5	3		6
2					1			
							8	
		5			6			3
					3		5	
	5	7		4		2		
	9				7			

INSANE - 155

	8			3			1	6
					6			
						2	7	
	6		9	5				
4		5						
		1		7		3	4	
		8	1					
	1	7			8			
			4	6		9		

INSANE - 156

	2		4					3
		3					7	
1								
		5				3		
3				7				
6		8		4			1	
7			2			1		
	1		6			4	8	
					5	2	6	

Solutions on Page: 122 & 123

INSANE - 157

		7						
5	8				4	2		6
		3			5		8	
	6				2			
8		9						
				7		5		
				5		3		
7					6	4		
4		2		3	8			

INSANE - 158

	4	2						
	6	8		3	5			
				6				4
	8					9		
		5			7			
	1		5					3
			9			6		1
	2	4	8				9	
		3				4		

INSANE - 159

			4		6	7		
	9					8	5	2
					3	5		1
	4							
	1			2			7	3
			8		5			
	8	7		3	1		6	
6							8	

INSANE - 160

		3			6	2	7	
	6						5	
				4			6	
					3		4	
		1		8		7		
				7		1		8
1	5	7		9				
6			2					4
	8							

INSANE - 161

	3		5	8		9		
	6		1			2		
2								
	5	6	3	4				
							5	
4	9	3	7			1		
	1							
								8
	8			5	3			2

INSANE - 162

	6				5			3
					1			2
	8					5		
								8
		1	7		6		9	
		2	3		8	7		
				8		1		6
		4				8		
9						2	7	

INSANE - 163

1			7		2			
		6	8		1	5		
		9		3				8
		7		5			3	
			9			1		
	7		3			2	8	
	4			8				6
2						4		

INSANE - 164

		8						6
		2						
	4	6	7	9			1	2
9	6		1			7		
		4	6					
				2	3	6		
				4				
1			3			5		
						8	7	

INSANE - 165

	8					1	7	9
			6					
					9		4	8
	6				1			
		4						5
			9	5		2		
	7		3	2				
	2		7				8	
		3	4		6			

INSANE - 166

4				8	9			
	1	3						
7						5	6	
					6			
2							7	6
1	6	9	8					
		4		5				8
				2		9	5	
				3	4			

INSANE - 167

6		2				9	5	
			6	4				2
			9					6
		5				1		
3				2		7		
								8
	1	6		5	9			
2			8				3	
4		8						

INSANE - 168

							6	
	9						3	1
7		5	4					
1				4	7			
				3				2
					5	8	7	
	3			7	2			
8	2	1				7		4
							1	

Solutions on Page: 123

INSANE - 169

8	7			3		6		
		6				1		
	9		8					2
		5		2			9	
2	3				4			
		9						7
	4							
				8	6	4		5
	6			1				

INSANE - 170

9	8							1
	2					8		5
	6				2			
		9						
			3			2	4	
	7			6		9	8	
4			8		9			
8						3		
		2		5		7		

INSANE - 171

	2	8		9				
	1	5	7	2		4		
3			1			9		
		2			4			
	4		5					3
							9	
				7		3		8
							1	
		9			8	5		

INSANE - 172

4					9	8		
							3	
5	9				4			
1		3				2		4
	4	9	8		5			
							5	
	8			7				
				2		6		
6						3	4	2

INSANE - 173

				6		3		
	7				8			
	6	2			9	1		5
			7					
							2	7
	4				3			
1	3			5				
	9		4				3	8
		7			2		9	

INSANE - 174

	1		5					
8	7	2		9			3	
			4		8		9	1
2					3			
			8			4		
						9		
5			6		9			
		3	2					7
		1					2	

INSANE - 175

2			8					
			7				8	
6		9			1			
					3	7		6
		4		6		9	5	
			9				4	
	3			8				2
		6						3
	9			3	5			

INSANE - 176

							1	6
	2	3				8		
1			8					
		6	7			9		
	5			2	9		7	8
						5		
		1	9					
7								
6			1		5	7	9	

INSANE - 177

					7			
3	6							
	1		9				6	
		6		5				4
	2	1				9		7
9			4					1
		5	6					9
				4		5		
			3			1		2

INSANE - 178

		5	8	9				
7	4				2		6	
								9
1				5		8	7	
		8		4		1		
9		3					2	6
			5					
		2	7	8	4			

INSANE - 179

1				6			2	
		8		1		6	5	
					9			
5							7	
3	6				8			
			2	7	5			4
	1		3				4	
					2		8	
		6		4				

INSANE - 180

		5	8			1	6	
	3	6					8	
						7		
	4			1		8		9
9				7			3	
			2					
	6							2
					9			
				8	3	6	1	7

Solutions on Page: 123

INSANE - 181

	6			2			5	4
		4				9		
					5			
		2		8	1			
								9
		3	6			4		
	7	9		4				8
	2			6				
		5			2	1		3

INSANE - 182

			3		9			
		9				8		
	7	8	6					1
	5	7						
	8				6			3
	6		4	2				
		6						
		4				3	1	8
5				1		2		

INSANE - 183

							1	
7	6		1				2	5
			3			9		
	8						7	
6		4						
9					8		4	2
	5	2	4	9		6		
		3		2				
				3				

INSANE - 184

		6				4		
	7						5	
2			1	6			8	
				1				
8				5				
	1		7		2	3		
	2		3					1
3						7	2	
4	8							3

INSANE - 185

4	6			5		7		
					7	2		
		8						1
9	2							
				9	3			
		6				4		
7		4		8		1		
	3					5		2
6		2			9			

INSANE - 186

	1	9	6		8			
6			3				2	
							3	
		7			2	8		
		5					4	
9		1						
		4						5
				4		1	8	
1	2		5				7	

INSANE - 187

6		8	2			5		
3	1							
			4				2	
	2				4		5	
7		3			6			
					3			
	3			8		7		
		1				6		
4				9		8		3

INSANE - 188

2		7			5	4		6
		5	1			2		
							3	8
	6	1		3				
			6	7				2
	7							
			4					
						9	6	7
				1	2	3		

INSANE - 189

1					7			
				9		4		1
				4		3	8	9
	6	8				2		
2		7					3	
			7					
3	4		2		8			
		9	3	1		5		

INSANE - 190

					2			4
					3	1		
			9	7				
7	1							5
	9		1					
5	3				8			9
3					7	4		
					5			
8	5	4					9	3

INSANE - 191

	7		6			3	1	
			4			5		
9			5	8				
								2
		1		6			4	5
8				2				
			8			6		3
		4					9	
	6			7	1			

INSANE - 192

	5	4		3				
			7				5	
8						7	6	
	1		2		7			
						9		
				1	4			
6								1
4	2				3			
	8			5	9	4		6

Solutions on Page: 124

INSANE - 193

		3					9	8
					1			
1	6	4			9			
	2						3	7
3	9		6					
		1	2		8			
		7			2			6
					5	7		
		2		9				

INSANE - 194

		6	7		5			
	7			3				
						3		8
	2		3	9				
		7	5	8	2			1
1	8							
				6	9	1		
2								
		3				5	6	

INSANE - 195

6			5					9
2				7	8	1		
	9		6		3			
					5			
7			3			4		
			2	9				
			4				1	5
					6			
	8		7			3	9	

INSANE - 196

	3		7					8
2						9		
5	7		3					2
1		7				6		
	2			7				5
	4		6					
	6				9			
9						3		
			5		6	1		

INSANE - 197

			8					
3	7							
	2						5	8
			9	1			4	
	6		2		4			
	8		3					
				5		7		9
7				3	9			
	3	6		8				5

INSANE - 198

							2	7
		2			3		8	
3		5					9	
1		3	2					
	6		5		8			
				9		6	3	
						1	6	
				2	5	8		
	1		8					

INSANE - 199

		3				2		
				7	9		3	6
7			1			5		
	9			4	5			
						3	4	
	1		6		8			
2							5	
				6		1		2
				1				7

INSANE - 200

		5	8					
	8			2				4
3	4						6	5
				6	7		5	
			4					
1					3			
4				9			7	8
		6					9	
		1	7			2		

INSANE - 201

					3			7
	6			5				
		5				3	9	
6		9			8		3	
3	4			7				
		8				6		
5		1		2				9
			1					
			8			2	4	

INSANE - 202

		1	7		3		5	
					9		7	
						3		8
2						1		
	7					4		
4		3						
					7		9	
5	6		3		2			
		8			1	5		4

INSANE - 203

		1	6					
			4			5		3
		2				8		
	6	8			5			
		5		9	2			8
						3		
8	5	6						
							8	4
2			7		6			9

INSANE - 204

		5			1		3	7
				7				2
				8	3			
	8	1				2		
		3	4	2				
					6			5
2				9		1		
4		7						
	5			1	7			

Solutions on Page: 124

EASY - 1

4	3	1	5	8	9	6	7	2
9	7	2	1	3	6	4	5	8
5	8	6	7	2	4	1	9	3
1	4	8	3	7	2	5	6	9
2	5	7	9	6	1	8	3	4
6	9	3	8	4	5	7	2	1
8	1	9	2	5	7	3	4	6
3	6	5	4	9	8	2	1	7
7	2	4	6	1	3	9	8	5

EASY - 2

1	8	3	7	9	6	4	2	5
4	7	6	5	1	2	3	8	9
2	9	5	4	3	8	6	7	1
3	1	7	2	6	5	9	4	8
6	2	8	1	4	9	5	3	7
9	5	4	8	7	3	1	6	2
8	6	2	9	5	4	7	1	3
7	4	9	3	2	1	8	5	6
5	3	1	6	8	7	2	9	4

EASY - 3

8	2	6	9	1	5	7	3	4
7	9	3	2	6	4	1	8	5
5	1	4	8	7	3	6	2	9
2	8	1	5	4	7	3	9	6
6	7	5	3	8	9	2	4	1
3	4	9	6	2	1	5	7	8
9	5	2	4	3	6	8	1	7
1	6	8	7	9	2	4	5	3
4	3	7	1	5	8	9	6	2

EASY - 4

6	5	1	7	8	2	4	9	3
7	3	2	5	9	4	8	6	1
9	4	8	3	1	6	7	2	5
1	7	6	8	2	5	9	3	4
2	9	4	1	3	7	5	8	6
3	8	5	6	4	9	2	1	7
8	2	7	4	6	3	1	5	9
5	1	3	9	7	8	6	4	2
4	6	9	2	5	1	3	7	8

EASY - 5

8	7	9	3	2	1	6	5	4
5	4	3	8	6	9	7	2	1
2	6	1	5	4	7	9	3	8
4	2	5	7	8	6	1	9	3
3	8	6	1	9	5	4	7	2
9	1	7	4	3	2	8	6	5
6	5	2	9	1	4	3	8	7
1	9	8	2	7	3	5	4	6
7	3	4	6	5	8	2	1	9

EASY - 6

4	2	1	9	3	7	6	5	8
3	9	5	8	4	6	7	2	1
7	6	8	2	1	5	3	9	4
6	8	9	4	5	2	1	3	7
1	3	4	7	8	9	5	6	2
2	5	7	3	6	1	4	8	9
9	7	6	1	2	3	8	4	5
5	4	2	6	7	8	9	1	3
8	1	3	5	9	4	2	7	6

EASY - 7

9	5	7	6	3	1	4	8	2
6	2	1	8	9	4	5	3	7
4	8	3	5	2	7	9	6	1
5	9	6	4	8	2	7	1	3
7	3	8	9	1	5	6	2	4
1	4	2	3	7	6	8	9	5
8	6	4	1	5	3	2	7	9
3	7	5	2	6	9	1	4	8
2	1	9	7	4	8	3	5	6

EASY - 8

8	3	6	9	5	7	1	2	4
7	2	4	8	1	3	9	5	6
9	5	1	4	2	6	7	3	8
6	7	2	1	4	8	3	9	5
3	1	9	7	6	5	4	8	2
4	8	5	2	3	9	6	7	1
1	4	8	3	9	2	5	6	7
2	6	3	5	7	1	8	4	9
5	9	7	6	8	4	2	1	3

EASY - 9

1	4	2	6	3	9	7	8	5
5	7	8	2	1	4	9	6	3
3	6	9	8	7	5	2	1	4
2	3	7	5	4	6	1	9	8
9	8	4	3	2	1	5	7	6
6	5	1	7	9	8	3	4	2
7	2	6	1	8	3	4	5	9
8	9	3	4	5	7	6	2	1
4	1	5	9	6	2	8	3	7

EASY - 10

5	8	1	6	7	9	4	2	3
3	4	9	1	8	2	7	6	5
7	2	6	4	5	3	8	9	1
8	6	5	7	2	4	1	3	9
2	3	4	8	9	1	5	7	6
9	1	7	5	3	6	2	8	4
6	5	2	3	1	8	9	4	7
4	7	8	9	6	5	3	1	2
1	9	3	2	4	7	6	5	8

EASY - 11

7	2	9	1	3	4	6	5	8
4	3	1	6	8	5	9	7	2
6	5	8	9	7	2	4	1	3
5	1	4	8	2	9	3	6	7
2	9	7	4	6	3	5	8	1
3	8	6	7	5	1	2	9	4
8	4	5	2	1	6	7	3	9
1	6	2	3	9	7	8	4	5
9	7	3	5	4	8	1	2	6

EASY - 12

5	1	6	3	2	4	9	8	7
4	8	7	6	9	1	5	2	3
2	3	9	5	8	7	4	1	6
8	5	3	7	1	9	2	6	4
6	4	1	2	5	8	3	7	9
9	7	2	4	3	6	1	5	8
7	2	8	9	4	5	6	3	1
3	6	4	1	7	2	8	9	5
1	9	5	8	6	3	7	4	2

EASY - 13

3	5	9	8	4	1	2	7	6
8	1	4	6	2	7	5	9	3
7	2	6	5	9	3	4	1	8
6	8	2	1	3	5	9	4	7
1	3	5	9	7	4	8	6	2
9	4	7	2	6	8	3	5	1
2	6	1	3	5	9	7	8	4
4	9	8	7	1	2	6	3	5
5	7	3	4	8	6	1	2	9

EASY - 14

9	8	1	3	5	4	6	7	2
6	4	5	7	1	2	3	9	8
7	2	3	8	9	6	5	1	4
8	1	4	5	7	9	2	3	6
2	7	6	1	4	3	8	5	9
3	5	9	6	2	8	7	4	1
5	6	2	4	3	1	9	8	7
1	9	7	2	8	5	4	6	3
4	3	8	9	6	7	1	2	5

EASY - 15

3	7	4	6	5	8	2	1	9
9	2	5	1	3	4	7	8	6
1	8	6	2	9	7	3	4	5
7	1	8	4	6	5	9	2	3
4	5	9	8	2	3	6	7	1
2	6	3	9	7	1	8	5	4
5	9	7	3	4	2	1	6	8
8	3	2	5	1	6	4	9	7
6	4	1	7	8	9	5	3	2

EASY - 16

3	4	5	9	1	2	8	6	7
7	6	9	5	4	8	2	3	1
8	2	1	6	7	3	4	5	9
9	8	6	4	5	7	3	1	2
4	3	7	8	2	1	6	9	5
1	5	2	3	9	6	7	8	4
6	7	4	1	3	9	5	2	8
5	9	8	2	6	4	1	7	3
2	1	3	7	8	5	9	4	6

EASY - 17

8	1	5	4	9	7	2	3	6
6	7	4	5	3	2	9	8	1
9	3	2	1	8	6	7	4	5
2	4	6	8	7	3	1	5	9
3	5	8	6	1	9	4	2	7
1	9	7	2	5	4	3	6	8
7	8	1	3	2	5	6	9	4
5	6	3	9	4	1	8	7	2
4	2	9	7	6	8	5	1	3

EASY - 18

1	4	8	9	7	5	3	6	2
6	7	5	1	2	3	4	9	8
9	2	3	8	4	6	5	7	1
2	3	6	7	8	9	1	4	5
8	5	4	3	6	1	9	2	7
7	1	9	4	5	2	6	8	3
3	6	2	5	9	8	7	1	4
5	8	7	6	1	4	2	3	9
4	9	1	2	3	7	8	5	6

EASY - 19

5	9	1	8	7	4	6	2	3
6	4	3	2	5	1	7	8	9
8	7	2	3	6	9	4	1	5
9	6	7	1	2	3	5	4	8
2	3	8	5	4	6	1	9	7
1	5	4	7	9	8	3	6	2
4	8	9	6	3	7	2	5	1
3	2	6	9	1	5	8	7	4
7	1	5	4	8	2	9	3	6

EASY - 20

3	8	9	4	5	7	6	2	1
5	2	1	9	3	6	4	7	8
4	7	6	8	1	2	5	9	3
7	1	4	2	8	5	9	3	6
6	5	2	1	9	3	7	8	4
9	3	8	7	6	4	2	1	5
1	6	7	5	2	8	3	4	9
2	9	5	3	4	1	8	6	7
8	4	3	6	7	9	1	5	2

EASY - 21

9	1	6	8	2	3	7	4	5
3	8	7	9	5	4	6	2	1
4	2	5	7	6	1	9	8	3
6	9	3	5	1	2	4	7	8
1	4	8	6	3	7	2	5	9
5	7	2	4	9	8	1	3	6
2	6	1	3	4	5	8	9	7
8	5	9	2	7	6	3	1	4
7	3	4	1	8	9	5	6	2

EASY - 22

5	6	2	4	7	8	3	1	9
9	4	1	3	2	6	7	8	5
3	7	8	9	5	1	6	2	4
6	2	9	7	1	3	5	4	8
1	3	7	5	8	4	9	6	2
4	8	5	2	6	9	1	7	3
8	9	4	6	3	7	2	5	1
7	5	3	1	4	2	8	9	6
2	1	6	8	9	5	4	3	7

EASY - 23

8	2	9	7	3	4	1	5	6
5	6	3	9	2	1	4	8	7
7	1	4	8	6	5	9	3	2
2	3	8	1	4	6	7	9	5
4	5	7	2	9	8	3	6	1
1	9	6	3	5	7	8	2	4
3	7	5	4	8	2	6	1	9
6	8	1	5	7	9	2	4	3
9	4	2	6	1	3	5	7	8

EASY - 24

9	5	2	4	1	6	7	3	8
6	8	1	9	3	7	4	2	5
3	7	4	5	8	2	6	1	9
7	6	8	1	5	4	3	9	2
2	4	9	3	6	8	1	5	7
5	1	3	7	2	9	8	4	6
1	3	6	2	7	5	9	8	4
8	9	5	6	4	1	2	7	3
4	2	7	8	9	3	5	6	1

EASY - 25

7	2	6	5	8	9	1	4	3
4	5	1	2	7	3	8	6	9
3	9	8	4	1	6	7	5	2
5	6	3	8	4	1	9	2	7
8	7	2	9	3	5	4	1	6
9	1	4	7	6	2	5	3	8
6	8	9	1	2	4	3	7	5
2	4	7	3	5	8	6	9	1
1	3	5	6	9	7	2	8	4

EASY - 26

6	7	1	3	5	4	9	8	2
4	8	5	9	6	2	7	1	3
2	3	9	8	1	7	4	6	5
1	4	7	2	8	9	5	3	6
5	6	2	1	7	3	8	9	4
8	9	3	6	4	5	1	2	7
3	5	6	7	9	8	2	4	1
9	1	4	5	2	6	3	7	8
7	2	8	4	3	1	6	5	9

EASY - 27

4	7	6	1	2	3	9	8	5
1	2	5	7	8	9	4	6	3
8	3	9	6	4	5	7	2	1
3	9	1	4	6	8	5	7	2
2	8	7	9	5	1	6	3	4
5	6	4	3	7	2	1	9	8
6	5	3	2	9	4	8	1	7
9	4	2	8	1	7	3	5	6
7	1	8	5	3	6	2	4	9

EASY - 28

6	2	5	1	4	9	7	8	3
8	9	4	2	7	3	5	1	6
1	7	3	5	8	6	9	2	4
2	5	6	7	9	8	4	3	1
9	1	7	3	2	4	8	6	5
3	4	8	6	1	5	2	7	9
4	6	1	8	5	2	3	9	7
5	3	2	9	6	7	1	4	8
7	8	9	4	3	1	6	5	2

EASY - 29

3	6	8	2	4	9	5	7	1
4	2	7	5	3	1	8	9	6
1	5	9	6	8	7	3	2	4
8	7	6	1	2	5	4	3	9
2	1	4	7	9	3	6	5	8
5	9	3	4	6	8	7	1	2
9	4	1	3	7	6	2	8	5
7	8	2	9	5	4	1	6	3
6	3	5	8	1	2	9	4	7

EASY - 30

5	9	6	8	1	4	7	3	2
3	7	8	6	2	9	4	1	5
4	1	2	3	7	5	8	6	9
1	3	7	9	4	6	5	2	8
8	2	4	5	3	1	6	9	7
6	5	9	7	8	2	1	4	3
7	6	1	2	5	3	9	8	4
2	4	5	1	9	8	3	7	6
9	8	3	4	6	7	2	5	1

EASY - 31

3	7	9	2	1	6	8	5	4
6	2	8	5	3	4	7	9	1
4	1	5	8	7	9	3	6	2
2	9	7	1	6	5	4	3	8
1	5	6	3	4	8	9	2	7
8	4	3	9	2	7	6	1	5
7	8	2	6	9	1	5	4	3
5	6	1	4	8	3	2	7	9
9	3	4	7	5	2	1	8	6

EASY - 32

1	4	5	7	2	6	3	8	9
7	2	3	4	8	9	5	1	6
8	6	9	3	5	1	4	2	7
5	9	4	6	7	8	2	3	1
3	8	6	9	1	2	7	5	4
2	7	1	5	4	3	6	9	8
4	5	8	2	9	7	1	6	3
9	3	2	1	6	4	8	7	5
6	1	7	8	3	5	9	4	2

EASY - 33

3	9	4	1	5	2	6	8	7
2	7	5	4	6	8	3	9	1
8	6	1	3	7	9	5	2	4
1	3	2	9	4	6	8	7	5
5	8	6	2	1	7	9	4	3
7	4	9	8	3	5	1	6	2
4	1	8	6	2	3	7	5	9
9	5	3	7	8	4	2	1	6
6	2	7	5	9	1	4	3	8

EASY - 34

7	4	8	2	1	5	6	9	3
3	6	9	8	4	7	1	5	2
2	1	5	9	6	3	8	7	4
1	3	7	5	8	9	4	2	6
5	9	4	6	2	1	7	3	8
8	2	6	7	3	4	9	1	5
9	5	3	4	7	6	2	8	1
6	7	2	1	5	8	3	4	9
4	8	1	3	9	2	5	6	7

EASY - 35

8	2	7	5	1	3	4	6	9
1	6	3	7	9	4	5	2	8
9	4	5	2	8	6	3	1	7
7	8	4	6	2	9	1	3	5
5	1	9	4	3	7	6	8	2
6	3	2	1	5	8	9	7	4
4	9	8	3	6	2	7	5	1
3	7	1	8	4	5	2	9	6
2	5	6	9	7	1	8	4	3

EASY - 36

9	3	2	8	5	4	1	7	6
4	6	5	7	3	1	2	8	9
8	7	1	2	6	9	3	5	4
5	4	3	1	9	6	8	2	7
7	1	6	4	2	8	5	9	3
2	9	8	3	7	5	4	6	1
3	5	9	6	1	2	7	4	8
6	8	7	5	4	3	9	1	2
1	2	4	9	8	7	6	3	5

EASY - 37

9	1	2	5	3	6	7	4	8
5	6	4	8	9	7	2	1	3
3	7	8	1	2	4	6	5	9
2	9	6	7	8	5	4	3	1
7	8	3	2	4	1	5	9	6
1	4	5	3	6	9	8	2	7
4	5	1	6	7	3	9	8	2
8	3	7	9	5	2	1	6	4
6	2	9	4	1	8	3	7	5

EASY - 38

8	7	1	9	4	3	6	5	2
5	2	6	1	7	8	3	9	4
4	9	3	6	2	5	7	8	1
9	5	2	7	8	6	4	1	3
3	4	8	2	5	1	9	7	6
6	1	7	3	9	4	5	2	8
2	8	9	4	3	7	1	6	5
1	3	5	8	6	9	2	4	7
7	6	4	5	1	2	8	3	9

EASY - 39

7	2	8	1	9	6	4	5	3
3	6	9	7	5	4	8	1	2
1	4	5	3	2	8	7	9	6
9	8	3	5	7	1	6	2	4
4	7	6	2	8	9	5	3	1
2	5	1	4	6	3	9	7	8
5	1	4	6	3	7	2	8	9
8	3	7	9	4	2	1	6	5
6	9	2	8	1	5	3	4	7

EASY - 40

3	9	1	4	6	8	7	2	5
8	4	6	5	7	2	3	1	9
2	7	5	9	1	3	4	8	6
4	3	8	6	5	1	2	9	7
5	1	2	7	3	9	8	6	4
9	6	7	2	8	4	5	3	1
1	8	4	3	9	7	6	5	2
6	2	9	8	4	5	1	7	3
7	5	3	1	2	6	9	4	8

EASY - 41

8	7	1	3	4	9	5	6	2
2	3	4	5	6	1	7	8	9
6	5	9	8	7	2	1	3	4
5	4	2	6	9	3	8	7	1
1	8	6	2	5	7	4	9	3
3	9	7	1	8	4	2	5	6
4	6	3	7	1	5	9	2	8
9	2	5	4	3	8	6	1	7
7	1	8	9	2	6	3	4	5

EASY - 42

5	9	6	8	1	7	2	4	3
4	7	2	9	6	3	8	1	5
3	1	8	4	2	5	9	6	7
6	3	7	1	9	8	4	5	2
9	5	1	2	7	4	6	3	8
8	2	4	3	5	6	1	7	9
2	4	9	5	3	1	7	8	6
7	8	5	6	4	2	3	9	1
1	6	3	7	8	9	5	2	4

EASY - 43

1	3	7	6	2	8	4	5	9
2	8	4	7	5	9	1	6	3
5	6	9	3	1	4	7	8	2
9	5	8	1	4	2	3	7	6
7	4	3	8	9	6	2	1	5
6	1	2	5	3	7	9	4	8
4	2	1	9	6	5	8	3	7
8	9	5	4	7	3	6	2	1
3	7	6	2	8	1	5	9	4

EASY - 44

2	8	1	5	6	9	7	3	4
4	5	6	7	3	2	9	8	1
3	7	9	4	1	8	2	5	6
8	1	7	2	4	6	3	9	5
6	2	5	9	8	3	1	4	7
9	4	3	1	5	7	8	6	2
7	9	8	6	2	5	4	1	3
5	3	4	8	7	1	6	2	9
1	6	2	3	9	4	5	7	8

EASY - 45

8	7	5	9	1	6	2	4	3
6	4	9	7	2	3	8	1	5
1	3	2	5	8	4	7	9	6
9	6	1	2	7	5	3	8	4
2	5	7	4	3	8	1	6	9
4	8	3	6	9	1	5	7	2
7	9	4	1	5	2	6	3	8
3	2	6	8	4	7	9	5	1
5	1	8	3	6	9	4	2	7

EASY - 46

4	7	1	2	5	8	3	9	6
9	8	3	7	1	6	5	4	2
5	2	6	3	9	4	7	1	8
8	9	7	5	3	1	2	6	4
2	1	5	4	6	9	8	7	3
6	3	4	8	2	7	1	5	9
7	5	2	9	4	3	6	8	1
3	6	9	1	8	5	4	2	7
1	4	8	6	7	2	9	3	5

EASY - 47

9	2	6	4	5	8	1	3	7
3	4	7	1	2	6	5	9	8
8	1	5	7	9	3	2	4	6
1	8	4	5	6	7	3	2	9
5	3	9	2	8	4	7	6	1
7	6	2	9	3	1	8	5	4
6	7	1	3	4	2	9	8	5
4	9	3	8	7	5	6	1	2
2	5	8	6	1	9	4	7	3

EASY - 48

4	3	6	9	8	2	5	1	7
8	7	9	1	4	5	3	6	2
5	1	2	3	6	7	8	4	9
1	8	5	6	7	4	9	2	3
9	4	7	8	2	3	6	5	1
2	6	3	5	9	1	4	7	8
7	2	8	4	3	6	1	9	5
3	5	4	7	1	9	2	8	6
6	9	1	2	5	8	7	3	4

EASY - 49

1	3	2	7	5	8	6	4	9
8	5	9	6	1	4	7	2	3
4	6	7	3	2	9	1	8	5
6	4	3	2	7	5	9	1	8
2	7	1	9	8	3	5	6	4
5	9	8	1	4	6	2	3	7
3	8	6	5	9	2	4	7	1
7	2	5	4	3	1	8	9	6
9	1	4	8	6	7	3	5	2

EASY - 50

4	3	2	1	7	5	9	6	8
6	7	9	4	3	8	1	2	5
8	5	1	6	2	9	3	4	7
1	2	7	9	4	6	5	8	3
5	9	4	7	8	3	2	1	6
3	6	8	2	5	1	7	9	4
7	8	6	3	9	2	4	5	1
9	4	5	8	1	7	6	3	2
2	1	3	5	6	4	8	7	9

EASY - 51

8	1	3	9	6	4	7	2	5
9	6	5	8	2	7	1	4	3
2	4	7	1	3	5	6	9	8
6	2	8	4	9	1	3	5	7
4	3	9	5	7	6	2	8	1
5	7	1	2	8	3	9	6	4
7	8	6	3	4	9	5	1	2
3	5	2	6	1	8	4	7	9
1	9	4	7	5	2	8	3	6

EASY - 52

5	7	1	4	6	2	3	8	9
3	4	6	5	8	9	1	7	2
2	8	9	1	3	7	5	6	4
1	9	3	7	2	6	8	4	5
7	6	2	8	5	4	9	1	3
4	5	8	9	1	3	7	2	6
6	1	4	3	9	8	2	5	7
8	3	7	2	4	5	6	9	1
9	2	5	6	7	1	4	3	8

EASY - 53

5	6	3	7	8	9	2	1	4
7	4	9	6	2	1	5	3	8
8	1	2	3	4	5	9	7	6
9	2	4	1	3	6	7	8	5
3	5	8	9	7	4	1	6	2
6	7	1	8	5	2	3	4	9
2	3	5	4	6	7	8	9	1
1	8	6	2	9	3	4	5	7
4	9	7	5	1	8	6	2	3

EASY - 54

9	6	4	3	8	5	1	7	2
5	3	7	9	1	2	4	6	8
8	1	2	7	6	4	5	9	3
7	9	1	6	5	8	2	3	4
4	2	3	1	9	7	8	5	6
6	8	5	4	2	3	7	1	9
1	4	8	5	3	6	9	2	7
3	7	9	2	4	1	6	8	5
2	5	6	8	7	9	3	4	1

EASY - 55

3	1	5	8	7	4	2	6	9
7	8	2	6	9	5	1	3	4
4	6	9	3	2	1	7	5	8
2	5	4	7	3	8	9	1	6
9	3	8	1	5	6	4	2	7
6	7	1	9	4	2	5	8	3
5	2	7	4	6	3	8	9	1
8	9	3	5	1	7	6	4	2
1	4	6	2	8	9	3	7	5

EASY - 56

9	2	4	1	8	5	6	7	3
3	6	5	7	2	9	1	4	8
7	1	8	4	6	3	9	5	2
2	9	6	8	1	4	7	3	5
8	5	3	9	7	6	4	2	1
4	7	1	3	5	2	8	9	6
1	4	7	2	3	8	5	6	9
6	8	2	5	9	7	3	1	4
5	3	9	6	4	1	2	8	7

EASY - 57

4	7	5	3	8	9	2	6	1
8	2	1	5	6	4	7	9	3
3	9	6	1	7	2	8	5	4
5	8	4	6	1	7	9	3	2
6	3	9	2	5	8	4	1	7
7	1	2	9	4	3	6	8	5
2	5	8	7	3	6	1	4	9
1	4	7	8	9	5	3	2	6
9	6	3	4	2	1	5	7	8

EASY - 58

5	2	4	8	6	7	1	3	9
3	6	9	4	1	5	7	8	2
1	7	8	3	9	2	5	6	4
2	4	7	9	5	8	3	1	6
6	8	3	1	7	4	9	2	5
9	1	5	6	2	3	8	4	7
8	5	6	2	3	9	4	7	1
7	3	2	5	4	1	6	9	8
4	9	1	7	8	6	2	5	3

EASY - 59

4	5	8	9	7	2	1	6	3
6	3	9	4	5	1	2	8	7
7	1	2	8	6	3	5	9	4
9	6	1	7	4	8	3	5	2
2	4	5	3	1	9	6	7	8
3	8	7	5	2	6	9	4	1
8	2	4	1	9	5	7	3	6
1	9	3	6	8	7	4	2	5
5	7	6	2	3	4	8	1	9

EASY - 60

2	4	7	9	1	5	8	6	3
9	8	3	6	4	7	1	2	5
6	5	1	2	8	3	7	9	4
4	7	8	3	5	2	9	1	6
5	6	2	8	9	1	4	3	7
3	1	9	4	7	6	5	8	2
1	3	5	7	2	9	6	4	8
8	9	6	5	3	4	2	7	1
7	2	4	1	6	8	3	5	9

EASY - 61

2	3	8	1	5	6	4	9	7
4	5	7	8	3	9	2	6	1
1	6	9	4	2	7	8	3	5
6	2	4	7	8	3	1	5	9
8	9	1	2	6	5	3	7	4
3	7	5	9	4	1	6	2	8
7	8	3	6	9	4	5	1	2
5	1	2	3	7	8	9	4	6
9	4	6	5	1	2	7	8	3

EASY - 62

9	8	4	6	1	3	5	2	7
7	5	1	2	8	4	6	3	9
2	3	6	5	7	9	8	4	1
4	2	5	3	9	7	1	6	8
3	7	9	8	6	1	2	5	4
6	1	8	4	5	2	9	7	3
1	6	3	7	2	8	4	9	5
8	4	2	9	3	5	7	1	6
5	9	7	1	4	6	3	8	2

EASY - 63

3	8	2	5	9	6	7	4	1
4	7	5	1	2	8	6	3	9
1	6	9	4	3	7	8	2	5
8	1	3	6	7	4	9	5	2
9	2	6	3	5	1	4	7	8
7	5	4	9	8	2	1	6	3
2	4	7	8	1	5	3	9	6
6	9	1	2	4	3	5	8	7
5	3	8	7	6	9	2	1	4

EASY - 64

8	5	7	6	9	3	2	1	4
4	3	6	8	1	2	9	7	5
9	1	2	5	7	4	8	6	3
6	4	5	9	3	8	1	2	7
3	7	9	2	5	1	6	4	8
2	8	1	7	4	6	3	5	9
1	2	3	4	8	7	5	9	6
7	9	8	1	6	5	4	3	2
5	6	4	3	2	9	7	8	1

EASY - 65

1	8	5	6	9	7	3	2	4
3	6	9	8	4	2	1	5	7
4	7	2	5	1	3	6	9	8
5	9	1	4	6	8	7	3	2
8	2	6	3	7	1	5	4	9
7	3	4	2	5	9	8	1	6
9	4	3	1	8	6	2	7	5
2	5	8	7	3	4	9	6	1
6	1	7	9	2	5	4	8	3

EASY - 66

2	4	8	7	3	9	1	5	6
1	9	6	8	4	5	3	2	7
3	5	7	6	2	1	9	8	4
5	6	1	2	8	4	7	3	9
7	3	2	1	9	6	8	4	5
4	8	9	5	7	3	2	6	1
9	7	4	3	6	2	5	1	8
8	1	3	4	5	7	6	9	2
6	2	5	9	1	8	4	7	3

EASY - 67

3	4	5	8	7	6	9	2	1
7	2	8	5	1	9	4	6	3
1	9	6	4	2	3	7	5	8
5	3	2	9	6	8	1	7	4
6	1	4	2	3	7	8	9	5
8	7	9	1	4	5	2	3	6
2	5	1	3	9	4	6	8	7
9	6	3	7	8	1	5	4	2
4	8	7	6	5	2	3	1	9

EASY - 68

9	4	5	6	2	8	3	1	7
2	3	6	9	1	7	5	4	8
1	7	8	5	4	3	2	6	9
8	9	3	1	7	4	6	5	2
4	5	2	8	9	6	1	7	3
6	1	7	2	3	5	9	8	4
7	8	9	3	5	1	4	2	6
5	2	4	7	6	9	8	3	1
3	6	1	4	8	2	7	9	5

EASY - 69

4	7	6	1	5	9	2	3	8
2	5	9	3	7	8	1	6	4
8	1	3	6	4	2	5	9	7
6	8	1	7	9	5	3	4	2
9	2	7	8	3	4	6	5	1
5	3	4	2	1	6	8	7	9
1	9	5	4	8	3	7	2	6
3	6	8	9	2	7	4	1	5
7	4	2	5	6	1	9	8	3

EASY - 70

9	7	8	3	4	1	5	6	2
4	5	1	7	6	2	8	3	9
3	2	6	5	8	9	7	1	4
7	8	2	4	9	6	1	5	3
6	9	5	2	1	3	4	8	7
1	3	4	8	5	7	9	2	6
2	4	3	1	7	8	6	9	5
5	1	9	6	2	4	3	7	8
8	6	7	9	3	5	2	4	1

EASY - 71

1	4	2	3	6	8	9	5	7
7	5	3	4	9	2	8	1	6
8	9	6	1	5	7	4	2	3
5	2	9	6	3	4	7	8	1
4	3	8	2	7	1	5	6	9
6	7	1	5	8	9	2	3	4
2	6	7	9	1	5	3	4	8
3	8	5	7	4	6	1	9	2
9	1	4	8	2	3	6	7	5

EASY - 72

3	5	8	9	7	6	4	2	1
1	7	9	5	2	4	3	6	8
2	6	4	8	3	1	5	9	7
4	8	2	1	9	5	6	7	3
5	3	7	4	6	2	1	8	9
6	9	1	3	8	7	2	4	5
9	2	6	7	1	3	8	5	4
8	4	3	2	5	9	7	1	6
7	1	5	6	4	8	9	3	2

EASY - 73

8	4	2	6	3	1	5	9	7
3	6	5	9	8	7	2	1	4
7	1	9	5	2	4	3	8	6
5	9	8	4	1	6	7	3	2
1	2	3	7	5	8	4	6	9
6	7	4	3	9	2	8	5	1
4	3	6	1	7	5	9	2	8
2	5	1	8	4	9	6	7	3
9	8	7	2	6	3	1	4	5

EASY - 74

8	2	9	1	5	3	6	4	7
7	1	4	8	9	6	2	3	5
5	3	6	7	4	2	8	1	9
6	4	1	2	3	5	7	9	8
2	7	3	4	8	9	5	6	1
9	5	8	6	1	7	3	2	4
1	6	7	9	2	8	4	5	3
4	8	5	3	6	1	9	7	2
3	9	2	5	7	4	1	8	6

EASY - 75

5	8	3	9	2	7	6	1	4
9	1	6	4	8	5	7	3	2
2	4	7	1	3	6	5	8	9
8	7	1	6	5	4	9	2	3
6	3	2	8	7	9	1	4	5
4	9	5	3	1	2	8	7	6
3	6	9	7	4	1	2	5	8
1	5	4	2	6	8	3	9	7
7	2	8	5	9	3	4	6	1

EASY - 76

9	7	4	5	3	6	8	2	1
8	2	1	7	9	4	3	5	6
5	3	6	1	2	8	4	9	7
2	1	5	6	4	3	9	7	8
4	6	7	8	5	9	2	1	3
3	9	8	2	1	7	6	4	5
1	8	9	3	7	2	5	6	4
6	5	2	4	8	1	7	3	9
7	4	3	9	6	5	1	8	2

EASY - 77

9	1	7	6	3	8	5	4	2
8	5	6	1	2	4	3	7	9
2	3	4	9	7	5	6	8	1
7	8	2	4	5	9	1	6	3
3	4	9	7	6	1	2	5	8
5	6	1	2	8	3	4	9	7
6	7	8	5	1	2	9	3	4
1	9	5	3	4	7	8	2	6
4	2	3	8	9	6	7	1	5

EASY - 78

6	9	5	3	4	2	8	7	1
4	8	1	5	7	9	6	3	2
2	7	3	8	1	6	4	9	5
5	2	6	4	8	3	7	1	9
1	3	7	9	6	5	2	8	4
9	4	8	7	2	1	5	6	3
7	1	9	6	5	4	3	2	8
3	6	4	2	9	8	1	5	7
8	5	2	1	3	7	9	4	6

EASY - 79

8	2	9	3	4	7	6	5	1
6	4	7	8	1	5	9	2	3
5	1	3	9	2	6	8	4	7
4	7	2	6	8	3	1	9	5
1	9	8	7	5	4	3	6	2
3	6	5	1	9	2	4	7	8
2	5	1	4	6	8	7	3	9
9	3	4	2	7	1	5	8	6
7	8	6	5	3	9	2	1	4

EASY - 80

5	8	6	7	1	2	9	4	3
4	9	3	6	5	8	2	1	7
7	1	2	9	4	3	8	5	6
8	3	1	4	6	7	5	9	2
2	4	9	1	3	5	6	7	8
6	7	5	2	8	9	1	3	4
3	5	7	8	9	6	4	2	1
1	2	8	5	7	4	3	6	9
9	6	4	3	2	1	7	8	5

EASY - 81

8	2	7	5	3	4	6	9	1
4	6	9	2	7	1	8	5	3
3	5	1	9	6	8	4	2	7
7	4	5	8	1	3	9	6	2
6	9	8	7	2	5	3	1	4
2	1	3	6	4	9	7	8	5
9	3	2	1	8	7	5	4	6
1	8	4	3	5	6	2	7	9
5	7	6	4	9	2	1	3	8

EASY - 82

4	7	5	1	6	2	8	9	3
1	2	9	4	3	8	5	7	6
6	8	3	9	5	7	2	4	1
9	4	7	5	8	1	6	3	2
5	3	8	2	4	6	9	1	7
2	1	6	3	7	9	4	8	5
7	5	1	8	2	4	3	6	9
8	6	2	7	9	3	1	5	4
3	9	4	6	1	5	7	2	8

EASY - 83

1	3	9	5	6	2	4	8	7
8	4	6	9	1	7	5	2	3
7	2	5	3	8	4	9	6	1
6	9	4	2	5	3	7	1	8
5	1	7	8	4	6	3	9	2
3	8	2	7	9	1	6	4	5
9	7	1	6	2	5	8	3	4
2	6	3	4	7	8	1	5	9
4	5	8	1	3	9	2	7	6

EASY - 84

9	7	1	2	8	5	4	3	6
2	8	6	1	4	3	7	5	9
3	5	4	9	6	7	8	2	1
6	4	5	3	9	2	1	7	8
8	1	9	7	5	6	2	4	3
7	2	3	8	1	4	9	6	5
4	9	7	5	3	1	6	8	2
5	6	8	4	2	9	3	1	7
1	3	2	6	7	8	5	9	4

EASY - 85

8	5	9	3	7	4	6	2	1
7	1	2	9	6	5	4	3	8
4	6	3	2	8	1	7	5	9
1	2	4	6	5	9	3	8	7
9	7	5	8	2	3	1	6	4
3	8	6	1	4	7	2	9	5
6	9	7	4	3	8	5	1	2
5	3	1	7	9	2	8	4	6
2	4	8	5	1	6	9	7	3

EASY - 86

6	7	9	3	2	5	8	1	4
4	5	2	1	8	6	3	9	7
8	1	3	4	7	9	2	5	6
7	8	4	5	1	2	9	6	3
1	9	5	6	3	7	4	2	8
2	3	6	8	9	4	5	7	1
5	6	7	2	4	3	1	8	9
3	2	8	9	6	1	7	4	5
9	4	1	7	5	8	6	3	2

EASY - 87

4	8	9	2	1	3	7	5	6
6	1	5	8	4	7	9	2	3
3	7	2	9	6	5	4	1	8
9	2	1	5	8	4	3	6	7
8	6	3	7	2	1	5	4	9
7	5	4	6	3	9	2	8	1
2	4	7	1	9	6	8	3	5
5	3	6	4	7	8	1	9	2
1	9	8	3	5	2	6	7	4

EASY - 88

3	6	7	9	5	4	1	2	8
2	1	4	6	8	7	5	3	9
9	5	8	3	2	1	4	7	6
8	9	2	5	1	6	3	4	7
4	7	5	2	9	3	6	8	1
6	3	1	7	4	8	2	9	5
1	8	3	4	6	9	7	5	2
7	2	9	1	3	5	8	6	4
5	4	6	8	7	2	9	1	3

EASY - 89

8	1	4	3	5	2	9	6	7
5	3	9	4	7	6	2	1	8
2	7	6	9	1	8	4	3	5
1	5	8	2	4	7	6	9	3
7	6	3	1	8	9	5	4	2
9	4	2	5	6	3	8	7	1
3	9	5	7	2	4	1	8	6
4	8	1	6	3	5	7	2	9
6	2	7	8	9	1	3	5	4

EASY - 90

4	1	6	9	5	7	3	2	8
8	5	7	6	3	2	9	4	1
9	2	3	4	1	8	7	6	5
1	7	9	2	4	3	8	5	6
6	8	2	7	9	5	4	1	3
3	4	5	1	8	6	2	7	9
7	6	1	8	2	9	5	3	4
2	3	8	5	6	4	1	9	7
5	9	4	3	7	1	6	8	2

EASY - 91

7	6	2	4	3	9	8	5	1
3	8	5	2	7	1	4	6	9
4	9	1	8	5	6	7	2	3
5	1	3	6	4	7	9	8	2
6	2	9	5	1	8	3	7	4
8	7	4	9	2	3	5	1	6
2	3	6	7	9	5	1	4	8
1	4	7	3	8	2	6	9	5
9	5	8	1	6	4	2	3	7

EASY - 92

1	5	9	4	3	2	7	6	8
7	3	4	5	6	8	1	2	9
2	8	6	7	1	9	3	4	5
6	9	8	3	2	7	5	1	4
5	4	1	9	8	6	2	3	7
3	2	7	1	5	4	8	9	6
4	7	2	8	9	1	6	5	3
9	6	3	2	7	5	4	8	1
8	1	5	6	4	3	9	7	2

EASY - 93

3	7	9	2	1	5	4	6	8
1	6	4	8	9	3	2	7	5
2	8	5	6	7	4	3	1	9
4	3	1	9	2	6	8	5	7
7	5	8	3	4	1	6	9	2
9	2	6	7	5	8	1	3	4
5	1	2	4	6	9	7	8	3
8	9	7	1	3	2	5	4	6
6	4	3	5	8	7	9	2	1

EASY - 94

9	1	7	2	4	3	8	5	6
2	8	4	1	5	6	7	3	9
3	5	6	8	9	7	4	2	1
8	4	1	3	2	5	9	6	7
7	6	3	4	8	9	2	1	5
5	2	9	7	6	1	3	4	8
4	3	5	6	7	8	1	9	2
1	9	8	5	3	2	6	7	4
6	7	2	9	1	4	5	8	3

EASY - 95

2	3	7	5	8	6	1	9	4
6	1	5	3	4	9	7	2	8
4	9	8	1	7	2	6	3	5
5	8	2	7	1	4	9	6	3
7	6	1	9	3	5	4	8	2
9	4	3	2	6	8	5	1	7
3	5	4	6	2	1	8	7	9
8	7	6	4	9	3	2	5	1
1	2	9	8	5	7	3	4	6

EASY - 96

4	8	2	3	1	5	7	6	9
3	7	9	4	8	6	2	5	1
1	5	6	2	7	9	3	8	4
7	6	1	5	9	2	4	3	8
8	4	5	7	3	1	6	9	2
2	9	3	8	6	4	1	7	5
5	3	4	6	2	8	9	1	7
6	1	8	9	4	7	5	2	3
9	2	7	1	5	3	8	4	6

EASY - 97

4	5	6	3	2	9	7	1	8
9	3	1	6	7	8	2	4	5
7	8	2	1	4	5	6	3	9
6	2	7	9	8	1	3	5	4
8	4	5	7	6	3	1	9	2
3	1	9	2	5	4	8	7	6
1	6	8	4	9	7	5	2	3
5	7	4	8	3	2	9	6	1
2	9	3	5	1	6	4	8	7

EASY - 98

5	6	2	9	8	3	1	4	7
4	1	9	2	6	7	8	5	3
8	7	3	1	4	5	6	9	2
9	5	4	7	3	1	2	8	6
3	2	6	8	9	4	5	7	1
7	8	1	5	2	6	9	3	4
1	9	8	4	7	2	3	6	5
6	4	5	3	1	8	7	2	9
2	3	7	6	5	9	4	1	8

EASY - 99

3	5	9	8	7	6	2	1	4
1	2	4	9	3	5	6	7	8
6	8	7	4	2	1	9	3	5
2	7	3	5	4	8	1	9	6
5	9	8	1	6	2	3	4	7
4	6	1	3	9	7	8	5	2
7	4	2	6	1	9	5	8	3
8	1	6	7	5	3	4	2	9
9	3	5	2	8	4	7	6	1

EASY - 100

5	6	9	2	1	8	7	4	3
1	8	3	9	7	4	2	6	5
4	7	2	3	6	5	9	8	1
8	2	7	1	4	6	5	3	9
3	5	6	8	9	7	4	1	2
9	1	4	5	3	2	8	7	6
6	4	1	7	2	9	3	5	8
7	9	8	6	5	3	1	2	4
2	3	5	4	8	1	6	9	7

EASY - 101

4	1	5	3	8	9	7	6	2
3	6	9	2	5	7	1	4	8
2	7	8	6	1	4	9	5	3
1	3	6	7	4	2	8	9	5
8	4	7	9	6	5	3	2	1
5	9	2	8	3	1	4	7	6
7	5	1	4	2	3	6	8	9
6	2	4	1	9	8	5	3	7
9	8	3	5	7	6	2	1	4

EASY - 102

4	6	1	2	3	7	8	9	5
8	3	2	1	5	9	4	6	7
9	7	5	4	8	6	2	1	3
2	1	9	3	7	8	6	5	4
5	4	3	6	1	2	7	8	9
6	8	7	9	4	5	1	3	2
3	2	6	7	9	1	5	4	8
1	9	8	5	2	4	3	7	6
7	5	4	8	6	3	9	2	1

EASY - 103

9	2	8	1	5	3	6	7	4
6	3	4	7	9	2	5	1	8
1	5	7	4	8	6	3	2	9
3	7	1	8	2	9	4	5	6
2	9	6	5	4	1	7	8	3
4	8	5	3	6	7	2	9	1
5	4	2	6	1	8	9	3	7
7	1	9	2	3	4	8	6	5
8	6	3	9	7	5	1	4	2

EASY - 104

7	9	3	4	2	1	8	5	6
2	4	8	3	5	6	7	1	9
5	1	6	7	9	8	3	2	4
8	7	1	5	4	9	6	3	2
4	2	5	6	1	3	9	7	8
3	6	9	8	7	2	5	4	1
1	5	4	9	6	7	2	8	3
9	3	7	2	8	4	1	6	5
6	8	2	1	3	5	4	9	7

EASY - 105

5	1	4	7	2	8	9	6	3
6	9	8	4	3	1	5	2	7
3	7	2	6	9	5	4	8	1
1	6	7	8	5	2	3	4	9
9	8	5	3	4	6	7	1	2
2	4	3	9	1	7	6	5	8
8	5	9	2	6	3	1	7	4
4	2	6	1	7	9	8	3	5
7	3	1	5	8	4	2	9	6

EASY - 106

2	1	5	6	7	8	4	9	3
3	4	9	1	2	5	7	6	8
7	8	6	3	9	4	1	2	5
5	6	3	2	1	7	8	4	9
1	2	7	8	4	9	5	3	6
4	9	8	5	3	6	2	7	1
9	5	4	7	8	3	6	1	2
8	7	1	9	6	2	3	5	4
6	3	2	4	5	1	9	8	7

EASY - 107

3	8	4	7	5	6	1	9	2
6	9	1	3	2	8	4	5	7
7	2	5	4	1	9	8	3	6
8	1	6	9	4	5	7	2	3
9	4	7	6	3	2	5	1	8
2	5	3	1	8	7	6	4	9
1	3	8	2	6	4	9	7	5
5	7	2	8	9	1	3	6	4
4	6	9	5	7	3	2	8	1

EASY - 108

7	9	2	8	3	4	1	5	6
3	1	4	6	9	5	2	7	8
8	5	6	7	2	1	4	9	3
5	3	9	1	8	6	7	4	2
6	4	7	9	5	2	8	3	1
2	8	1	3	4	7	5	6	9
9	6	5	4	1	8	3	2	7
4	7	8	2	6	3	9	1	5
1	2	3	5	7	9	6	8	4

EASY - 109

8	1	2	7	9	3	6	4	5
6	5	3	2	4	8	1	7	9
9	4	7	1	6	5	8	3	2
4	7	9	6	1	2	3	5	8
2	8	1	3	5	7	4	9	6
5	3	6	9	8	4	7	2	1
7	2	8	5	3	1	9	6	4
3	6	4	8	2	9	5	1	7
1	9	5	4	7	6	2	8	3

EASY - 110

2	9	6	5	1	7	8	4	3
7	5	4	6	3	8	9	1	2
8	3	1	2	9	4	6	5	7
3	6	8	4	7	2	1	9	5
9	7	2	8	5	1	3	6	4
1	4	5	9	6	3	7	2	8
4	2	3	1	8	6	5	7	9
6	8	9	7	2	5	4	3	1
5	1	7	3	4	9	2	8	6

EASY - 111

2	1	8	5	6	7	4	3	9
5	4	6	9	8	3	1	2	7
3	7	9	2	4	1	6	8	5
6	3	5	4	2	9	7	1	8
7	2	4	6	1	8	9	5	3
8	9	1	3	7	5	2	4	6
1	6	7	8	5	4	3	9	2
4	8	3	7	9	2	5	6	1
9	5	2	1	3	6	8	7	4

EASY - 112

2	1	7	4	8	5	6	9	3
6	9	4	3	7	2	5	8	1
8	3	5	9	6	1	7	2	4
5	4	6	2	1	7	8	3	9
9	2	8	6	5	3	1	4	7
3	7	1	8	4	9	2	5	6
4	5	3	1	2	6	9	7	8
1	8	2	7	9	4	3	6	5
7	6	9	5	3	8	4	1	2

EASY - 113

4	9	3	6	1	8	7	5	2
5	8	6	2	3	7	1	9	4
7	2	1	9	4	5	8	3	6
1	4	7	3	5	6	2	8	9
6	5	2	7	8	9	4	1	3
8	3	9	1	2	4	6	7	5
3	1	5	4	7	2	9	6	8
2	6	8	5	9	1	3	4	7
9	7	4	8	6	3	5	2	1

EASY - 114

2	6	1	5	4	8	3	7	9
3	7	8	6	9	1	4	5	2
9	4	5	2	3	7	6	8	1
6	5	2	1	8	9	7	4	3
8	3	7	4	2	5	9	1	6
1	9	4	3	7	6	5	2	8
4	8	6	7	1	3	2	9	5
7	1	3	9	5	2	8	6	4
5	2	9	8	6	4	1	3	7

EASY - 115

8	1	4	7	9	6	2	3	5
6	7	9	5	2	3	1	4	8
2	3	5	1	8	4	9	7	6
9	8	3	2	1	5	7	6	4
4	6	2	8	3	7	5	1	9
1	5	7	4	6	9	8	2	3
3	2	6	9	5	1	4	8	7
7	9	1	3	4	8	6	5	2
5	4	8	6	7	2	3	9	1

EASY - 116

3	8	2	6	1	5	7	4	9
1	9	5	8	7	4	3	2	6
6	7	4	2	3	9	5	8	1
4	3	9	7	8	2	1	6	5
2	1	7	9	5	6	8	3	4
8	5	6	1	4	3	9	7	2
7	6	3	5	2	1	4	9	8
9	4	1	3	6	8	2	5	7
5	2	8	4	9	7	6	1	3

EASY - 117

6	2	8	7	1	9	3	4	5
1	7	4	5	8	3	6	2	9
5	3	9	2	4	6	8	1	7
3	5	2	8	6	1	9	7	4
7	8	6	9	5	4	1	3	2
9	4	1	3	2	7	5	8	6
4	9	5	1	7	8	2	6	3
2	1	7	6	3	5	4	9	8
8	6	3	4	9	2	7	5	1

EASY - 118

9	3	4	5	1	6	2	7	8
6	2	1	3	7	8	4	9	5
7	8	5	2	4	9	1	3	6
1	6	3	9	8	7	5	2	4
5	9	2	6	3	4	8	1	7
8	4	7	1	5	2	3	6	9
2	7	8	4	9	3	6	5	1
4	5	6	7	2	1	9	8	3
3	1	9	8	6	5	7	4	2

EASY - 119

8	9	2	7	3	5	4	6	1
1	4	6	2	9	8	3	5	7
7	5	3	6	4	1	8	9	2
9	1	5	3	2	4	7	8	6
4	6	7	5	8	9	2	1	3
3	2	8	1	7	6	9	4	5
6	3	4	8	1	7	5	2	9
5	7	9	4	6	2	1	3	8
2	8	1	9	5	3	6	7	4

EASY - 120

1	4	3	9	6	7	8	5	2
8	9	2	3	5	1	6	7	4
6	7	5	8	4	2	9	1	3
3	6	1	2	8	5	7	4	9
9	8	4	6	7	3	1	2	5
2	5	7	4	1	9	3	8	6
7	2	9	1	3	4	5	6	8
5	3	6	7	2	8	4	9	1
4	1	8	5	9	6	2	3	7

EASY - 121

9	8	3	7	2	6	5	1	4
7	2	5	4	8	1	3	6	9
4	1	6	3	9	5	7	2	8
3	9	2	8	1	7	6	4	5
5	7	1	6	4	9	8	3	2
6	4	8	5	3	2	1	9	7
1	6	9	2	5	8	4	7	3
2	5	4	1	7	3	9	8	6
8	3	7	9	6	4	2	5	1

EASY - 122

2	8	9	3	4	7	6	5	1
6	7	1	2	8	5	9	4	3
3	4	5	1	9	6	2	8	7
1	2	4	8	3	9	7	6	5
9	3	6	7	5	1	8	2	4
7	5	8	6	2	4	1	3	9
8	9	7	4	6	3	5	1	2
4	1	2	5	7	8	3	9	6
5	6	3	9	1	2	4	7	8

EASY - 123

7	1	6	5	8	4	2	3	9
9	5	4	2	3	1	8	7	6
3	8	2	9	6	7	1	4	5
5	4	8	6	2	9	3	1	7
1	6	7	3	4	8	5	9	2
2	3	9	1	7	5	4	6	8
4	9	3	7	5	2	6	8	1
8	2	1	4	9	6	7	5	3
6	7	5	8	1	3	9	2	4

EASY - 124

9	4	7	5	3	8	2	1	6
6	1	3	4	9	2	5	7	8
5	8	2	1	6	7	4	9	3
2	6	9	8	4	3	7	5	1
1	7	4	9	5	6	3	8	2
3	5	8	2	7	1	9	6	4
4	9	6	3	8	5	1	2	7
7	2	5	6	1	4	8	3	9
8	3	1	7	2	9	6	4	5

EASY - 125

3	4	2	8	7	6	5	1	9
9	8	6	1	5	2	7	4	3
1	5	7	9	4	3	6	2	8
4	3	9	6	1	5	2	8	7
2	6	1	7	3	8	4	9	5
8	7	5	4	2	9	1	3	6
6	9	4	5	8	1	3	7	2
5	1	3	2	9	7	8	6	4
7	2	8	3	6	4	9	5	1

EASY - 126

9	2	3	7	6	4	8	5	1
1	7	4	2	5	8	9	6	3
5	6	8	1	9	3	2	7	4
3	5	9	4	8	6	1	2	7
7	4	1	5	3	2	6	8	9
6	8	2	9	7	1	3	4	5
4	9	6	3	2	7	5	1	8
8	1	5	6	4	9	7	3	2
2	3	7	8	1	5	4	9	6

EASY - 127

1	4	5	2	9	6	7	3	8
7	8	9	3	1	4	6	2	5
6	3	2	8	5	7	9	1	4
2	5	4	7	3	1	8	6	9
8	7	1	4	6	9	3	5	2
3	9	6	5	2	8	1	4	7
9	1	8	6	4	5	2	7	3
4	6	3	9	7	2	5	8	1
5	2	7	1	8	3	4	9	6

EASY - 128

1	7	9	3	5	8	2	6	4
8	6	4	1	9	2	5	7	3
2	3	5	6	4	7	8	9	1
6	9	3	5	2	4	7	1	8
4	5	2	8	7	1	9	3	6
7	8	1	9	6	3	4	5	2
9	2	8	7	3	6	1	4	5
3	4	7	2	1	5	6	8	9
5	1	6	4	8	9	3	2	7

EASY - 129

4	6	9	2	5	1	7	8	3
1	2	7	9	3	8	5	6	4
5	8	3	6	4	7	1	9	2
7	3	6	8	9	2	4	1	5
8	5	2	3	1	4	6	7	9
9	1	4	5	7	6	2	3	8
3	9	1	4	6	5	8	2	7
2	7	5	1	8	3	9	4	6
6	4	8	7	2	9	3	5	1

EASY - 130

3	1	2	6	8	4	9	7	5
9	5	7	2	3	1	8	6	4
8	6	4	7	5	9	3	2	1
2	7	3	9	1	6	5	4	8
6	9	1	8	4	5	7	3	2
4	8	5	3	2	7	1	9	6
5	2	8	4	9	3	6	1	7
1	3	6	5	7	2	4	8	9
7	4	9	1	6	8	2	5	3

EASY - 131

6	7	8	2	1	5	4	3	9
2	9	1	3	8	4	7	6	5
4	5	3	6	9	7	2	8	1
5	4	7	1	6	3	9	2	8
9	8	6	5	4	2	1	7	3
1	3	2	9	7	8	6	5	4
7	2	5	4	3	9	8	1	6
8	1	9	7	5	6	3	4	2
3	6	4	8	2	1	5	9	7

EASY - 132

4	8	1	2	7	9	3	5	6
3	6	5	8	1	4	7	2	9
7	9	2	3	6	5	4	1	8
5	1	6	4	3	2	9	8	7
8	7	9	6	5	1	2	4	3
2	3	4	9	8	7	1	6	5
1	5	3	7	2	6	8	9	4
6	4	7	1	9	8	5	3	2
9	2	8	5	4	3	6	7	1

EASY - 133

3	4	1	9	7	2	5	8	6
6	7	5	8	4	1	9	3	2
2	9	8	3	5	6	1	4	7
7	3	4	1	2	5	6	9	8
5	6	2	7	8	9	4	1	3
8	1	9	4	6	3	2	7	5
4	8	6	2	9	7	3	5	1
1	2	7	5	3	4	8	6	9
9	5	3	6	1	8	7	2	4

EASY - 134

5	1	3	6	7	2	4	8	9
2	4	9	3	1	8	5	6	7
7	6	8	4	5	9	1	2	3
6	5	1	9	2	3	8	7	4
8	7	4	1	6	5	3	9	2
9	3	2	7	8	4	6	1	5
3	8	5	2	9	6	7	4	1
1	2	6	5	4	7	9	3	8
4	9	7	8	3	1	2	5	6

EASY - 135

7	5	9	8	1	6	2	4	3
6	4	3	9	7	2	8	1	5
8	2	1	5	4	3	9	6	7
2	3	5	4	8	7	1	9	6
9	6	7	3	2	1	5	8	4
1	8	4	6	5	9	7	3	2
4	9	8	2	6	5	3	7	1
3	1	2	7	9	4	6	5	8
5	7	6	1	3	8	4	2	9

EASY - 136

8	9	3	2	4	1	5	6	7
6	7	1	9	3	5	8	2	4
5	4	2	6	8	7	3	9	1
3	6	7	1	5	9	2	4	8
9	8	4	7	2	3	1	5	6
2	1	5	8	6	4	9	7	3
4	3	6	5	9	8	7	1	2
1	2	9	3	7	6	4	8	5
7	5	8	4	1	2	6	3	9

EASY - 137

6	9	4	8	2	5	3	1	7
7	1	3	9	6	4	5	2	8
8	2	5	7	3	1	9	6	4
4	5	9	1	8	7	2	3	6
2	7	8	3	9	6	1	4	5
3	6	1	4	5	2	7	8	9
9	4	7	6	1	3	8	5	2
5	3	6	2	7	8	4	9	1
1	8	2	5	4	9	6	7	3

EASY - 138

2	4	7	3	6	9	1	5	8
6	8	1	2	7	5	3	9	4
5	3	9	4	8	1	6	7	2
1	7	6	5	3	4	8	2	9
8	9	5	6	2	7	4	3	1
3	2	4	1	9	8	7	6	5
4	1	2	7	5	6	9	8	3
7	5	8	9	4	3	2	1	6
9	6	3	8	1	2	5	4	7

EASY - 139

3	2	1	4	8	7	6	9	5
6	9	8	3	5	2	4	1	7
4	7	5	6	1	9	2	3	8
7	1	2	8	9	4	3	5	6
5	3	9	7	6	1	8	2	4
8	4	6	2	3	5	1	7	9
2	5	4	1	7	6	9	8	3
1	8	7	9	4	3	5	6	2
9	6	3	5	2	8	7	4	1

EASY - 140

1	4	3	9	6	7	8	5	2
7	2	5	1	8	3	4	6	9
9	8	6	2	5	4	1	7	3
5	1	9	8	4	2	7	3	6
4	7	8	6	3	9	5	2	1
6	3	2	7	1	5	9	4	8
3	5	1	4	2	8	6	9	7
8	9	4	3	7	6	2	1	5
2	6	7	5	9	1	3	8	4

EASY - 141

1	4	7	8	9	2	3	6	5
5	6	3	4	1	7	8	9	2
9	8	2	6	3	5	4	7	1
3	1	6	7	8	9	5	2	4
7	2	5	1	6	4	9	8	3
8	9	4	5	2	3	7	1	6
4	3	8	2	7	1	6	5	9
2	7	9	3	5	6	1	4	8
6	5	1	9	4	8	2	3	7

EASY - 142

1	3	7	9	2	6	8	4	5
5	6	4	8	1	7	9	3	2
9	8	2	4	3	5	7	6	1
6	9	1	2	5	4	3	7	8
2	4	3	7	6	8	5	1	9
7	5	8	3	9	1	4	2	6
8	7	9	6	4	2	1	5	3
4	2	5	1	8	3	6	9	7
3	1	6	5	7	9	2	8	4

EASY - 143

1	4	5	7	3	2	8	9	6
8	2	3	5	9	6	1	7	4
9	6	7	1	8	4	5	2	3
2	5	1	4	7	8	3	6	9
6	9	4	3	2	1	7	5	8
7	3	8	6	5	9	2	4	1
4	7	2	9	1	3	6	8	5
3	8	9	2	6	5	4	1	7
5	1	6	8	4	7	9	3	2

EASY - 144

7	3	9	1	6	2	8	5	4
5	1	6	8	7	4	9	2	3
4	8	2	9	3	5	6	1	7
8	6	3	4	5	9	1	7	2
1	5	4	6	2	7	3	8	9
2	9	7	3	1	8	5	4	6
6	4	1	7	8	3	2	9	5
9	2	8	5	4	6	7	3	1
3	7	5	2	9	1	4	6	8

EASY - 145

1	3	7	6	2	8	4	5	9
2	6	8	4	9	5	7	3	1
9	5	4	7	3	1	6	8	2
4	2	6	9	8	7	5	1	3
8	9	3	5	1	6	2	7	4
5	7	1	3	4	2	8	9	6
6	4	9	8	7	3	1	2	5
7	1	5	2	6	9	3	4	8
3	8	2	1	5	4	9	6	7

EASY - 146

5	6	9	4	1	2	8	3	7
2	7	8	3	9	6	1	5	4
3	1	4	8	5	7	2	6	9
9	2	6	5	4	1	7	8	3
4	3	1	6	7	8	5	9	2
8	5	7	2	3	9	6	4	1
6	4	5	7	2	3	9	1	8
1	8	2	9	6	4	3	7	5
7	9	3	1	8	5	4	2	6

EASY - 147

3	5	2	7	9	8	4	6	1
6	1	7	3	4	5	8	2	9
8	9	4	1	6	2	3	7	5
5	7	8	9	2	3	1	4	6
2	6	3	8	1	4	9	5	7
1	4	9	5	7	6	2	8	3
7	8	5	4	3	1	6	9	2
4	2	1	6	5	9	7	3	8
9	3	6	2	8	7	5	1	4

EASY - 148

3	6	7	9	5	4	1	2	8
1	5	9	2	6	8	3	4	7
2	4	8	3	7	1	6	9	5
7	1	2	5	4	6	9	8	3
9	3	5	1	8	2	4	7	6
4	8	6	7	3	9	2	5	1
6	2	1	8	9	5	7	3	4
8	9	3	4	1	7	5	6	2
5	7	4	6	2	3	8	1	9

EASY - 149

4	2	1	7	3	9	5	8	6
7	5	9	4	8	6	1	3	2
8	6	3	2	5	1	7	9	4
2	7	5	8	9	3	4	6	1
9	1	6	5	4	2	8	7	3
3	8	4	1	6	7	2	5	9
5	3	2	9	1	8	6	4	7
6	4	7	3	2	5	9	1	8
1	9	8	6	7	4	3	2	5

EASY - 150

4	8	1	3	7	9	5	2	6
5	6	3	2	4	1	7	8	9
2	7	9	6	5	8	3	1	4
3	1	7	8	9	6	2	4	5
8	2	6	4	3	5	9	7	1
9	5	4	1	2	7	6	3	8
6	9	2	7	1	4	8	5	3
1	3	5	9	8	2	4	6	7
7	4	8	5	6	3	1	9	2

EASY - 151

1	9	8	6	3	5	7	4	2
3	2	4	8	1	7	9	6	5
6	5	7	4	9	2	1	3	8
4	8	3	7	6	9	5	2	1
7	1	5	2	8	3	6	9	4
2	6	9	1	5	4	8	7	3
8	3	2	9	7	1	4	5	6
9	4	6	5	2	8	3	1	7
5	7	1	3	4	6	2	8	9

EASY - 152

1	2	9	5	7	6	3	8	4
4	5	7	3	1	8	6	9	2
6	8	3	9	4	2	1	5	7
9	7	5	8	6	3	4	2	1
8	3	1	2	9	4	5	7	6
2	6	4	7	5	1	9	3	8
5	4	8	6	3	7	2	1	9
3	1	2	4	8	9	7	6	5
7	9	6	1	2	5	8	4	3

EASY - 153

4	5	3	8	1	6	7	9	2
6	9	2	3	5	7	1	4	8
7	1	8	2	4	9	5	3	6
1	3	5	9	6	8	4	2	7
8	7	4	5	3	2	9	6	1
9	2	6	1	7	4	3	8	5
2	6	1	7	9	3	8	5	4
5	8	9	4	2	1	6	7	3
3	4	7	6	8	5	2	1	9

EASY - 154

1	7	2	3	4	8	9	6	5
8	9	4	6	5	2	7	3	1
5	3	6	9	1	7	4	8	2
9	1	8	4	2	5	6	7	3
2	5	3	7	8	6	1	4	9
6	4	7	1	9	3	2	5	8
4	8	9	5	6	1	3	2	7
7	2	1	8	3	4	5	9	6
3	6	5	2	7	9	8	1	4

EASY - 155

2	1	7	9	5	6	4	8	3
9	6	4	2	8	3	1	7	5
8	3	5	1	4	7	6	9	2
5	8	1	4	7	2	3	6	9
7	4	3	5	6	9	2	1	8
6	9	2	3	1	8	7	5	4
3	5	8	7	2	1	9	4	6
1	2	6	8	9	4	5	3	7
4	7	9	6	3	5	8	2	1

EASY - 156

3	2	8	6	1	9	7	5	4
5	1	4	8	7	2	3	6	9
6	9	7	3	4	5	2	8	1
2	8	6	1	9	4	5	7	3
1	4	5	7	6	3	8	9	2
7	3	9	2	5	8	4	1	6
9	7	2	5	3	6	1	4	8
8	6	1	4	2	7	9	3	5
4	5	3	9	8	1	6	2	7

EASY - 157

8	9	4	2	1	3	6	7	5
3	6	7	5	4	9	1	2	8
5	1	2	8	7	6	9	4	3
9	5	1	6	8	7	4	3	2
7	2	6	1	3	4	5	8	9
4	8	3	9	5	2	7	1	6
6	4	8	7	2	5	3	9	1
2	7	5	3	9	1	8	6	4
1	3	9	4	6	8	2	5	7

EASY - 158

6	9	3	2	7	8	4	1	5
7	1	5	6	4	9	3	2	8
2	4	8	3	5	1	6	7	9
5	8	9	4	3	7	1	6	2
1	6	4	5	9	2	8	3	7
3	2	7	8	1	6	9	5	4
9	7	6	1	8	5	2	4	3
8	3	2	7	6	4	5	9	1
4	5	1	9	2	3	7	8	6

EASY - 159

8	6	1	9	2	7	5	3	4
5	7	9	4	1	3	6	8	2
3	4	2	5	8	6	9	1	7
1	9	5	2	4	8	3	7	6
6	8	7	3	5	1	4	2	9
4	2	3	6	7	9	1	5	8
7	5	6	8	3	4	2	9	1
2	1	4	7	9	5	8	6	3
9	3	8	1	6	2	7	4	5

EASY - 160

1	6	8	9	4	3	5	7	2
3	4	5	8	7	2	1	6	9
7	9	2	6	5	1	3	8	4
2	8	7	4	9	5	6	1	3
9	3	6	7	1	8	4	2	5
4	5	1	2	3	6	8	9	7
6	1	4	3	2	7	9	5	8
5	7	3	1	8	9	2	4	6
8	2	9	5	6	4	7	3	1

EASY - 161

4	5	1	3	6	7	9	8	2
2	7	6	8	9	1	4	3	5
8	3	9	4	5	2	1	6	7
3	1	7	5	2	4	8	9	6
5	4	8	6	7	9	2	1	3
9	6	2	1	8	3	7	5	4
7	2	3	9	1	6	5	4	8
1	8	4	7	3	5	6	2	9
6	9	5	2	4	8	3	7	1

EASY - 162

6	1	3	2	5	9	8	4	7
2	8	4	3	1	7	5	6	9
9	7	5	8	4	6	2	3	1
4	6	8	1	9	3	7	2	5
5	2	1	4	7	8	6	9	3
7	3	9	5	6	2	4	1	8
3	9	6	7	2	5	1	8	4
8	4	7	6	3	1	9	5	2
1	5	2	9	8	4	3	7	6

EASY - 163

7	6	5	3	9	4	2	1	8
9	3	4	1	8	2	5	6	7
2	1	8	5	6	7	3	4	9
4	8	9	2	5	3	6	7	1
6	5	7	9	4	1	8	2	3
3	2	1	6	7	8	9	5	4
1	9	3	4	2	6	7	8	5
8	4	2	7	3	5	1	9	6
5	7	6	8	1	9	4	3	2

EASY - 164

4	5	6	2	8	7	1	9	3
9	1	3	6	5	4	7	2	8
7	8	2	1	9	3	4	5	6
2	7	9	8	1	6	3	4	5
5	4	8	3	7	9	6	1	2
3	6	1	4	2	5	9	8	7
6	9	4	5	3	8	2	7	1
8	2	7	9	6	1	5	3	4
1	3	5	7	4	2	8	6	9

EASY - 165

2	7	9	8	3	4	1	5	6
3	1	5	6	9	7	8	4	2
8	4	6	2	5	1	3	7	9
9	3	2	4	1	8	5	6	7
4	8	7	9	6	5	2	1	3
6	5	1	3	7	2	4	9	8
5	9	8	7	4	3	6	2	1
1	6	3	5	2	9	7	8	4
7	2	4	1	8	6	9	3	5

EASY - 166

9	8	3	6	4	2	7	1	5
2	4	6	7	1	5	9	8	3
1	5	7	3	9	8	2	4	6
6	7	9	2	3	4	8	5	1
8	3	1	5	7	9	4	6	2
5	2	4	1	8	6	3	9	7
4	1	5	9	2	7	6	3	8
7	6	8	4	5	3	1	2	9
3	9	2	8	6	1	5	7	4

EASY - 167

1	7	3	6	8	2	9	4	5
4	8	6	5	7	9	3	1	2
5	2	9	4	1	3	8	7	6
9	5	2	3	6	7	4	8	1
7	4	1	9	2	8	6	5	3
6	3	8	1	5	4	2	9	7
8	6	7	2	9	1	5	3	4
2	9	4	7	3	5	1	6	8
3	1	5	8	4	6	7	2	9

EASY - 168

6	7	8	3	2	4	9	1	5
4	5	2	8	9	1	7	3	6
9	1	3	5	6	7	4	2	8
1	8	6	9	7	3	2	5	4
7	3	5	4	1	2	8	6	9
2	4	9	6	5	8	3	7	1
5	9	7	2	4	6	1	8	3
8	2	4	1	3	5	6	9	7
3	6	1	7	8	9	5	4	2

EASY - 169

8	2	4	6	7	3	5	1	9
9	5	7	1	4	8	2	3	6
6	3	1	5	2	9	8	7	4
2	8	9	3	1	5	6	4	7
4	6	5	9	8	7	3	2	1
1	7	3	4	6	2	9	8	5
5	4	6	2	3	1	7	9	8
3	9	8	7	5	4	1	6	2
7	1	2	8	9	6	4	5	3

EASY - 170

2	1	8	5	3	6	4	9	7
7	5	4	1	9	8	2	6	3
3	9	6	4	2	7	1	5	8
8	4	7	6	1	5	9	3	2
1	3	2	7	8	9	5	4	6
5	6	9	2	4	3	8	7	1
4	2	3	9	7	1	6	8	5
9	8	5	3	6	2	7	1	4
6	7	1	8	5	4	3	2	9

EASY - 171

7	3	6	2	8	4	9	1	5
4	1	5	3	6	9	7	2	8
2	9	8	7	5	1	6	3	4
5	4	3	6	7	8	1	9	2
8	6	1	5	9	2	4	7	3
9	2	7	1	4	3	8	5	6
1	8	4	9	3	5	2	6	7
3	7	2	4	1	6	5	8	9
6	5	9	8	2	7	3	4	1

EASY - 172

9	4	5	7	1	6	3	2	8
7	8	2	3	4	5	1	6	9
1	3	6	8	2	9	5	4	7
5	1	9	6	3	4	7	8	2
3	2	7	5	9	8	4	1	6
4	6	8	1	7	2	9	5	3
8	7	3	4	6	1	2	9	5
6	9	4	2	5	7	8	3	1
2	5	1	9	8	3	6	7	4

EASY - 173

1	6	8	7	5	2	4	9	3
7	2	9	6	4	3	1	8	5
4	5	3	1	8	9	6	7	2
2	7	1	5	3	8	9	6	4
5	9	4	2	6	7	8	3	1
3	8	6	9	1	4	5	2	7
8	4	2	3	9	5	7	1	6
6	3	5	8	7	1	2	4	9
9	1	7	4	2	6	3	5	8

EASY - 174

8	3	9	2	4	7	5	1	6
7	6	2	1	3	5	8	9	4
1	5	4	6	9	8	3	7	2
2	1	8	9	5	3	4	6	7
4	7	3	8	2	6	9	5	1
6	9	5	4	7	1	2	3	8
5	8	7	3	1	2	6	4	9
9	2	1	5	6	4	7	8	3
3	4	6	7	8	9	1	2	5

EASY - 175

3	5	1	6	7	9	4	2	8
4	8	9	1	2	5	7	3	6
6	7	2	4	8	3	9	1	5
8	2	3	7	9	4	5	6	1
9	4	5	8	6	1	2	7	3
1	6	7	5	3	2	8	9	4
2	1	4	3	5	7	6	8	9
5	9	6	2	1	8	3	4	7
7	3	8	9	4	6	1	5	2

EASY - 176

4	9	6	8	5	1	7	2	3
5	7	3	4	6	2	9	8	1
1	2	8	3	9	7	4	6	5
8	4	1	2	3	6	5	9	7
3	6	9	5	7	8	2	1	4
7	5	2	1	4	9	6	3	8
9	1	7	6	8	4	3	5	2
6	8	5	7	2	3	1	4	9
2	3	4	9	1	5	8	7	6

EASY - 177

8	9	2	7	3	5	1	6	4
3	6	1	2	8	4	5	7	9
4	7	5	1	6	9	3	2	8
5	4	3	6	2	8	9	1	7
2	8	9	4	7	1	6	3	5
6	1	7	9	5	3	8	4	2
7	2	8	5	1	6	4	9	3
9	5	6	3	4	2	7	8	1
1	3	4	8	9	7	2	5	6

EASY - 178

8	9	6	5	2	4	7	3	1
7	2	1	3	9	8	4	5	6
5	4	3	6	7	1	8	2	9
1	3	2	8	5	6	9	7	4
6	5	9	7	4	2	3	1	8
4	7	8	1	3	9	2	6	5
2	8	7	9	6	5	1	4	3
9	6	4	2	1	3	5	8	7
3	1	5	4	8	7	6	9	2

EASY - 179

9	1	4	7	6	2	3	8	5
3	7	8	9	5	4	2	6	1
6	5	2	3	8	1	9	7	4
8	4	5	6	9	7	1	3	2
7	3	9	1	2	8	4	5	6
1	2	6	5	4	3	8	9	7
5	8	7	4	1	9	6	2	3
2	6	1	8	3	5	7	4	9
4	9	3	2	7	6	5	1	8

EASY - 180

2	7	5	4	3	6	1	9	8
9	6	3	8	2	1	5	4	7
1	8	4	5	9	7	3	6	2
4	9	7	2	5	3	8	1	6
8	3	6	7	1	9	2	5	4
5	2	1	6	8	4	9	7	3
7	1	2	9	6	8	4	3	5
6	5	9	3	4	2	7	8	1
3	4	8	1	7	5	6	2	9

EASY - 181

3	5	9	8	4	1	2	7	6
1	4	8	2	7	6	5	3	9
2	6	7	5	9	3	1	4	8
4	8	2	6	1	9	7	5	3
7	9	3	4	5	8	6	1	2
6	1	5	3	2	7	8	9	4
5	2	6	1	3	4	9	8	7
8	7	4	9	6	5	3	2	1
9	3	1	7	8	2	4	6	5

EASY - 182

2	7	8	4	9	1	6	3	5
1	3	6	7	5	2	4	9	8
4	5	9	3	6	8	2	7	1
3	9	1	6	2	4	5	8	7
8	2	7	9	1	5	3	4	6
5	6	4	8	7	3	9	1	2
7	1	5	2	4	9	8	6	3
9	8	2	1	3	6	7	5	4
6	4	3	5	8	7	1	2	9

EASY - 183

8	3	5	7	2	1	4	6	9
7	2	1	6	9	4	8	3	5
6	9	4	5	8	3	2	1	7
2	7	3	4	5	6	1	9	8
1	5	6	8	7	9	3	2	4
4	8	9	3	1	2	7	5	6
3	6	2	9	4	7	5	8	1
5	1	7	2	6	8	9	4	3
9	4	8	1	3	5	6	7	2

EASY - 184

1	6	2	8	3	4	9	5	7
9	7	3	1	2	5	8	6	4
4	5	8	6	9	7	3	1	2
2	8	4	5	1	3	6	7	9
5	9	7	2	8	6	4	3	1
3	1	6	7	4	9	5	2	8
8	3	1	9	5	2	7	4	6
6	2	5	4	7	8	1	9	3
7	4	9	3	6	1	2	8	5

EASY - 185

7	2	5	3	1	6	8	9	4
8	4	6	2	7	9	1	5	3
9	1	3	8	4	5	6	7	2
5	6	2	9	3	4	7	8	1
3	9	1	7	6	8	2	4	5
4	7	8	1	5	2	9	3	6
1	8	7	4	2	3	5	6	9
6	3	9	5	8	1	4	2	7
2	5	4	6	9	7	3	1	8

EASY - 186

5	9	3	4	2	7	6	8	1
8	4	7	9	1	6	5	3	2
2	1	6	3	8	5	9	4	7
3	5	8	6	7	4	1	2	9
9	2	4	8	5	1	7	6	3
6	7	1	2	9	3	4	5	8
4	8	2	7	6	9	3	1	5
7	3	5	1	4	8	2	9	6
1	6	9	5	3	2	8	7	4

EASY - 187

7	5	1	2	9	4	6	8	3
4	3	8	5	1	6	9	2	7
9	2	6	7	8	3	5	4	1
6	4	3	8	5	1	7	9	2
2	7	5	6	3	9	4	1	8
8	1	9	4	2	7	3	6	5
1	6	7	3	4	2	8	5	9
5	9	4	1	7	8	2	3	6
3	8	2	9	6	5	1	7	4

EASY - 188

3	6	2	7	8	4	9	1	5
7	4	1	5	3	9	8	6	2
8	9	5	6	2	1	4	3	7
1	2	7	4	9	8	6	5	3
6	5	8	3	1	7	2	9	4
4	3	9	2	6	5	1	7	8
2	7	4	9	5	6	3	8	1
9	8	3	1	7	2	5	4	6
5	1	6	8	4	3	7	2	9

EASY - 189

7	5	1	3	2	8	6	9	4
4	2	8	7	6	9	3	5	1
3	6	9	4	5	1	2	7	8
8	7	2	1	9	6	4	3	5
5	1	6	2	3	4	7	8	9
9	3	4	8	7	5	1	6	2
2	8	5	6	4	3	9	1	7
6	9	7	5	1	2	8	4	3
1	4	3	9	8	7	5	2	6

EASY - 190

5	1	2	4	8	3	7	6	9
6	8	7	2	1	9	3	4	5
3	9	4	6	7	5	8	1	2
4	5	8	9	3	6	2	7	1
1	3	6	7	2	4	5	9	8
2	7	9	1	5	8	4	3	6
8	4	1	3	6	2	9	5	7
7	2	3	5	9	1	6	8	4
9	6	5	8	4	7	1	2	3

EASY - 191

2	9	6	7	8	5	4	1	3
7	8	1	4	3	9	6	2	5
5	3	4	6	1	2	7	9	8
6	2	8	3	4	7	1	5	9
1	7	5	8	9	6	3	4	2
3	4	9	5	2	1	8	6	7
9	6	7	1	5	3	2	8	4
8	5	3	2	6	4	9	7	1
4	1	2	9	7	8	5	3	6

EASY - 192

5	7	4	6	1	8	2	3	9
2	8	9	3	5	7	6	4	1
3	1	6	9	2	4	8	5	7
7	9	8	5	4	6	3	1	2
6	4	3	2	8	1	7	9	5
1	5	2	7	3	9	4	8	6
9	2	5	4	6	3	1	7	8
4	6	1	8	7	5	9	2	3
8	3	7	1	9	2	5	6	4

EASY - 193

1	4	5	7	2	6	3	8	9
7	3	9	1	4	8	2	5	6
2	8	6	3	9	5	7	1	4
3	9	7	4	6	1	8	2	5
4	1	8	5	7	2	6	9	3
5	6	2	9	8	3	1	4	7
8	2	3	6	5	9	4	7	1
9	7	1	8	3	4	5	6	2
6	5	4	2	1	7	9	3	8

EASY - 194

8	2	4	3	1	5	6	9	7
7	9	5	8	6	2	4	1	3
3	1	6	4	9	7	8	5	2
5	4	7	1	2	3	9	6	8
1	8	9	7	4	6	3	2	5
6	3	2	9	5	8	7	4	1
4	5	8	2	7	9	1	3	6
2	7	1	6	3	4	5	8	9
9	6	3	5	8	1	2	7	4

EASY - 195

9	3	6	4	1	5	2	7	8
2	4	5	9	8	7	3	6	1
7	8	1	2	6	3	4	5	9
4	6	9	8	2	1	5	3	7
3	1	8	5	7	4	6	9	2
5	2	7	3	9	6	1	8	4
6	5	2	7	4	8	9	1	3
1	7	4	6	3	9	8	2	5
8	9	3	1	5	2	7	4	6

EASY - 196

5	6	3	4	9	7	1	2	8
1	8	2	6	5	3	7	4	9
9	7	4	1	8	2	6	3	5
8	4	7	5	6	1	3	9	2
2	1	9	7	3	4	5	8	6
6	3	5	9	2	8	4	1	7
3	9	6	2	4	5	8	7	1
7	2	8	3	1	6	9	5	4
4	5	1	8	7	9	2	6	3

EASY - 197

3	5	2	1	4	8	7	9	6
9	7	1	2	5	6	3	4	8
4	6	8	7	9	3	1	5	2
1	9	3	6	7	5	8	2	4
5	8	7	4	3	2	9	6	1
6	2	4	9	8	1	5	7	3
7	1	5	8	6	4	2	3	9
2	3	6	5	1	9	4	8	7
8	4	9	3	2	7	6	1	5

EASY - 198

3	1	7	6	4	9	8	5	2
9	2	5	7	3	8	1	4	6
6	8	4	1	2	5	9	7	3
8	6	9	5	7	1	3	2	4
2	7	1	4	9	3	5	6	8
4	5	3	8	6	2	7	1	9
1	9	2	3	5	4	6	8	7
7	4	8	9	1	6	2	3	5
5	3	6	2	8	7	4	9	1

EASY - 199

2	4	6	1	9	8	7	5	3
1	7	3	6	2	5	4	9	8
9	5	8	3	4	7	6	2	1
6	8	7	9	1	3	2	4	5
5	3	2	8	6	4	9	1	7
4	1	9	7	5	2	3	8	6
3	6	1	2	8	9	5	7	4
7	2	5	4	3	1	8	6	9
8	9	4	5	7	6	1	3	2

EASY - 200

7	8	9	1	3	6	2	5	4
5	4	2	8	7	9	3	1	6
3	1	6	4	5	2	7	8	9
1	2	5	9	8	3	6	4	7
8	9	4	2	6	7	1	3	5
6	3	7	5	4	1	8	9	2
9	5	1	7	2	8	4	6	3
2	6	8	3	9	4	5	7	1
4	7	3	6	1	5	9	2	8

EASY - 201

7	1	8	9	2	4	3	6	5
2	6	9	8	3	5	7	1	4
4	5	3	1	7	6	9	2	8
3	9	4	7	1	2	8	5	6
5	7	1	3	6	8	2	4	9
8	2	6	4	5	9	1	3	7
6	3	2	5	8	7	4	9	1
1	4	7	6	9	3	5	8	2
9	8	5	2	4	1	6	7	3

EASY - 202

6	9	7	5	1	2	8	4	3
5	1	4	9	8	3	6	2	7
8	2	3	7	6	4	9	5	1
7	6	9	4	2	8	3	1	5
2	5	1	6	3	7	4	8	9
3	4	8	1	9	5	2	7	6
4	8	6	3	7	1	5	9	2
9	7	5	2	4	6	1	3	8
1	3	2	8	5	9	7	6	4

EASY - 203

7	3	4	5	9	6	2	1	8
2	6	5	1	8	3	4	9	7
1	8	9	4	7	2	5	6	3
9	2	3	7	4	8	1	5	6
6	5	1	3	2	9	7	8	4
4	7	8	6	5	1	9	3	2
3	9	6	2	1	4	8	7	5
8	4	7	9	6	5	3	2	1
5	1	2	8	3	7	6	4	9

EASY - 204

6	3	1	4	5	2	9	7	8
9	4	2	1	8	7	5	6	3
5	8	7	3	6	9	4	2	1
7	2	9	6	1	4	8	3	5
8	6	3	9	2	5	7	1	4
1	5	4	8	7	3	2	9	6
3	9	8	2	4	6	1	5	7
2	1	5	7	3	8	6	4	9
4	7	6	5	9	1	3	8	2

INTERMEDIATE - 1

3	2	9	4	5	1	8	6	7
1	7	8	2	9	6	4	5	3
4	5	6	7	8	3	2	9	1
9	6	7	1	3	8	5	2	4
2	4	5	9	6	7	1	3	8
8	3	1	5	4	2	9	7	6
6	8	2	3	1	9	7	4	5
7	1	4	6	2	5	3	8	9
5	9	3	8	7	4	6	1	2

INTERMEDIATE - 2

1	5	2	6	3	4	7	8	9
9	7	3	2	1	8	6	5	4
6	4	8	9	5	7	1	3	2
5	3	4	8	6	1	2	9	7
8	9	6	5	7	2	3	4	1
2	1	7	4	9	3	5	6	8
4	8	1	3	2	6	9	7	5
7	6	9	1	4	5	8	2	3
3	2	5	7	8	9	4	1	6

INTERMEDIATE - 3

6	8	3	2	7	4	5	1	9
7	2	5	3	9	1	6	8	4
9	1	4	6	8	5	2	7	3
5	9	6	7	3	8	1	4	2
3	4	8	9	1	2	7	5	6
2	7	1	4	5	6	3	9	8
8	5	2	1	6	9	4	3	7
4	3	9	5	2	7	8	6	1
1	6	7	8	4	3	9	2	5

INTERMEDIATE - 4

1	7	4	5	8	9	2	3	6
5	9	6	2	1	3	8	7	4
8	2	3	4	6	7	9	1	5
7	1	2	3	5	4	6	9	8
9	4	5	6	7	8	1	2	3
6	3	8	1	9	2	4	5	7
2	5	9	8	3	6	7	4	1
4	8	1	7	2	5	3	6	9
3	6	7	9	4	1	5	8	2

INTERMEDIATE - 5

6	5	4	8	7	2	9	1	3
7	1	2	3	4	9	5	6	8
8	9	3	5	1	6	4	7	2
9	8	1	6	5	7	2	3	4
2	3	7	9	8	4	1	5	6
4	6	5	1	2	3	8	9	7
5	7	6	2	9	8	3	4	1
1	4	8	7	3	5	6	2	9
3	2	9	4	6	1	7	8	5

INTERMEDIATE - 6

8	3	1	4	5	9	6	2	7
2	5	4	7	3	6	1	9	8
6	9	7	8	1	2	3	5	4
5	8	3	6	2	4	7	1	9
4	1	2	9	7	3	8	6	5
7	6	9	1	8	5	2	4	3
9	2	6	3	4	7	5	8	1
3	4	8	5	6	1	9	7	2
1	7	5	2	9	8	4	3	6

INTERMEDIATE - 7

5	1	4	8	3	9	6	2	7
6	7	2	1	4	5	9	8	3
9	3	8	2	6	7	4	5	1
4	8	6	9	1	3	2	7	5
3	2	5	4	7	8	1	6	9
1	9	7	5	2	6	3	4	8
7	6	1	3	5	4	8	9	2
2	5	9	6	8	1	7	3	4
8	4	3	7	9	2	5	1	6

INTERMEDIATE - 8

6	5	8	9	7	1	2	4	3
4	2	7	6	8	3	5	9	1
1	3	9	2	4	5	8	7	6
8	1	3	4	6	7	9	2	5
2	9	5	1	3	8	4	6	7
7	4	6	5	9	2	1	3	8
5	6	4	7	1	9	3	8	2
9	8	1	3	2	6	7	5	4
3	7	2	8	5	4	6	1	9

INTERMEDIATE - 9

9	3	4	6	2	8	5	1	7
8	7	1	5	4	9	3	6	2
6	2	5	3	1	7	4	9	8
3	6	2	4	7	1	8	5	9
4	5	8	9	6	2	1	7	3
1	9	7	8	3	5	2	4	6
7	4	3	1	8	6	9	2	5
5	8	6	2	9	4	7	3	1
2	1	9	7	5	3	6	8	4

INTERMEDIATE - 10

4	7	9	6	8	3	1	2	5
5	1	3	7	9	2	8	6	4
2	8	6	1	5	4	7	9	3
7	5	8	3	4	6	9	1	2
6	2	4	5	1	9	3	8	7
3	9	1	8	2	7	5	4	6
9	6	7	2	3	8	4	5	1
8	3	5	4	6	1	2	7	9
1	4	2	9	7	5	6	3	8

INTERMEDIATE - 11

9	1	2	5	3	6	7	4	8
5	6	8	4	9	7	2	1	3
3	7	4	1	2	8	9	6	5
7	8	3	6	5	4	1	9	2
6	5	9	7	1	2	8	3	4
2	4	1	3	8	9	6	5	7
8	2	6	9	4	5	3	7	1
1	9	5	2	7	3	4	8	6
4	3	7	8	6	1	5	2	9

INTERMEDIATE - 12

9	1	7	8	3	6	4	2	5
5	2	6	7	1	4	3	8	9
3	4	8	9	2	5	7	1	6
6	5	9	2	7	1	8	4	3
8	3	2	6	4	9	5	7	1
4	7	1	5	8	3	6	9	2
7	6	5	1	9	8	2	3	4
1	8	3	4	6	2	9	5	7
2	9	4	3	5	7	1	6	8

INTERMEDIATE - 13

9	1	4	5	7	8	6	3	2
7	2	3	9	6	4	8	5	1
8	5	6	1	2	3	7	4	9
2	3	9	8	5	7	1	6	4
1	6	8	2	4	9	5	7	3
5	4	7	6	3	1	9	2	8
4	9	5	7	1	2	3	8	6
6	8	2	3	9	5	4	1	7
3	7	1	4	8	6	2	9	5

INTERMEDIATE - 14

6	1	4	2	9	5	8	7	3
2	7	9	4	3	8	6	5	1
8	5	3	6	7	1	4	9	2
7	8	5	3	4	9	1	2	6
3	2	1	5	6	7	9	8	4
4	9	6	1	8	2	5	3	7
1	4	7	8	5	3	2	6	9
9	6	8	7	2	4	3	1	5
5	3	2	9	1	6	7	4	8

INTERMEDIATE - 15

8	1	5	3	9	4	6	7	2
6	9	2	7	1	5	4	8	3
3	7	4	2	6	8	1	9	5
5	2	3	4	7	6	8	1	9
1	6	9	5	8	2	7	3	4
4	8	7	1	3	9	2	5	6
9	3	1	6	4	7	5	2	8
2	4	8	9	5	1	3	6	7
7	5	6	8	2	3	9	4	1

INTERMEDIATE - 16

2	3	6	8	1	4	9	5	7
4	1	7	2	9	5	6	8	3
5	8	9	6	7	3	1	2	4
8	2	4	7	3	6	5	9	1
7	9	3	5	8	1	4	6	2
1	6	5	4	2	9	7	3	8
9	5	8	1	4	2	3	7	6
6	7	1	3	5	8	2	4	9
3	4	2	9	6	7	8	1	5

INTERMEDIATE - 17

5	9	6	1	7	8	2	4	3
4	2	7	6	5	3	1	9	8
8	1	3	9	2	4	5	6	7
1	7	8	3	9	2	6	5	4
9	4	5	8	6	1	7	3	2
6	3	2	7	4	5	9	8	1
2	8	4	5	1	6	3	7	9
7	6	1	4	3	9	8	2	5
3	5	9	2	8	7	4	1	6

INTERMEDIATE - 18

3	8	4	1	5	9	6	2	7
9	2	5	3	7	6	1	8	4
6	1	7	4	8	2	9	5	3
1	7	3	8	9	4	5	6	2
8	9	2	6	3	5	7	4	1
4	5	6	2	1	7	3	9	8
2	3	1	9	6	8	4	7	5
7	4	9	5	2	3	8	1	6
5	6	8	7	4	1	2	3	9

INTERMEDIATE - 19

6	7	2	4	9	8	1	3	5
9	4	8	3	1	5	7	2	6
1	3	5	7	6	2	4	8	9
5	6	9	1	3	7	8	4	2
3	2	4	8	5	6	9	7	1
8	1	7	2	4	9	6	5	3
2	9	6	5	7	4	3	1	8
4	5	1	6	8	3	2	9	7
7	8	3	9	2	1	5	6	4

INTERMEDIATE - 20

3	8	1	5	6	4	2	9	7
4	5	7	3	2	9	6	1	8
9	6	2	1	7	8	5	3	4
7	9	5	8	1	2	4	6	3
6	3	8	9	4	7	1	5	2
1	2	4	6	5	3	7	8	9
2	1	9	4	3	5	8	7	6
8	4	6	7	9	1	3	2	5
5	7	3	2	8	6	9	4	1

INTERMEDIATE - 21

6	9	5	2	4	3	7	1	8
3	2	8	1	5	7	4	9	6
1	7	4	8	9	6	3	5	2
7	3	2	6	1	9	8	4	5
5	1	6	7	8	4	9	2	3
8	4	9	3	2	5	1	6	7
2	5	3	4	7	1	6	8	9
9	6	1	5	3	8	2	7	4
4	8	7	9	6	2	5	3	1

INTERMEDIATE - 22

4	6	9	2	3	8	5	1	7
2	8	5	9	1	7	4	3	6
1	3	7	5	6	4	2	9	8
5	4	3	6	9	1	8	7	2
8	7	1	4	5	2	3	6	9
9	2	6	8	7	3	1	5	4
7	1	8	3	2	6	9	4	5
3	9	4	7	8	5	6	2	1
6	5	2	1	4	9	7	8	3

INTERMEDIATE - 23

8	7	2	3	1	9	6	4	5
9	1	5	4	7	6	8	3	2
3	4	6	8	2	5	1	7	9
7	3	9	1	5	4	2	6	8
1	2	4	6	3	8	5	9	7
6	5	8	2	9	7	4	1	3
5	9	1	7	6	2	3	8	4
4	6	7	5	8	3	9	2	1
2	8	3	9	4	1	7	5	6

INTERMEDIATE - 24

6	1	5	9	3	8	2	4	7
7	4	2	1	5	6	9	8	3
8	9	3	7	2	4	6	5	1
9	8	1	4	7	2	5	3	6
5	3	4	8	6	1	7	9	2
2	7	6	3	9	5	4	1	8
4	6	8	2	1	9	3	7	5
1	2	7	5	4	3	8	6	9
3	5	9	6	8	7	1	2	4

INTERMEDIATE - 25

4	9	1	6	3	7	8	5	2
8	2	7	1	9	5	4	3	6
5	6	3	2	8	4	9	7	1
9	1	6	8	2	3	7	4	5
7	4	8	9	5	1	6	2	3
2	3	5	7	4	6	1	9	8
1	8	2	3	7	9	5	6	4
6	7	4	5	1	2	3	8	9
3	5	9	4	6	8	2	1	7

INTERMEDIATE - 26

3	4	6	8	2	5	9	1	7
1	5	9	6	3	7	2	8	4
8	2	7	1	4	9	6	5	3
4	8	1	3	7	6	5	2	9
2	6	3	5	9	8	7	4	1
9	7	5	2	1	4	3	6	8
5	3	4	7	6	1	8	9	2
6	9	2	4	8	3	1	7	5
7	1	8	9	5	2	4	3	6

INTERMEDIATE - 27

4	8	1	3	2	7	5	6	9
3	2	5	6	4	9	1	8	7
7	6	9	5	8	1	4	2	3
1	4	2	7	9	8	6	3	5
9	3	7	1	6	5	8	4	2
8	5	6	4	3	2	9	7	1
5	1	3	8	7	4	2	9	6
6	9	8	2	5	3	7	1	4
2	7	4	9	1	6	3	5	8

INTERMEDIATE - 28

2	7	6	1	5	9	8	3	4
5	3	8	2	7	4	1	6	9
1	4	9	6	3	8	2	7	5
4	2	1	3	6	7	9	5	8
6	5	3	9	8	2	7	4	1
9	8	7	4	1	5	3	2	6
3	9	2	5	4	1	6	8	7
7	1	5	8	2	6	4	9	3
8	6	4	7	9	3	5	1	2

INTERMEDIATE - 29

9	6	3	1	7	2	5	4	8
5	4	8	6	3	9	2	1	7
1	7	2	5	4	8	6	9	3
3	8	5	7	6	1	9	2	4
7	9	4	2	8	5	3	6	1
6	2	1	3	9	4	8	7	5
2	1	9	8	5	7	4	3	6
4	5	6	9	1	3	7	8	2
8	3	7	4	2	6	1	5	9

INTERMEDIATE - 30

6	5	4	7	3	8	2	9	1
2	3	7	5	1	9	8	4	6
1	8	9	2	6	4	3	5	7
7	9	1	3	8	2	5	6	4
5	6	8	9	4	1	7	3	2
4	2	3	6	7	5	9	1	8
9	7	5	1	2	6	4	8	3
8	1	2	4	9	3	6	7	5
3	4	6	8	5	7	1	2	9

INTERMEDIATE - 31

2	6	3	5	8	1	4	7	9
9	7	4	3	2	6	5	8	1
5	1	8	9	4	7	6	3	2
7	9	6	1	3	5	2	4	8
1	4	2	7	6	8	9	5	3
8	3	5	4	9	2	7	1	6
3	8	7	6	5	9	1	2	4
6	2	1	8	7	4	3	9	5
4	5	9	2	1	3	8	6	7

INTERMEDIATE - 32

5	2	9	7	1	8	4	3	6
8	3	7	9	4	6	2	5	1
1	6	4	3	5	2	8	9	7
3	9	1	4	6	7	5	8	2
7	8	2	1	9	5	6	4	3
4	5	6	2	8	3	1	7	9
2	1	5	8	7	9	3	6	4
6	7	3	5	2	4	9	1	8
9	4	8	6	3	1	7	2	5

INTERMEDIATE - 33

3	2	5	6	7	1	4	8	9
7	4	1	2	9	8	5	3	6
6	8	9	4	5	3	1	2	7
9	3	2	8	6	5	7	1	4
1	7	4	9	3	2	8	6	5
5	6	8	7	1	4	3	9	2
2	9	3	5	8	7	6	4	1
4	1	7	3	2	6	9	5	8
8	5	6	1	4	9	2	7	3

INTERMEDIATE - 34

7	3	6	4	9	2	8	5	1
8	2	1	6	7	5	3	4	9
5	4	9	3	1	8	6	7	2
6	1	7	5	8	4	9	2	3
2	5	4	1	3	9	7	6	8
3	9	8	2	6	7	4	1	5
4	6	2	8	5	3	1	9	7
9	8	5	7	4	1	2	3	6
1	7	3	9	2	6	5	8	4

INTERMEDIATE - 35

5	7	9	8	4	1	6	3	2
6	8	2	3	7	9	4	1	5
1	4	3	2	6	5	8	7	9
8	3	7	9	1	4	5	2	6
9	6	5	7	8	2	3	4	1
2	1	4	5	3	6	9	8	7
7	2	8	6	5	3	1	9	4
3	5	1	4	9	7	2	6	8
4	9	6	1	2	8	7	5	3

INTERMEDIATE - 36

6	8	5	3	1	9	2	4	7
4	7	1	5	2	8	3	9	6
9	3	2	6	7	4	8	5	1
8	1	9	4	3	2	7	6	5
2	4	6	8	5	7	9	1	3
3	5	7	9	6	1	4	8	2
1	2	4	7	8	5	6	3	9
5	9	3	2	4	6	1	7	8
7	6	8	1	9	3	5	2	4

INTERMEDIATE - 37

5	7	8	9	3	4	1	6	2
9	6	4	2	1	5	7	8	3
3	1	2	8	6	7	4	5	9
6	9	1	7	8	2	5	3	4
8	5	3	4	9	1	6	2	7
4	2	7	6	5	3	9	1	8
7	4	5	1	2	8	3	9	6
1	8	9	3	4	6	2	7	5
2	3	6	5	7	9	8	4	1

INTERMEDIATE - 38

7	9	6	4	1	3	8	5	2
1	2	4	7	5	8	9	3	6
8	5	3	2	9	6	4	1	7
4	7	1	5	8	9	6	2	3
9	8	5	6	3	2	7	4	1
3	6	2	1	7	4	5	8	9
5	4	7	9	2	1	3	6	8
2	3	9	8	6	5	1	7	4
6	1	8	3	4	7	2	9	5

INTERMEDIATE - 39

5	8	1	4	3	6	7	2	9
4	2	6	8	7	9	5	3	1
3	9	7	5	1	2	6	4	8
6	1	5	9	2	3	4	8	7
7	4	2	6	8	5	9	1	3
9	3	8	7	4	1	2	5	6
1	7	9	2	5	8	3	6	4
8	5	4	3	6	7	1	9	2
2	6	3	1	9	4	8	7	5

INTERMEDIATE - 40

1	6	7	3	8	9	4	5	2
5	8	4	2	7	1	3	6	9
3	9	2	4	6	5	8	1	7
8	2	1	7	9	4	5	3	6
9	3	5	8	1	6	7	2	4
7	4	6	5	2	3	1	9	8
6	1	8	9	3	7	2	4	5
4	7	3	6	5	2	9	8	1
2	5	9	1	4	8	6	7	3

INTERMEDIATE - 41

5	8	9	7	3	4	1	2	6
4	1	3	9	2	6	8	5	7
6	7	2	1	8	5	9	4	3
1	4	5	3	6	9	7	8	2
3	2	7	5	4	8	6	9	1
9	6	8	2	7	1	5	3	4
8	5	6	4	1	2	3	7	9
7	9	4	6	5	3	2	1	8
2	3	1	8	9	7	4	6	5

INTERMEDIATE - 42

1	8	6	9	4	3	7	5	2
5	2	4	1	7	8	6	9	3
9	3	7	2	6	5	4	1	8
6	5	3	7	1	4	2	8	9
4	9	8	3	5	2	1	7	6
7	1	2	8	9	6	3	4	5
2	4	9	5	3	1	8	6	7
8	6	5	4	2	7	9	3	1
3	7	1	6	8	9	5	2	4

INTERMEDIATE - 43

4	6	1	8	3	2	9	5	7
5	3	8	1	7	9	2	4	6
9	2	7	4	6	5	3	1	8
6	4	3	9	1	7	8	2	5
7	1	9	5	2	8	6	3	4
2	8	5	6	4	3	1	7	9
1	5	4	2	8	6	7	9	3
8	7	2	3	9	4	5	6	1
3	9	6	7	5	1	4	8	2

INTERMEDIATE - 44

7	3	8	2	6	4	9	5	1
6	2	4	9	1	5	8	3	7
5	1	9	8	3	7	6	4	2
4	7	5	1	8	3	2	6	9
2	9	1	7	4	6	3	8	5
3	8	6	5	2	9	1	7	4
9	5	3	6	7	2	4	1	8
8	6	2	4	5	1	7	9	3
1	4	7	3	9	8	5	2	6

INTERMEDIATE - 45

2	1	7	8	9	5	6	4	3
9	5	4	3	2	6	1	7	8
8	6	3	7	4	1	2	5	9
7	2	6	1	5	9	8	3	4
3	8	1	4	6	7	9	2	5
5	4	9	2	8	3	7	1	6
6	3	8	5	7	2	4	9	1
4	7	5	9	1	8	3	6	2
1	9	2	6	3	4	5	8	7

INTERMEDIATE - 46

2	3	7	4	9	1	6	8	5
5	6	8	2	7	3	9	4	1
9	1	4	5	6	8	2	3	7
1	7	6	3	5	9	4	2	8
8	9	2	1	4	6	7	5	3
3	4	5	8	2	7	1	9	6
4	2	3	6	1	5	8	7	9
6	5	9	7	8	2	3	1	4
7	8	1	9	3	4	5	6	2

INTERMEDIATE - 47

3	2	4	5	7	6	1	9	8
9	1	7	3	8	4	2	6	5
5	6	8	2	1	9	3	7	4
4	5	1	6	9	7	8	2	3
2	9	6	1	3	8	5	4	7
7	8	3	4	2	5	9	1	6
1	7	9	8	4	3	6	5	2
6	3	2	7	5	1	4	8	9
8	4	5	9	6	2	7	3	1

INTERMEDIATE - 48

9	1	8	4	6	5	2	3	7
6	5	3	1	2	7	8	4	9
7	2	4	3	8	9	6	1	5
3	4	7	6	5	1	9	2	8
8	9	2	7	4	3	1	5	6
1	6	5	2	9	8	4	7	3
2	8	6	5	7	4	3	9	1
5	3	9	8	1	2	7	6	4
4	7	1	9	3	6	5	8	2

INTERMEDIATE - 49

4	3	9	1	2	7	8	5	6
6	5	2	4	9	8	3	7	1
7	8	1	5	3	6	4	2	9
8	7	5	6	4	2	9	1	3
3	1	4	9	7	5	2	6	8
9	2	6	8	1	3	5	4	7
1	9	7	2	8	4	6	3	5
5	4	3	7	6	9	1	8	2
2	6	8	3	5	1	7	9	4

INTERMEDIATE - 50

9	3	5	2	8	6	1	7	4
2	1	7	5	9	4	8	6	3
4	8	6	3	7	1	2	9	5
8	4	1	6	3	5	9	2	7
7	9	2	4	1	8	5	3	6
5	6	3	7	2	9	4	8	1
1	2	4	8	6	3	7	5	9
6	5	8	9	4	7	3	1	2
3	7	9	1	5	2	6	4	8

INTERMEDIATE - 51

2	9	3	5	6	4	1	8	7
8	1	5	7	2	9	4	6	3
7	4	6	8	3	1	9	2	5
5	7	1	4	8	3	6	9	2
3	8	2	1	9	6	5	7	4
9	6	4	2	5	7	8	3	1
4	3	9	6	7	5	2	1	8
1	2	7	9	4	8	3	5	6
6	5	8	3	1	2	7	4	9

INTERMEDIATE - 52

6	9	4	8	2	5	3	1	7
8	7	3	9	4	1	6	5	2
5	2	1	6	3	7	4	8	9
2	5	8	3	7	4	1	9	6
4	6	7	2	1	9	8	3	5
1	3	9	5	8	6	2	7	4
9	1	5	4	6	8	7	2	3
3	8	6	7	9	2	5	4	1
7	4	2	1	5	3	9	6	8

INTERMEDIATE - 53

2	4	8	6	7	1	3	9	5
6	5	1	2	3	9	8	7	4
9	3	7	8	4	5	6	1	2
8	7	5	4	2	3	1	6	9
4	2	9	1	6	7	5	8	3
1	6	3	9	5	8	2	4	7
5	8	6	3	9	4	7	2	1
7	9	2	5	1	6	4	3	8
3	1	4	7	8	2	9	5	6

INTERMEDIATE - 54

9	7	2	1	5	6	4	8	3
6	3	4	7	8	2	9	1	5
1	5	8	9	3	4	2	7	6
2	1	6	3	7	9	5	4	8
4	8	5	6	2	1	7	3	9
7	9	3	8	4	5	6	2	1
8	4	9	2	6	3	1	5	7
5	6	7	4	1	8	3	9	2
3	2	1	5	9	7	8	6	4

INTERMEDIATE - 55

8	7	4	6	9	3	1	2	5
2	3	1	7	4	5	9	6	8
5	6	9	1	8	2	4	7	3
4	1	7	5	6	9	8	3	2
9	5	8	3	2	4	6	1	7
3	2	6	8	1	7	5	4	9
1	4	3	2	5	8	7	9	6
7	9	5	4	3	6	2	8	1
6	8	2	9	7	1	3	5	4

INTERMEDIATE - 56

8	3	5	7	6	2	4	9	1
1	2	6	8	4	9	5	7	3
9	7	4	1	3	5	6	2	8
5	1	8	6	2	4	9	3	7
2	4	9	5	7	3	8	1	6
3	6	7	9	8	1	2	5	4
4	8	1	2	5	7	3	6	9
7	5	3	4	9	6	1	8	2
6	9	2	3	1	8	7	4	5

INTERMEDIATE - 57

2	4	8	7	3	9	1	5	6
1	9	6	8	5	4	3	2	7
3	5	7	6	2	1	9	8	4
5	6	1	9	4	2	7	3	8
7	8	4	1	6	3	2	9	5
9	3	2	5	8	7	6	4	1
8	1	9	3	7	5	4	6	2
4	7	5	2	9	6	8	1	3
6	2	3	4	1	8	5	7	9

INTERMEDIATE - 58

1	7	9	3	5	8	2	6	4
6	8	5	1	2	4	9	7	3
4	2	3	6	9	7	8	5	1
3	5	7	9	4	1	6	2	8
2	9	4	7	8	6	1	3	5
8	1	6	2	3	5	4	9	7
5	4	2	8	6	3	7	1	9
9	3	1	4	7	2	5	8	6
7	6	8	5	1	9	3	4	2

INTERMEDIATE - 59

3	2	4	5	7	6	1	9	8
9	1	7	3	8	4	6	5	2
8	6	5	1	2	9	3	7	4
6	8	2	9	4	5	7	3	1
7	9	3	6	1	8	4	2	5
4	5	1	2	3	7	8	6	9
2	7	8	4	9	3	5	1	6
1	4	6	7	5	2	9	8	3
5	3	9	8	6	1	2	4	7

INTERMEDIATE - 60

8	5	4	1	3	9	2	7	6
1	9	2	6	5	7	4	3	8
7	3	6	8	2	4	5	1	9
5	4	8	7	1	2	9	6	3
6	1	3	4	9	5	7	8	2
2	7	9	3	8	6	1	5	4
4	8	5	2	7	3	6	9	1
3	2	7	9	6	1	8	4	5
9	6	1	5	4	8	3	2	7

INTERMEDIATE - 61

4	1	6	8	9	5	7	3	2
5	9	8	7	2	3	6	1	4
3	2	7	6	4	1	5	9	8
8	4	2	1	5	7	9	6	3
6	5	1	3	8	9	4	2	7
7	3	9	2	6	4	1	8	5
9	8	5	4	1	2	3	7	6
2	7	4	9	3	6	8	5	1
1	6	3	5	7	8	2	4	9

INTERMEDIATE - 62

3	5	1	7	2	6	4	9	8
2	6	4	3	8	9	5	1	7
8	9	7	1	5	4	6	2	3
4	8	2	6	7	3	9	5	1
9	3	5	4	1	2	8	7	6
7	1	6	5	9	8	3	4	2
5	2	3	9	6	1	7	8	4
1	4	9	8	3	7	2	6	5
6	7	8	2	4	5	1	3	9

INTERMEDIATE - 63

2	7	3	9	4	6	5	8	1
4	1	6	3	5	8	9	2	7
5	8	9	7	2	1	3	6	4
1	3	2	8	6	9	4	7	5
7	4	5	1	3	2	6	9	8
9	6	8	4	7	5	1	3	2
3	5	1	6	8	7	2	4	9
8	9	4	2	1	3	7	5	6
6	2	7	5	9	4	8	1	3

INTERMEDIATE - 64

1	5	3	2	4	6	8	7	9
9	2	4	5	8	7	1	6	3
6	7	8	1	9	3	5	2	4
2	9	7	4	5	8	6	3	1
3	8	1	9	6	2	7	4	5
5	4	6	3	7	1	9	8	2
4	6	2	7	1	5	3	9	8
8	3	5	6	2	9	4	1	7
7	1	9	8	3	4	2	5	6

INTERMEDIATE - 65

7	5	1	4	3	9	6	8	2
3	8	9	2	7	6	1	5	4
6	2	4	5	8	1	7	9	3
9	7	6	8	5	2	3	4	1
5	4	8	6	1	3	9	2	7
1	3	2	9	4	7	8	6	5
8	1	5	3	9	4	2	7	6
2	9	3	7	6	5	4	1	8
4	6	7	1	2	8	5	3	9

INTERMEDIATE - 66

2	8	3	6	5	4	7	9	1
1	4	7	2	3	9	5	8	6
5	9	6	7	1	8	4	3	2
7	2	1	5	9	6	8	4	3
3	6	4	1	8	7	2	5	9
8	5	9	3	4	2	1	6	7
4	1	5	9	2	3	6	7	8
6	3	8	4	7	1	9	2	5
9	7	2	8	6	5	3	1	4

INTERMEDIATE - 67

2	9	4	3	1	5	8	7	6
1	6	3	9	7	8	2	5	4
8	7	5	4	2	6	1	9	3
4	1	8	6	9	2	5	3	7
7	2	6	1	5	3	9	4	8
5	3	9	7	8	4	6	2	1
6	5	7	2	3	1	4	8	9
3	8	1	5	4	9	7	6	2
9	4	2	8	6	7	3	1	5

INTERMEDIATE - 68

1	9	7	2	8	5	6	4	3
8	6	2	4	7	3	5	9	1
4	3	5	6	1	9	8	7	2
3	8	9	5	6	7	2	1	4
7	5	1	3	2	4	9	8	6
6	2	4	1	9	8	7	3	5
2	4	8	9	3	6	1	5	7
9	1	3	7	5	2	4	6	8
5	7	6	8	4	1	3	2	9

INTERMEDIATE - 69

5	7	1	3	4	8	9	2	6
2	6	3	5	1	9	8	7	4
9	4	8	2	6	7	3	5	1
4	1	2	7	5	3	6	8	9
7	8	9	1	2	6	5	4	3
6	3	5	9	8	4	7	1	2
1	5	6	8	9	2	4	3	7
3	2	4	6	7	5	1	9	8
8	9	7	4	3	1	2	6	5

INTERMEDIATE - 70

6	9	2	8	1	5	4	7	3
7	4	3	9	2	6	8	1	5
5	1	8	3	7	4	2	9	6
8	2	6	7	5	9	3	4	1
9	3	4	2	6	1	7	5	8
1	5	7	4	3	8	9	6	2
4	7	1	6	8	2	5	3	9
2	6	9	5	4	3	1	8	7
3	8	5	1	9	7	6	2	4

INTERMEDIATE - 71

9	3	7	4	6	5	8	2	1
5	1	8	3	2	7	9	4	6
4	6	2	9	8	1	3	7	5
3	5	1	8	9	2	4	6	7
6	8	4	1	7	3	5	9	2
7	2	9	6	5	4	1	8	3
2	4	5	7	1	9	6	3	8
8	7	3	5	4	6	2	1	9
1	9	6	2	3	8	7	5	4

INTERMEDIATE - 72

5	6	1	8	7	9	4	3	2
4	9	8	6	2	3	1	7	5
2	7	3	1	4	5	6	9	8
7	3	4	9	5	6	2	8	1
9	1	2	3	8	4	5	6	7
8	5	6	7	1	2	9	4	3
1	4	7	5	6	8	3	2	9
6	8	9	2	3	1	7	5	4
3	2	5	4	9	7	8	1	6

INTERMEDIATE - 73

5	6	1	8	9	4	3	7	2
2	8	3	6	5	7	9	4	1
9	7	4	1	2	3	5	8	6
8	5	6	9	4	2	7	1	3
4	9	2	3	7	1	6	5	8
1	3	7	5	6	8	2	9	4
7	2	5	4	1	6	8	3	9
3	1	9	2	8	5	4	6	7
6	4	8	7	3	9	1	2	5

INTERMEDIATE - 74

7	2	5	4	3	1	9	6	8
8	9	3	2	6	7	4	1	5
6	1	4	9	8	5	3	2	7
9	4	8	7	1	2	5	3	6
2	5	6	8	9	3	7	4	1
1	3	7	6	5	4	2	8	9
5	8	2	1	4	9	6	7	3
4	6	9	3	7	8	1	5	2
3	7	1	5	2	6	8	9	4

INTERMEDIATE - 75

6	1	2	9	4	5	7	3	8
4	9	7	6	8	3	5	1	2
5	3	8	7	2	1	4	6	9
2	6	9	3	1	7	8	5	4
8	4	5	2	9	6	3	7	1
1	7	3	8	5	4	9	2	6
9	2	6	5	7	8	1	4	3
7	8	1	4	3	2	6	9	5
3	5	4	1	6	9	2	8	7

INTERMEDIATE - 76

3	2	8	4	1	5	7	6	9
4	7	1	3	9	6	5	8	2
6	5	9	7	8	2	1	4	3
2	3	7	9	4	8	6	1	5
9	8	4	6	5	1	2	3	7
1	6	5	2	7	3	8	9	4
7	9	2	8	6	4	3	5	1
8	1	3	5	2	9	4	7	6
5	4	6	1	3	7	9	2	8

INTERMEDIATE - 77

5	8	2	6	9	7	1	4	3
4	7	9	8	3	1	2	6	5
6	1	3	5	2	4	8	9	7
7	9	1	3	6	5	4	8	2
2	5	4	7	1	8	9	3	6
8	3	6	2	4	9	5	7	1
9	4	5	1	7	3	6	2	8
3	6	8	9	5	2	7	1	4
1	2	7	4	8	6	3	5	9

INTERMEDIATE - 78

1	3	9	2	6	4	7	8	5
8	2	7	1	9	5	4	3	6
6	4	5	8	7	3	9	2	1
5	7	2	9	1	8	3	6	4
3	6	4	7	5	2	8	1	9
9	1	8	3	4	6	2	5	7
2	5	6	4	8	9	1	7	3
4	8	1	6	3	7	5	9	2
7	9	3	5	2	1	6	4	8

INTERMEDIATE - 79

8	3	5	7	2	1	4	6	9
6	7	2	3	4	9	1	5	8
1	4	9	8	5	6	7	2	3
5	8	7	4	3	2	6	9	1
4	1	3	9	6	5	8	7	2
2	9	6	1	7	8	3	4	5
9	5	4	6	1	3	2	8	7
7	2	1	5	8	4	9	3	6
3	6	8	2	9	7	5	1	4

INTERMEDIATE - 80

4	3	5	1	2	7	6	9	8
7	1	2	6	8	9	4	3	5
8	9	6	3	5	4	2	7	1
6	5	9	7	1	3	8	4	2
2	8	3	4	6	5	9	1	7
1	7	4	2	9	8	5	6	3
9	4	7	5	3	2	1	8	6
3	2	1	8	4	6	7	5	9
5	6	8	9	7	1	3	2	4

INTERMEDIATE - 81

4	5	3	8	1	9	6	2	7
6	1	9	2	4	7	3	5	8
2	8	7	3	5	6	9	4	1
5	9	1	7	8	2	4	3	6
8	6	4	5	9	3	7	1	2
3	7	2	4	6	1	8	9	5
1	3	5	9	7	8	2	6	4
9	4	8	6	2	5	1	7	3
7	2	6	1	3	4	5	8	9

INTERMEDIATE - 82

8	5	9	2	7	6	4	1	3
6	7	1	4	3	9	5	8	2
3	4	2	5	8	1	7	9	6
5	6	4	8	9	2	3	7	1
9	3	8	7	1	5	2	6	4
1	2	7	6	4	3	9	5	8
4	8	6	9	2	7	1	3	5
2	9	3	1	5	8	6	4	7
7	1	5	3	6	4	8	2	9

INTERMEDIATE - 83

2	1	9	8	5	7	4	6	3
6	4	5	9	2	3	7	8	1
7	8	3	4	6	1	9	2	5
5	6	4	7	9	8	3	1	2
3	9	1	2	4	5	8	7	6
8	7	2	1	3	6	5	9	4
1	2	8	5	7	4	6	3	9
9	5	6	3	8	2	1	4	7
4	3	7	6	1	9	2	5	8

INTERMEDIATE - 84

4	7	8	5	3	1	9	2	6
2	3	5	4	6	9	1	7	8
1	9	6	2	7	8	3	4	5
3	1	7	6	4	5	2	8	9
8	4	9	1	2	7	5	6	3
5	6	2	8	9	3	7	1	4
6	2	1	3	5	4	8	9	7
9	5	4	7	8	2	6	3	1
7	8	3	9	1	6	4	5	2

INTERMEDIATE - 85

5	4	8	9	3	7	6	2	1
7	2	6	4	1	8	3	5	9
9	3	1	6	2	5	8	7	4
4	8	3	5	7	6	1	9	2
6	7	2	3	9	1	4	8	5
1	9	5	8	4	2	7	3	6
2	6	9	7	8	4	5	1	3
8	1	4	2	5	3	9	6	7
3	5	7	1	6	9	2	4	8

INTERMEDIATE - 86

4	9	1	3	5	2	6	7	8
2	8	5	6	4	7	9	3	1
6	3	7	9	8	1	2	5	4
5	6	3	4	7	9	8	1	2
1	2	4	8	3	5	7	6	9
8	7	9	2	1	6	5	4	3
7	1	8	5	2	4	3	9	6
3	5	6	1	9	8	4	2	7
9	4	2	7	6	3	1	8	5

INTERMEDIATE - 87

4	3	7	9	8	5	2	6	1
9	2	6	1	4	3	5	7	8
5	8	1	7	2	6	4	3	9
8	6	9	5	7	2	3	1	4
2	5	4	3	1	9	7	8	6
1	7	3	8	6	4	9	5	2
7	4	2	6	3	8	1	9	5
3	9	8	2	5	1	6	4	7
6	1	5	4	9	7	8	2	3

INTERMEDIATE - 88

9	1	4	8	5	6	3	2	7
3	7	6	1	9	2	4	8	5
2	8	5	3	7	4	6	9	1
5	6	2	7	8	3	9	1	4
8	9	3	4	6	1	5	7	2
1	4	7	9	2	5	8	3	6
7	2	8	5	4	9	1	6	3
6	5	1	2	3	8	7	4	9
4	3	9	6	1	7	2	5	8

INTERMEDIATE - 89

3	2	6	8	5	4	7	1	9
1	4	5	9	7	3	8	2	6
7	9	8	2	1	6	3	4	5
6	3	4	5	8	2	9	7	1
5	1	7	3	6	9	4	8	2
9	8	2	7	4	1	5	6	3
8	5	1	6	3	7	2	9	4
2	6	3	4	9	8	1	5	7
4	7	9	1	2	5	6	3	8

INTERMEDIATE - 90

4	3	7	2	8	9	1	5	6
9	8	5	4	1	6	7	3	2
1	6	2	3	5	7	8	9	4
8	2	3	9	7	1	6	4	5
6	4	1	8	3	5	9	2	7
7	5	9	6	2	4	3	1	8
3	1	6	7	4	2	5	8	9
2	7	8	5	9	3	4	6	1
5	9	4	1	6	8	2	7	3

INTERMEDIATE - 91

2	7	4	5	1	6	9	3	8
3	1	9	4	7	8	5	6	2
6	8	5	3	2	9	7	1	4
1	3	2	9	4	7	8	5	6
7	4	6	1	8	5	3	2	9
9	5	8	6	3	2	1	4	7
4	6	3	8	9	1	2	7	5
8	2	1	7	5	4	6	9	3
5	9	7	2	6	3	4	8	1

INTERMEDIATE - 92

4	7	2	9	8	3	6	1	5
8	3	9	1	5	6	4	7	2
1	6	5	4	2	7	8	3	9
3	9	7	2	6	5	1	8	4
5	8	4	3	1	9	7	2	6
2	1	6	7	4	8	5	9	3
7	2	1	6	3	4	9	5	8
6	5	3	8	9	1	2	4	7
9	4	8	5	7	2	3	6	1

INTERMEDIATE - 93

5	4	2	1	3	8	6	9	7
8	9	1	4	7	6	3	5	2
6	3	7	2	5	9	4	8	1
4	1	5	3	6	7	9	2	8
3	7	8	9	4	2	5	1	6
2	6	9	5	8	1	7	3	4
1	8	4	6	9	5	2	7	3
9	2	6	7	1	3	8	4	5
7	5	3	8	2	4	1	6	9

INTERMEDIATE - 94

8	1	6	9	3	4	2	5	7
7	5	2	1	8	6	9	4	3
9	3	4	2	5	7	1	8	6
6	7	8	5	9	1	4	3	2
1	2	3	6	4	8	7	9	5
4	9	5	7	2	3	6	1	8
3	4	7	8	1	2	5	6	9
5	6	1	3	7	9	8	2	4
2	8	9	4	6	5	3	7	1

INTERMEDIATE - 95

7	6	1	5	9	8	3	4	2
9	8	2	4	6	3	1	5	7
4	3	5	2	7	1	9	8	6
5	1	6	3	2	4	7	9	8
2	9	3	1	8	7	5	6	4
8	7	4	9	5	6	2	1	3
3	2	9	6	4	5	8	7	1
1	4	7	8	3	9	6	2	5
6	5	8	7	1	2	4	3	9

INTERMEDIATE - 96

3	8	7	6	4	2	9	1	5
1	9	5	3	8	7	6	2	4
4	2	6	9	5	1	3	7	8
8	7	3	5	1	4	2	6	9
6	1	4	7	2	9	5	8	3
9	5	2	8	6	3	7	4	1
7	6	9	4	3	8	1	5	2
5	4	1	2	9	6	8	3	7
2	3	8	1	7	5	4	9	6

INTERMEDIATE - 97

8	4	6	5	2	9	3	1	7
9	7	3	8	1	6	4	5	2
1	5	2	4	7	3	9	8	6
4	8	5	7	3	1	6	2	9
3	1	9	6	8	2	5	7	4
6	2	7	9	4	5	8	3	1
2	3	8	1	9	4	7	6	5
7	6	4	2	5	8	1	9	3
5	9	1	3	6	7	2	4	8

INTERMEDIATE - 98

4	9	1	2	8	7	6	5	3
7	2	8	5	6	3	1	9	4
5	6	3	9	1	4	8	2	7
1	5	6	4	3	9	7	8	2
3	4	2	6	7	8	9	1	5
8	7	9	1	2	5	4	3	6
6	8	5	7	9	2	3	4	1
9	1	4	3	5	6	2	7	8
2	3	7	8	4	1	5	6	9

INTERMEDIATE - 99

5	8	9	7	3	4	2	1	6
7	3	6	9	1	2	4	8	5
1	4	2	8	5	6	3	9	7
6	5	7	1	8	3	9	4	2
9	2	3	4	6	5	1	7	8
4	1	8	2	7	9	5	6	3
3	6	4	5	9	7	8	2	1
2	7	1	3	4	8	6	5	9
8	9	5	6	2	1	7	3	4

INTERMEDIATE - 100

5	4	3	2	8	6	9	7	1
2	7	8	1	5	9	4	3	6
9	6	1	4	7	3	5	2	8
7	8	2	9	4	1	6	5	3
1	3	6	8	2	5	7	9	4
4	5	9	3	6	7	8	1	2
8	9	5	6	1	2	3	4	7
6	1	7	5	3	4	2	8	9
3	2	4	7	9	8	1	6	5

INTERMEDIATE - 101

6	5	3	8	4	2	9	1	7
4	7	1	9	3	6	8	2	5
9	8	2	1	5	7	3	4	6
8	3	7	2	1	9	6	5	4
1	9	6	5	8	4	7	3	2
2	4	5	7	6	3	1	9	8
7	6	4	3	9	5	2	8	1
5	1	9	6	2	8	4	7	3
3	2	8	4	7	1	5	6	9

INTERMEDIATE - 102

7	3	4	9	1	5	6	8	2
9	8	5	4	6	2	3	1	7
1	2	6	8	7	3	9	5	4
5	7	3	1	2	9	8	4	6
2	9	8	3	4	6	1	7	5
4	6	1	7	5	8	2	3	9
6	1	7	2	8	4	5	9	3
3	4	2	5	9	1	7	6	8
8	5	9	6	3	7	4	2	1

INTERMEDIATE - 103

1	3	2	8	5	9	6	7	4
6	5	9	7	1	4	3	2	8
7	8	4	2	3	6	5	9	1
5	6	3	4	9	2	1	8	7
2	9	8	3	7	1	4	6	5
4	1	7	5	6	8	2	3	9
9	7	1	6	2	5	8	4	3
8	2	5	9	4	3	7	1	6
3	4	6	1	8	7	9	5	2

INTERMEDIATE - 104

4	2	6	1	3	9	5	8	7
5	7	9	6	4	8	1	2	3
8	1	3	7	2	5	6	4	9
7	8	2	3	9	6	4	1	5
1	3	4	2	5	7	9	6	8
9	6	5	8	1	4	7	3	2
2	5	7	4	8	1	3	9	6
6	4	8	9	7	3	2	5	1
3	9	1	5	6	2	8	7	4

INTERMEDIATE - 105

6	2	3	5	1	9	7	8	4
9	4	1	6	7	8	2	3	5
7	8	5	2	3	4	9	1	6
2	7	8	3	4	1	5	6	9
3	9	4	8	5	6	1	7	2
5	1	6	7	9	2	3	4	8
4	6	7	1	2	5	8	9	3
8	3	2	9	6	7	4	5	1
1	5	9	4	8	3	6	2	7

INTERMEDIATE - 106

7	2	4	5	9	3	6	1	8
8	9	6	7	4	1	2	3	5
3	5	1	6	2	8	9	7	4
9	4	2	8	1	6	3	5	7
5	8	3	9	7	2	4	6	1
1	6	7	4	3	5	8	2	9
2	7	8	3	5	4	1	9	6
4	3	9	1	6	7	5	8	2
6	1	5	2	8	9	7	4	3

INTERMEDIATE - 107

9	8	7	5	2	6	1	3	4
5	6	2	4	1	3	9	7	8
4	3	1	7	9	8	2	5	6
3	5	8	2	4	7	6	1	9
2	7	6	9	3	1	8	4	5
1	9	4	6	8	5	7	2	3
7	4	5	8	6	2	3	9	1
8	2	3	1	5	9	4	6	7
6	1	9	3	7	4	5	8	2

INTERMEDIATE - 108

2	3	8	1	4	5	9	7	6
7	9	6	3	2	8	4	5	1
1	4	5	9	6	7	8	3	2
5	8	9	7	3	2	6	1	4
6	2	3	5	1	4	7	8	9
4	7	1	6	8	9	5	2	3
9	6	7	2	5	3	1	4	8
8	1	2	4	7	6	3	9	5
3	5	4	8	9	1	2	6	7

INTERMEDIATE - 109

1	7	8	4	9	3	2	6	5
4	6	9	2	7	5	3	8	1
3	5	2	8	6	1	4	9	7
7	2	5	3	1	9	6	4	8
8	9	3	7	4	6	1	5	2
6	4	1	5	2	8	9	7	3
5	3	4	9	8	2	7	1	6
2	1	7	6	5	4	8	3	9
9	8	6	1	3	7	5	2	4

INTERMEDIATE - 110

2	4	8	1	5	6	7	9	3
1	9	3	2	4	7	6	8	5
5	6	7	9	3	8	2	1	4
6	2	1	5	7	3	8	4	9
3	5	9	8	2	4	1	6	7
7	8	4	6	9	1	5	3	2
4	3	6	7	1	5	9	2	8
8	7	2	3	6	9	4	5	1
9	1	5	4	8	2	3	7	6

INTERMEDIATE - 111

5	1	2	4	9	3	7	8	6
3	8	4	7	5	6	1	2	9
7	9	6	1	8	2	3	4	5
4	2	7	3	1	9	5	6	8
6	5	8	2	7	4	9	3	1
1	3	9	8	6	5	2	7	4
2	7	1	5	4	8	6	9	3
9	4	5	6	3	7	8	1	2
8	6	3	9	2	1	4	5	7

INTERMEDIATE - 112

6	2	5	4	1	8	3	7	9
7	4	8	2	9	3	6	5	1
1	9	3	6	7	5	8	2	4
8	1	6	5	2	7	9	4	3
5	7	9	8	3	4	1	6	2
2	3	4	1	6	9	5	8	7
4	8	1	9	5	2	7	3	6
9	5	7	3	4	6	2	1	8
3	6	2	7	8	1	4	9	5

INTERMEDIATE - 113

1	6	2	5	4	8	3	9	7
9	5	3	1	6	7	2	8	4
8	4	7	3	2	9	5	1	6
4	7	6	9	8	2	1	3	5
5	3	8	7	1	4	9	6	2
2	1	9	6	5	3	7	4	8
6	9	4	2	3	5	8	7	1
7	8	5	4	9	1	6	2	3
3	2	1	8	7	6	4	5	9

INTERMEDIATE - 114

5	4	2	7	3	1	8	9	6
3	6	9	8	2	4	7	1	5
7	1	8	5	9	6	4	2	3
6	3	4	9	7	2	1	5	8
8	9	7	3	1	5	6	4	2
1	2	5	6	4	8	3	7	9
4	7	6	2	8	9	5	3	1
2	8	1	4	5	3	9	6	7
9	5	3	1	6	7	2	8	4

INTERMEDIATE - 115

5	8	4	9	1	3	6	2	7
7	3	9	6	5	2	8	4	1
1	6	2	8	4	7	5	3	9
4	2	5	1	8	6	7	9	3
3	1	7	5	2	9	4	8	6
6	9	8	3	7	4	2	1	5
2	4	1	7	9	5	3	6	8
8	5	6	2	3	1	9	7	4
9	7	3	4	6	8	1	5	2

INTERMEDIATE - 116

5	8	2	1	9	7	6	4	3
1	3	6	5	8	4	2	7	9
7	9	4	3	2	6	1	8	5
4	6	9	2	3	1	8	5	7
8	2	7	4	6	5	3	9	1
3	1	5	9	7	8	4	2	6
9	7	1	6	4	2	5	3	8
2	5	8	7	1	3	9	6	4
6	4	3	8	5	9	7	1	2

INTERMEDIATE - 117

3	5	6	8	1	4	2	7	9
7	4	2	3	6	9	1	5	8
1	9	8	5	7	2	6	3	4
6	2	4	9	5	7	3	8	1
8	3	9	6	4	1	5	2	7
5	7	1	2	3	8	9	4	6
9	1	3	4	8	5	7	6	2
4	6	7	1	2	3	8	9	5
2	8	5	7	9	6	4	1	3

INTERMEDIATE - 118

3	6	7	2	1	4	8	9	5
1	2	8	3	5	9	6	4	7
5	9	4	8	7	6	1	2	3
8	5	9	1	2	7	4	3	6
7	1	6	9	4	3	2	5	8
4	3	2	6	8	5	9	7	1
9	7	1	4	3	8	5	6	2
2	4	5	7	6	1	3	8	9
6	8	3	5	9	2	7	1	4

INTERMEDIATE - 119

8	2	7	4	9	1	5	6	3
9	1	6	3	5	8	4	7	2
3	4	5	6	7	2	8	9	1
7	5	8	1	4	3	6	2	9
1	6	2	7	8	9	3	4	5
4	3	9	5	2	6	7	1	8
2	8	4	9	3	7	1	5	6
6	7	3	2	1	5	9	8	4
5	9	1	8	6	4	2	3	7

INTERMEDIATE - 120

2	5	3	4	7	8	6	1	9
6	1	4	3	9	5	8	2	7
8	7	9	6	1	2	5	4	3
4	3	7	5	8	9	2	6	1
5	2	8	1	6	3	7	9	4
9	6	1	7	2	4	3	5	8
7	9	5	8	4	6	1	3	2
3	8	2	9	5	1	4	7	6
1	4	6	2	3	7	9	8	5

INTERMEDIATE - 121

4	2	9	7	3	5	1	6	8
5	8	1	4	2	6	9	7	3
3	7	6	1	8	9	5	4	2
7	3	4	5	9	8	2	1	6
1	5	2	6	7	3	8	9	4
6	9	8	2	4	1	3	5	7
8	1	5	3	6	4	7	2	9
9	4	7	8	5	2	6	3	1
2	6	3	9	1	7	4	8	5

INTERMEDIATE - 122

9	3	7	1	8	6	4	2	5
2	4	6	7	9	5	3	1	8
8	5	1	4	3	2	7	9	6
4	6	5	3	2	9	8	7	1
1	2	3	8	4	7	6	5	9
7	8	9	5	6	1	2	4	3
6	7	2	9	1	3	5	8	4
3	9	4	2	5	8	1	6	7
5	1	8	6	7	4	9	3	2

INTERMEDIATE - 123

8	5	6	9	1	7	2	4	3
7	9	3	8	2	4	1	6	5
2	4	1	3	5	6	7	9	8
3	7	4	6	9	1	8	5	2
1	6	8	5	3	2	9	7	4
5	2	9	7	4	8	6	3	1
4	8	5	2	6	9	3	1	7
9	1	7	4	8	3	5	2	6
6	3	2	1	7	5	4	8	9

INTERMEDIATE - 124

3	5	1	6	7	9	4	2	8
4	8	9	1	2	5	7	3	6
6	7	2	4	8	3	9	1	5
8	2	3	7	1	4	5	6	9
7	9	5	3	6	8	1	4	2
1	4	6	9	5	2	3	8	7
5	6	4	2	9	1	8	7	3
9	1	7	8	3	6	2	5	4
2	3	8	5	4	7	6	9	1

INTERMEDIATE - 125

6	2	1	4	8	9	5	3	7
4	3	9	6	7	5	2	1	8
5	8	7	3	1	2	4	6	9
8	1	4	2	5	7	3	9	6
7	5	2	9	3	6	8	4	1
9	6	3	8	4	1	7	2	5
2	4	6	7	9	8	1	5	3
1	9	8	5	2	3	6	7	4
3	7	5	1	6	4	9	8	2

INTERMEDIATE - 126

7	2	3	4	1	6	9	8	5
5	9	6	8	7	2	1	4	3
8	1	4	9	3	5	6	7	2
6	8	1	5	4	7	2	3	9
3	5	7	2	8	9	4	6	1
2	4	9	3	6	1	7	5	8
1	7	2	6	5	3	8	9	4
9	3	8	7	2	4	5	1	6
4	6	5	1	9	8	3	2	7

INTERMEDIATE - 127

8	1	7	4	2	3	5	9	6
6	9	3	7	1	5	4	2	8
5	4	2	8	9	6	1	7	3
2	6	1	5	4	8	9	3	7
9	3	4	1	7	2	6	8	5
7	5	8	6	3	9	2	1	4
3	2	6	9	8	4	7	5	1
1	8	5	2	6	7	3	4	9
4	7	9	3	5	1	8	6	2

INTERMEDIATE - 128

7	4	8	1	6	5	3	9	2
2	5	6	4	9	3	1	8	7
1	9	3	8	7	2	5	6	4
6	2	9	5	4	7	8	3	1
4	7	5	3	8	1	6	2	9
8	3	1	9	2	6	4	7	5
5	8	4	2	3	9	7	1	6
9	1	7	6	5	8	2	4	3
3	6	2	7	1	4	9	5	8

INTERMEDIATE - 129

1	6	9	3	7	5	4	2	8
2	5	3	8	1	4	7	9	6
7	4	8	6	2	9	1	5	3
6	8	7	9	4	2	5	3	1
3	2	1	7	5	8	9	6	4
5	9	4	1	3	6	2	8	7
9	1	5	4	6	3	8	7	2
4	3	2	5	8	7	6	1	9
8	7	6	2	9	1	3	4	5

INTERMEDIATE - 130

3	2	5	9	8	1	6	4	7
7	4	1	5	3	6	9	2	8
9	6	8	2	4	7	1	3	5
4	9	2	3	1	8	7	5	6
5	8	6	4	7	2	3	9	1
1	7	3	6	9	5	4	8	2
8	1	9	7	2	3	5	6	4
2	5	4	1	6	9	8	7	3
6	3	7	8	5	4	2	1	9

INTERMEDIATE - 131

6	5	2	8	7	1	9	4	3
4	9	3	2	6	5	8	1	7
1	7	8	9	4	3	5	2	6
8	4	7	3	1	2	6	5	9
3	1	5	4	9	6	2	7	8
2	6	9	5	8	7	1	3	4
9	8	1	7	5	4	3	6	2
5	2	4	6	3	9	7	8	1
7	3	6	1	2	8	4	9	5

INTERMEDIATE - 132

6	8	4	7	2	3	1	5	9
3	7	9	4	5	1	6	8	2
5	1	2	9	6	8	7	3	4
9	2	1	8	4	7	3	6	5
4	3	6	2	1	5	9	7	8
8	5	7	3	9	6	2	4	1
1	9	3	5	7	4	8	2	6
7	6	5	1	8	2	4	9	3
2	4	8	6	3	9	5	1	7

INTERMEDIATE - 133

1	7	2	8	4	6	9	3	5
8	6	9	7	3	5	2	4	1
4	3	5	1	9	2	8	6	7
6	4	3	5	1	9	7	2	8
7	5	8	3	2	4	1	9	6
9	2	1	6	7	8	4	5	3
5	1	4	2	6	7	3	8	9
3	9	6	4	8	1	5	7	2
2	8	7	9	5	3	6	1	4

INTERMEDIATE - 134

4	7	5	6	1	8	9	3	2
6	2	8	3	5	9	1	4	7
9	3	1	2	7	4	8	5	6
2	6	4	1	9	5	7	8	3
5	9	3	7	8	2	6	1	4
8	1	7	4	3	6	2	9	5
1	4	2	8	6	3	5	7	9
7	5	6	9	4	1	3	2	8
3	8	9	5	2	7	4	6	1

INTERMEDIATE - 135

9	7	3	8	4	2	5	6	1
4	8	6	5	3	1	7	2	9
1	2	5	6	7	9	8	4	3
2	1	4	7	8	3	6	9	5
7	3	9	1	5	6	4	8	2
6	5	8	2	9	4	1	3	7
5	9	7	4	2	8	3	1	6
3	4	1	9	6	7	2	5	8
8	6	2	3	1	5	9	7	4

INTERMEDIATE - 136

6	1	9	4	8	3	2	7	5
8	3	5	9	7	2	6	1	4
7	4	2	1	5	6	9	8	3
5	7	3	6	4	1	8	2	9
1	9	6	2	3	8	4	5	7
4	2	8	7	9	5	3	6	1
3	5	7	8	2	9	1	4	6
2	6	4	3	1	7	5	9	8
9	8	1	5	6	4	7	3	2

INTERMEDIATE - 137

6	3	1	7	9	5	8	2	4
8	7	9	3	4	2	6	5	1
2	5	4	8	1	6	9	7	3
3	8	5	1	2	9	7	4	6
1	2	7	4	6	8	3	9	5
4	9	6	5	7	3	2	1	8
5	4	8	2	3	7	1	6	9
7	6	3	9	5	1	4	8	2
9	1	2	6	8	4	5	3	7

INTERMEDIATE - 138

8	1	3	6	9	4	5	7	2
5	9	4	3	2	7	8	6	1
7	6	2	5	8	1	3	4	9
3	7	6	2	4	9	1	8	5
1	5	8	7	3	6	2	9	4
2	4	9	8	1	5	6	3	7
6	3	5	4	7	2	9	1	8
9	2	7	1	6	8	4	5	3
4	8	1	9	5	3	7	2	6

INTERMEDIATE - 139

7	6	5	9	8	3	2	1	4
1	4	2	7	5	6	3	9	8
8	3	9	1	4	2	5	6	7
3	7	1	6	9	5	4	8	2
6	9	4	8	2	7	1	5	3
5	2	8	4	3	1	6	7	9
2	5	7	3	6	9	8	4	1
4	1	3	5	7	8	9	2	6
9	8	6	2	1	4	7	3	5

INTERMEDIATE - 140

2	1	6	3	4	7	9	5	8
5	4	3	6	8	9	7	1	2
7	8	9	2	5	1	6	4	3
3	5	8	1	6	4	2	9	7
9	7	1	5	3	2	4	8	6
4	6	2	9	7	8	1	3	5
1	3	5	7	9	6	8	2	4
6	9	4	8	2	5	3	7	1
8	2	7	4	1	3	5	6	9

INTERMEDIATE - 141

7	5	2	1	6	3	8	9	4
6	1	4	9	2	8	7	5	3
8	3	9	7	5	4	2	6	1
9	2	3	5	7	6	1	4	8
1	7	6	4	8	9	3	2	5
5	4	8	3	1	2	9	7	6
4	6	1	8	9	7	5	3	2
2	8	7	6	3	5	4	1	9
3	9	5	2	4	1	6	8	7

INTERMEDIATE - 142

8	1	4	6	5	7	2	3	9
3	2	5	4	9	1	8	6	7
7	9	6	8	2	3	1	4	5
4	6	7	1	8	2	9	5	3
5	8	9	3	4	6	7	1	2
1	3	2	9	7	5	6	8	4
2	4	3	7	1	8	5	9	6
6	7	8	5	3	9	4	2	1
9	5	1	2	6	4	3	7	8

INTERMEDIATE - 143

9	6	8	7	1	4	5	2	3
1	5	7	2	9	3	6	4	8
4	3	2	8	6	5	9	7	1
6	9	4	3	2	8	1	5	7
3	7	1	4	5	9	2	8	6
2	8	5	1	7	6	3	9	4
7	4	6	5	3	2	8	1	9
5	1	9	6	8	7	4	3	2
8	2	3	9	4	1	7	6	5

INTERMEDIATE - 144

8	9	7	3	6	4	2	1	5
4	5	6	7	2	1	8	9	3
2	3	1	9	8	5	7	6	4
3	7	8	1	5	2	6	4	9
9	2	4	8	3	6	1	5	7
6	1	5	4	7	9	3	2	8
7	6	2	5	9	3	4	8	1
1	8	9	2	4	7	5	3	6
5	4	3	6	1	8	9	7	2

INTERMEDIATE - 145

2	3	1	6	8	7	4	9	5
9	6	7	3	5	4	8	2	1
4	8	5	9	1	2	6	7	3
1	7	8	4	9	6	3	5	2
3	4	6	2	7	5	9	1	8
5	2	9	1	3	8	7	4	6
6	5	2	8	4	9	1	3	7
8	1	4	7	2	3	5	6	9
7	9	3	5	6	1	2	8	4

INTERMEDIATE - 146

9	3	5	2	4	8	6	7	1
2	4	7	1	6	9	3	8	5
8	6	1	7	5	3	9	4	2
7	2	6	9	3	4	1	5	8
3	9	4	5	8	1	7	2	6
5	1	8	6	2	7	4	3	9
1	8	2	4	7	6	5	9	3
4	5	9	3	1	2	8	6	7
6	7	3	8	9	5	2	1	4

INTERMEDIATE - 147

8	9	4	5	7	1	2	6	3
5	6	3	9	2	4	8	1	7
2	7	1	8	3	6	9	4	5
6	3	7	4	8	9	1	5	2
1	5	9	2	6	3	7	8	4
4	8	2	1	5	7	6	3	9
7	2	6	3	1	5	4	9	8
9	1	5	7	4	8	3	2	6
3	4	8	6	9	2	5	7	1

INTERMEDIATE - 148

7	3	6	8	5	4	2	1	9
4	9	8	3	2	1	5	7	6
5	1	2	7	9	6	8	3	4
2	7	5	6	3	9	4	8	1
1	8	9	5	4	2	3	6	7
3	6	4	1	8	7	9	5	2
6	2	3	4	1	8	7	9	5
8	4	7	9	6	5	1	2	3
9	5	1	2	7	3	6	4	8

INTERMEDIATE - 149

2	6	1	5	7	4	3	8	9
9	5	8	2	6	3	7	4	1
3	7	4	1	8	9	2	5	6
6	1	9	4	3	7	5	2	8
5	3	2	8	9	1	4	6	7
8	4	7	6	5	2	1	9	3
4	9	3	7	2	8	6	1	5
1	8	5	3	4	6	9	7	2
7	2	6	9	1	5	8	3	4

INTERMEDIATE - 150

9	1	5	6	7	4	8	2	3
8	6	3	5	1	2	9	7	4
4	7	2	8	3	9	1	5	6
3	4	7	1	2	5	6	9	8
2	9	1	4	8	6	5	3	7
5	8	6	3	9	7	4	1	2
6	3	9	2	5	8	7	4	1
7	2	8	9	4	1	3	6	5
1	5	4	7	6	3	2	8	9

INTERMEDIATE - 151

8	1	3	6	7	9	4	5	2
6	5	4	3	2	1	9	8	7
7	2	9	5	4	8	3	6	1
2	7	1	9	6	3	8	4	5
9	8	5	4	1	7	6	2	3
4	3	6	8	5	2	7	1	9
3	4	7	2	8	5	1	9	6
5	9	8	1	3	6	2	7	4
1	6	2	7	9	4	5	3	8

INTERMEDIATE - 152

3	5	1	4	8	7	6	9	2
6	9	2	5	1	3	7	8	4
8	4	7	6	2	9	3	1	5
9	2	6	8	5	4	1	3	7
4	3	5	7	9	1	8	2	6
1	7	8	3	6	2	4	5	9
2	6	9	1	7	8	5	4	3
5	8	3	9	4	6	2	7	1
7	1	4	2	3	5	9	6	8

INTERMEDIATE - 153

3	1	6	8	7	5	2	4	9
7	2	5	3	9	4	6	1	8
4	9	8	6	2	1	5	3	7
8	4	2	1	3	7	9	6	5
9	5	3	2	8	6	4	7	1
6	7	1	5	4	9	8	2	3
5	6	9	4	1	3	7	8	2
2	3	7	9	6	8	1	5	4
1	8	4	7	5	2	3	9	6

INTERMEDIATE - 154

1	6	5	7	3	9	4	2	8
9	3	4	8	1	2	6	5	7
2	8	7	4	6	5	3	9	1
7	4	8	6	2	3	9	1	5
6	9	1	5	4	7	8	3	2
3	5	2	9	8	1	7	6	4
5	2	6	3	7	8	1	4	9
8	1	3	2	9	4	5	7	6
4	7	9	1	5	6	2	8	3

INTERMEDIATE - 155

6	3	5	8	1	4	7	9	2
4	7	8	3	2	9	1	6	5
2	1	9	6	7	5	8	3	4
1	2	4	5	6	3	9	7	8
7	9	6	4	8	1	5	2	3
5	8	3	2	9	7	6	4	1
3	5	1	9	4	6	2	8	7
9	4	2	7	5	8	3	1	6
8	6	7	1	3	2	4	5	9

INTERMEDIATE - 156

9	3	6	4	1	5	2	7	8
2	4	5	9	7	8	3	6	1
8	1	7	3	6	2	9	5	4
6	8	4	2	9	7	1	3	5
5	9	2	6	3	1	8	4	7
3	7	1	5	8	4	6	2	9
1	2	3	7	5	9	4	8	6
4	5	8	1	2	6	7	9	3
7	6	9	8	4	3	5	1	2

INTERMEDIATE - 157

9	7	5	1	2	3	4	8	6
3	1	2	8	6	4	9	7	5
8	4	6	5	9	7	1	3	2
1	9	3	4	5	8	6	2	7
4	5	7	2	1	6	8	9	3
6	2	8	3	7	9	5	4	1
5	6	9	7	8	2	3	1	4
7	8	4	6	3	1	2	5	9
2	3	1	9	4	5	7	6	8

INTERMEDIATE - 158

4	5	6	1	7	3	2	9	8
9	1	8	4	6	2	5	3	7
3	7	2	8	9	5	6	1	4
6	9	7	3	5	4	1	8	2
5	2	4	7	1	8	9	6	3
1	8	3	6	2	9	7	4	5
8	6	9	2	3	7	4	5	1
2	3	5	9	4	1	8	7	6
7	4	1	5	8	6	3	2	9

INTERMEDIATE - 159

3	9	7	5	4	8	2	1	6
2	8	4	6	9	1	5	3	7
6	5	1	3	7	2	8	9	4
7	4	5	1	8	9	3	6	2
9	1	2	7	6	3	4	5	8
8	3	6	2	5	4	1	7	9
4	7	8	9	1	5	6	2	3
5	2	9	4	3	6	7	8	1
1	6	3	8	2	7	9	4	5

INTERMEDIATE - 160

7	5	3	6	1	9	4	2	8
1	4	8	2	5	7	6	3	9
9	6	2	3	8	4	7	5	1
2	1	9	8	3	6	5	4	7
5	8	4	9	7	2	3	1	6
3	7	6	1	4	5	9	8	2
6	9	5	4	2	1	8	7	3
8	2	7	5	9	3	1	6	4
4	3	1	7	6	8	2	9	5

INTERMEDIATE - 161

4	5	3	8	9	7	6	2	1
6	9	2	5	4	1	8	3	7
8	1	7	3	2	6	4	9	5
9	7	1	6	3	4	2	5	8
3	4	6	2	5	8	1	7	9
2	8	5	1	7	9	3	6	4
1	3	8	7	6	5	9	4	2
7	2	9	4	8	3	5	1	6
5	6	4	9	1	2	7	8	3

INTERMEDIATE - 162

9	1	8	4	5	3	2	7	6
2	3	6	7	8	1	4	5	9
4	5	7	2	9	6	3	1	8
8	4	9	3	1	5	6	2	7
1	2	5	6	7	4	9	8	3
7	6	3	9	2	8	5	4	1
5	7	4	1	6	9	8	3	2
3	9	1	8	4	2	7	6	5
6	8	2	5	3	7	1	9	4

INTERMEDIATE - 163

2	4	5	6	9	1	7	8	3
6	1	3	5	8	7	4	2	9
9	8	7	4	3	2	1	5	6
7	6	9	8	4	5	2	3	1
3	2	8	7	1	9	6	4	5
1	5	4	3	2	6	8	9	7
4	3	1	9	7	8	5	6	2
8	7	6	2	5	3	9	1	4
5	9	2	1	6	4	3	7	8

INTERMEDIATE - 164

7	8	5	4	1	6	3	2	9
1	6	9	2	8	3	5	7	4
3	4	2	7	9	5	1	8	6
6	1	3	8	7	2	4	9	5
2	5	7	3	4	9	6	1	8
4	9	8	5	6	1	7	3	2
5	7	1	6	2	8	9	4	3
8	3	4	9	5	7	2	6	1
9	2	6	1	3	4	8	5	7

INTERMEDIATE - 165

8	2	3	7	9	4	5	6	1
1	5	9	2	6	8	7	3	4
4	6	7	5	1	3	8	2	9
3	7	2	4	5	9	1	8	6
5	1	6	3	8	7	9	4	2
9	4	8	6	2	1	3	5	7
7	3	1	8	4	2	6	9	5
6	9	4	1	3	5	2	7	8
2	8	5	9	7	6	4	1	3

INTERMEDIATE - 166

7	1	5	2	8	4	6	3	9
3	4	9	1	6	5	2	8	7
2	8	6	3	9	7	1	4	5
5	6	3	4	2	9	7	1	8
4	7	2	6	1	8	5	9	3
1	9	8	7	5	3	4	2	6
8	3	1	5	7	2	9	6	4
9	2	7	8	4	6	3	5	1
6	5	4	9	3	1	8	7	2

INTERMEDIATE - 167

9	7	3	8	4	2	1	5	6
5	1	4	7	9	6	2	3	8
8	6	2	1	3	5	7	4	9
2	4	6	9	7	3	8	1	5
7	8	1	5	2	4	9	6	3
3	9	5	6	8	1	4	7	2
1	3	8	2	6	7	5	9	4
6	2	7	4	5	9	3	8	1
4	5	9	3	1	8	6	2	7

INTERMEDIATE - 168

8	1	5	7	3	6	4	2	9
2	3	7	9	4	1	6	5	8
9	6	4	2	5	8	1	7	3
4	5	6	1	2	9	3	8	7
3	2	9	4	8	7	5	6	1
1	7	8	5	6	3	9	4	2
5	4	3	8	1	2	7	9	6
7	8	1	6	9	4	2	3	5
6	9	2	3	7	5	8	1	4

INTERMEDIATE - 169

3	4	9	2	5	6	1	8	7
2	7	1	3	9	8	4	5	6
8	5	6	4	7	1	2	9	3
9	1	8	5	3	2	7	6	4
4	3	5	8	6	7	9	1	2
7	6	2	1	4	9	5	3	8
5	9	3	6	2	4	8	7	1
1	2	7	9	8	3	6	4	5
6	8	4	7	1	5	3	2	9

INTERMEDIATE - 170

7	1	2	9	3	4	8	5	6
4	9	8	7	6	5	1	3	2
6	3	5	8	1	2	7	4	9
1	7	3	2	8	9	5	6	4
8	2	6	5	4	3	9	7	1
5	4	9	6	7	1	3	2	8
3	5	4	1	9	6	2	8	7
9	6	7	3	2	8	4	1	5
2	8	1	4	5	7	6	9	3

INTERMEDIATE - 171

1	3	2	9	7	6	5	8	4
7	4	5	8	1	3	9	6	2
8	6	9	4	2	5	3	7	1
2	5	8	1	3	7	6	4	9
3	9	6	5	4	2	7	1	8
4	1	7	6	9	8	2	5	3
5	8	1	2	6	9	4	3	7
9	7	4	3	5	1	8	2	6
6	2	3	7	8	4	1	9	5

INTERMEDIATE - 172

3	6	4	9	5	8	7	2	1
9	2	5	1	6	7	8	3	4
1	7	8	2	4	3	5	6	9
7	9	3	4	2	6	1	5	8
4	1	6	3	8	5	9	7	2
5	8	2	7	1	9	3	4	6
2	5	1	8	7	4	6	9	3
6	4	9	5	3	1	2	8	7
8	3	7	6	9	2	4	1	5

INTERMEDIATE - 173

6	8	3	9	5	4	1	2	7
4	7	9	6	2	1	5	8	3
2	1	5	7	3	8	4	6	9
7	4	2	1	6	3	9	5	8
8	9	1	4	7	5	6	3	2
5	3	6	8	9	2	7	1	4
1	6	8	2	4	7	3	9	5
3	2	7	5	1	9	8	4	6
9	5	4	3	8	6	2	7	1

INTERMEDIATE - 174

6	4	1	3	2	7	8	9	5
3	5	2	8	6	9	7	4	1
7	9	8	5	4	1	2	3	6
9	6	7	2	5	8	3	1	4
8	1	3	9	7	4	5	6	2
5	2	4	6	1	3	9	8	7
1	3	5	4	9	2	6	7	8
2	7	9	1	8	6	4	5	3
4	8	6	7	3	5	1	2	9

INTERMEDIATE - 175

7	3	9	1	6	2	8	5	4
5	1	6	8	7	4	3	9	2
2	8	4	3	5	9	7	6	1
9	5	7	6	2	3	1	4	8
6	4	3	5	8	1	2	7	9
1	2	8	9	4	7	5	3	6
8	7	2	4	3	6	9	1	5
4	9	5	7	1	8	6	2	3
3	6	1	2	9	5	4	8	7

INTERMEDIATE - 176

4	9	8	1	6	3	5	2	7
6	7	5	2	4	8	1	3	9
1	2	3	7	5	9	4	6	8
3	5	6	4	8	2	9	7	1
8	1	9	6	7	5	2	4	3
2	4	7	3	9	1	8	5	6
5	3	4	8	1	6	7	9	2
7	8	2	9	3	4	6	1	5
9	6	1	5	2	7	3	8	4

INTERMEDIATE - 177

3	1	6	8	7	5	2	4	9
7	2	5	3	9	4	1	6	8
9	4	8	2	1	6	3	5	7
1	5	7	4	6	8	9	3	2
6	8	9	1	3	2	4	7	5
4	3	2	9	5	7	8	1	6
5	9	3	7	8	1	6	2	4
2	7	1	6	4	9	5	8	3
8	6	4	5	2	3	7	9	1

INTERMEDIATE - 178

7	3	9	8	2	6	1	5	4
5	8	1	3	9	4	6	7	2
6	2	4	7	1	5	8	3	9
1	6	2	5	8	7	4	9	3
4	7	3	2	6	9	5	1	8
9	5	8	4	3	1	7	2	6
3	9	6	1	5	8	2	4	7
8	4	5	9	7	2	3	6	1
2	1	7	6	4	3	9	8	5

INTERMEDIATE - 179

2	9	8	6	7	1	4	3	5
7	4	5	8	9	3	2	6	1
6	1	3	2	4	5	8	9	7
1	8	6	5	2	7	3	4	9
4	5	9	1	3	6	7	2	8
3	7	2	4	8	9	1	5	6
8	3	1	9	6	4	5	7	2
5	6	4	7	1	2	9	8	3
9	2	7	3	5	8	6	1	4

INTERMEDIATE - 180

2	4	3	8	5	1	9	7	6
8	5	6	4	9	7	2	1	3
9	1	7	2	3	6	5	8	4
6	9	1	3	4	2	8	5	7
7	2	8	9	6	5	3	4	1
4	3	5	1	7	8	6	2	9
1	7	2	6	8	3	4	9	5
5	6	9	7	2	4	1	3	8
3	8	4	5	1	9	7	6	2

INTERMEDIATE - 181

1	5	9	4	2	3	7	8	6
8	4	3	9	6	7	5	2	1
2	7	6	1	8	5	9	4	3
7	1	4	5	3	9	2	6	8
6	9	8	7	1	2	3	5	4
3	2	5	8	4	6	1	7	9
9	8	1	2	7	4	6	3	5
5	3	2	6	9	8	4	1	7
4	6	7	3	5	1	8	9	2

INTERMEDIATE - 182

6	5	1	7	8	2	4	3	9
7	3	2	5	9	4	6	1	8
4	9	8	3	6	1	2	5	7
1	4	7	9	3	5	8	6	2
8	6	3	4	2	7	5	9	1
5	2	9	6	1	8	3	7	4
2	7	4	1	5	6	9	8	3
3	8	6	2	7	9	1	4	5
9	1	5	8	4	3	7	2	6

INTERMEDIATE - 183

8	5	7	6	9	3	2	1	4
6	1	2	5	4	7	3	8	9
3	9	4	2	8	1	5	6	7
4	2	8	3	1	9	6	7	5
1	7	3	8	6	5	4	9	2
9	6	5	4	7	2	8	3	1
5	4	9	7	3	6	1	2	8
2	3	1	9	5	8	7	4	6
7	8	6	1	2	4	9	5	3

INTERMEDIATE - 184

7	1	4	5	6	8	3	9	2
2	9	8	7	1	3	6	5	4
3	6	5	4	2	9	7	1	8
4	8	2	1	7	5	9	6	3
9	3	7	6	8	4	1	2	5
6	5	1	9	3	2	8	4	7
1	4	3	2	9	7	5	8	6
8	2	9	3	5	6	4	7	1
5	7	6	8	4	1	2	3	9

INTERMEDIATE - 185

8	4	7	3	9	1	2	6	5
6	2	5	8	7	4	9	1	3
9	3	1	2	5	6	4	7	8
1	9	4	6	8	3	5	2	7
5	6	8	9	2	7	1	3	4
2	7	3	4	1	5	8	9	6
7	8	6	1	4	9	3	5	2
3	1	2	5	6	8	7	4	9
4	5	9	7	3	2	6	8	1

INTERMEDIATE - 186

5	2	1	3	8	9	4	7	6
9	8	4	5	6	7	3	1	2
3	7	6	1	4	2	9	8	5
6	4	3	7	1	5	8	2	9
2	5	8	4	9	3	7	6	1
1	9	7	8	2	6	5	3	4
8	3	2	9	5	1	6	4	7
7	1	9	6	3	4	2	5	8
4	6	5	2	7	8	1	9	3

INTERMEDIATE - 187

1	2	9	8	6	4	5	7	3
3	4	8	1	5	7	2	9	6
6	5	7	9	2	3	4	1	8
9	3	4	2	7	6	8	5	1
7	6	5	3	8	1	9	4	2
2	8	1	5	4	9	6	3	7
8	7	2	4	1	5	3	6	9
5	1	3	6	9	2	7	8	4
4	9	6	7	3	8	1	2	5

INTERMEDIATE - 188

1	4	3	7	2	6	5	9	8
2	6	7	8	9	5	3	4	1
8	5	9	3	4	1	2	6	7
4	7	1	9	5	3	8	2	6
5	2	8	4	6	7	9	1	3
3	9	6	2	1	8	4	7	5
9	1	5	6	3	2	7	8	4
7	3	4	1	8	9	6	5	2
6	8	2	5	7	4	1	3	9

INTERMEDIATE - 189

5	1	3	4	9	8	7	6	2
4	2	7	5	1	6	9	8	3
9	6	8	3	7	2	1	5	4
7	8	1	2	5	9	4	3	6
3	4	9	6	8	1	5	2	7
2	5	6	7	3	4	8	1	9
6	3	5	8	4	7	2	9	1
1	7	2	9	6	5	3	4	8
8	9	4	1	2	3	6	7	5

INTERMEDIATE - 190

2	7	4	6	3	9	5	1	8
5	1	3	4	2	8	6	9	7
6	8	9	1	7	5	3	4	2
1	6	5	7	9	4	2	8	3
7	9	2	5	8	3	1	6	4
4	3	8	2	6	1	7	5	9
8	2	6	9	1	7	4	3	5
3	4	1	8	5	2	9	7	6
9	5	7	3	4	6	8	2	1

INTERMEDIATE - 191

8	7	3	6	9	4	5	1	2
2	4	9	8	5	1	6	3	7
6	5	1	2	3	7	4	8	9
4	6	2	1	8	3	7	9	5
7	1	5	4	6	9	8	2	3
9	3	8	5	7	2	1	4	6
5	2	7	9	1	8	3	6	4
3	8	4	7	2	6	9	5	1
1	9	6	3	4	5	2	7	8

INTERMEDIATE - 192

7	6	5	9	8	3	2	1	4
1	4	9	5	6	2	3	8	7
8	3	2	4	1	7	6	9	5
6	2	1	7	9	4	8	5	3
4	9	3	8	5	1	7	2	6
5	8	7	3	2	6	1	4	9
9	7	6	1	4	8	5	3	2
3	5	8	2	7	9	4	6	1
2	1	4	6	3	5	9	7	8

INTERMEDIATE - 193

7	1	9	6	3	2	4	8	5
3	4	8	1	5	7	6	9	2
5	6	2	8	4	9	7	3	1
4	3	6	7	1	5	9	2	8
9	2	7	3	6	8	5	1	4
1	8	5	9	2	4	3	7	6
8	7	4	2	9	6	1	5	3
6	9	1	5	8	3	2	4	7
2	5	3	4	7	1	8	6	9

INTERMEDIATE - 194

1	5	8	3	9	4	7	2	6
9	4	3	6	7	2	8	5	1
6	7	2	8	1	5	9	4	3
5	1	4	7	2	9	6	3	8
3	2	9	1	8	6	4	7	5
7	8	6	4	5	3	1	9	2
8	9	1	2	3	7	5	6	4
2	6	7	5	4	8	3	1	9
4	3	5	9	6	1	2	8	7

INTERMEDIATE - 195

3	5	2	8	4	6	7	1	9
4	7	6	2	9	1	8	5	3
1	9	8	5	7	3	6	4	2
9	3	7	6	5	8	1	2	4
2	6	4	9	1	7	5	3	8
5	8	1	3	2	4	9	6	7
7	2	9	4	6	5	3	8	1
6	1	3	7	8	2	4	9	5
8	4	5	1	3	9	2	7	6

INTERMEDIATE - 196

7	3	6	4	8	9	5	2	1
9	2	1	6	5	7	4	3	8
8	5	4	2	3	1	6	7	9
5	9	7	8	1	6	2	4	3
6	4	3	9	2	5	1	8	7
1	8	2	7	4	3	9	5	6
4	1	8	3	6	2	7	9	5
3	6	9	5	7	4	8	1	2
2	7	5	1	9	8	3	6	4

INTERMEDIATE - 197

9	8	1	4	6	2	5	7	3
5	6	2	8	7	3	1	9	4
7	3	4	9	5	1	8	2	6
3	1	5	2	4	6	7	8	9
6	2	9	3	8	7	4	5	1
8	4	7	5	1	9	3	6	2
1	7	3	6	2	8	9	4	5
2	5	8	1	9	4	6	3	7
4	9	6	7	3	5	2	1	8

INTERMEDIATE - 198

6	7	8	4	9	1	5	3	2
1	5	2	3	6	8	4	9	7
9	4	3	5	2	7	8	1	6
2	9	4	1	3	5	6	7	8
7	3	6	8	4	2	9	5	1
8	1	5	9	7	6	3	2	4
3	2	7	6	8	9	1	4	5
5	6	9	7	1	4	2	8	3
4	8	1	2	5	3	7	6	9

INTERMEDIATE - 199

9	2	1	4	6	3	5	7	8
7	6	3	9	8	5	2	4	1
4	8	5	7	1	2	6	3	9
6	5	7	3	4	9	1	8	2
3	9	2	8	5	1	7	6	4
8	1	4	6	2	7	3	9	5
5	3	8	1	7	4	9	2	6
1	4	9	2	3	6	8	5	7
2	7	6	5	9	8	4	1	3

INTERMEDIATE - 200

5	3	4	7	1	6	2	8	9
7	2	1	8	3	9	4	5	6
6	8	9	5	2	4	7	1	3
9	6	7	1	8	3	5	2	4
3	1	8	2	4	5	6	9	7
2	4	5	9	6	7	8	3	1
4	7	2	3	5	1	9	6	8
8	9	3	6	7	2	1	4	5
1	5	6	4	9	8	3	7	2

INTERMEDIATE - 201

2	4	6	1	3	8	7	9	5
5	8	3	7	6	9	2	4	1
7	1	9	5	2	4	8	6	3
9	7	1	8	4	6	3	5	2
4	5	8	3	1	2	6	7	9
6	3	2	9	7	5	1	8	4
8	6	5	2	9	1	4	3	7
3	2	4	6	5	7	9	1	8
1	9	7	4	8	3	5	2	6

INTERMEDIATE - 202

5	6	4	9	1	7	8	3	2
3	2	9	6	4	8	7	1	5
1	7	8	5	3	2	9	6	4
8	3	2	4	5	6	1	9	7
9	4	1	2	7	3	6	5	8
6	5	7	1	8	9	2	4	3
2	1	3	8	6	4	5	7	9
4	9	6	7	2	5	3	8	1
7	8	5	3	9	1	4	2	6

INTERMEDIATE - 203

4	1	7	2	8	9	6	3	5
6	8	3	1	5	7	9	2	4
2	9	5	6	4	3	8	7	1
8	7	4	3	1	5	2	6	9
9	6	1	7	2	4	3	5	8
3	5	2	9	6	8	4	1	7
1	3	9	8	7	6	5	4	2
7	4	6	5	9	2	1	8	3
5	2	8	4	3	1	7	9	6

INTERMEDIATE - 204

8	1	7	4	9	2	3	5	6
5	2	9	3	7	6	1	4	8
6	3	4	1	5	8	2	7	9
9	7	3	6	2	4	5	8	1
1	6	8	9	3	5	4	2	7
4	5	2	7	8	1	9	6	3
2	4	6	8	1	9	7	3	5
7	8	1	5	4	3	6	9	2
3	9	5	2	6	7	8	1	4

HARD - 1

9	8	6	1	5	2	7	4	3
3	4	7	8	6	9	2	1	5
1	5	2	4	7	3	6	9	8
2	6	8	5	4	1	3	7	9
4	9	3	2	8	7	5	6	1
5	7	1	3	9	6	8	2	4
8	1	5	6	2	4	9	3	7
7	2	4	9	3	5	1	8	6
6	3	9	7	1	8	4	5	2

HARD - 2

4	3	5	2	8	6	9	7	1
7	8	1	5	4	9	3	2	6
9	2	6	3	7	1	4	5	8
8	9	7	4	6	3	2	1	5
1	5	2	8	9	7	6	4	3
6	4	3	1	2	5	7	8	9
5	6	8	7	3	2	1	9	4
2	1	9	6	5	4	8	3	7
3	7	4	9	1	8	5	6	2

HARD - 3

3	4	1	9	7	8	2	5	6
2	8	7	3	6	5	1	9	4
9	6	5	2	1	4	3	8	7
8	9	6	1	4	2	5	7	3
1	7	2	5	9	3	4	6	8
5	3	4	6	8	7	9	1	2
4	1	9	7	2	6	8	3	5
6	5	8	4	3	1	7	2	9
7	2	3	8	5	9	6	4	1

HARD - 4

1	8	4	2	9	3	7	6	5
9	5	2	7	6	4	1	8	3
7	6	3	1	5	8	4	2	9
5	9	6	3	7	2	8	4	1
2	1	7	8	4	9	3	5	6
4	3	8	5	1	6	9	7	2
3	2	5	9	8	7	6	1	4
6	7	1	4	3	5	2	9	8
8	4	9	6	2	1	5	3	7

HARD - 5

7	9	8	6	3	2	5	1	4
1	3	6	9	4	5	2	7	8
2	4	5	7	8	1	9	6	3
4	5	2	1	6	3	8	9	7
8	6	9	4	5	7	1	3	2
3	7	1	8	2	9	4	5	6
9	1	4	3	7	8	6	2	5
5	8	7	2	1	6	3	4	9
6	2	3	5	9	4	7	8	1

HARD - 6

8	9	4	2	1	3	7	6	5
7	1	5	8	6	9	3	4	2
2	3	6	5	4	7	9	1	8
5	2	3	4	7	1	6	8	9
4	8	9	3	5	6	2	7	1
6	7	1	9	2	8	4	5	3
9	5	7	6	8	2	1	3	4
3	6	8	1	9	4	5	2	7
1	4	2	7	3	5	8	9	6

HARD - 7

3	7	2	6	4	8	5	9	1
8	5	4	3	1	9	7	6	2
6	9	1	5	7	2	3	4	8
5	3	6	8	2	7	9	1	4
2	4	8	1	9	3	6	5	7
9	1	7	4	6	5	2	8	3
4	8	3	7	5	6	1	2	9
7	6	9	2	8	1	4	3	5
1	2	5	9	3	4	8	7	6

HARD - 8

7	1	2	3	5	4	9	6	8
8	9	6	7	1	2	3	5	4
3	4	5	8	6	9	7	2	1
5	2	8	1	4	7	6	9	3
9	7	1	2	3	6	4	8	5
4	6	3	9	8	5	2	1	7
1	3	9	6	7	8	5	4	2
6	5	7	4	2	1	8	3	9
2	8	4	5	9	3	1	7	6

HARD - 9

1	9	8	6	3	5	7	4	2
2	3	5	1	4	7	8	6	9
7	6	4	2	8	9	3	1	5
8	2	6	9	5	4	1	3	7
9	4	7	3	2	1	5	8	6
3	5	1	7	6	8	9	2	4
6	1	9	8	7	2	4	5	3
5	7	2	4	1	3	6	9	8
4	8	3	5	9	6	2	7	1

HARD - 10

6	4	1	2	3	7	8	9	5
7	5	2	6	9	8	3	1	4
8	3	9	4	1	5	2	7	6
4	2	7	9	8	1	5	6	3
5	9	6	7	2	3	4	8	1
1	8	3	5	6	4	7	2	9
9	6	5	3	7	2	1	4	8
3	7	8	1	4	6	9	5	2
2	1	4	8	5	9	6	3	7

HARD - 11

4	7	9	6	5	2	3	8	1
1	5	6	4	3	8	7	9	2
3	8	2	9	1	7	5	4	6
8	4	5	1	9	3	2	6	7
7	6	1	8	2	5	4	3	9
2	9	3	7	6	4	1	5	8
9	1	7	3	4	6	8	2	5
6	2	4	5	8	1	9	7	3
5	3	8	2	7	9	6	1	4

HARD - 12

8	7	5	6	2	4	1	3	9
1	9	3	8	5	7	6	2	4
4	6	2	9	3	1	7	5	8
6	3	9	1	4	2	8	7	5
5	4	8	3	7	6	9	1	2
2	1	7	5	8	9	4	6	3
7	8	6	2	9	3	5	4	1
9	2	1	4	6	5	3	8	7
3	5	4	7	1	8	2	9	6

HARD - 13

4	1	5	8	3	9	6	2	7
6	7	2	1	4	5	9	8	3
9	3	8	2	6	7	4	5	1
5	8	4	9	1	3	2	7	6
2	9	1	4	7	6	8	3	5
7	6	3	5	8	2	1	4	9
1	4	7	6	5	8	3	9	2
3	2	6	7	9	4	5	1	8
8	5	9	3	2	1	7	6	4

HARD - 14

9	1	5	7	8	4	2	3	6
6	7	4	5	2	3	1	9	8
2	3	8	6	1	9	4	7	5
3	8	6	1	4	5	7	2	9
1	4	7	9	6	2	8	5	3
5	2	9	8	3	7	6	1	4
4	6	3	2	5	1	9	8	7
8	9	2	3	7	6	5	4	1
7	5	1	4	9	8	3	6	2

HARD - 15

9	8	2	3	7	4	6	5	1
3	7	4	5	6	1	2	9	8
5	1	6	8	2	9	4	3	7
7	4	1	6	8	5	3	2	9
2	3	5	4	9	7	8	1	6
8	6	9	2	1	3	5	7	4
4	5	7	1	3	8	9	6	2
6	9	8	7	5	2	1	4	3
1	2	3	9	4	6	7	8	5

HARD - 16

5	6	4	9	3	1	7	8	2
8	1	9	6	2	7	5	3	4
3	7	2	8	4	5	1	6	9
7	5	1	2	8	9	6	4	3
6	9	3	1	5	4	8	2	7
2	4	8	7	6	3	9	5	1
1	2	5	4	9	8	3	7	6
9	3	6	5	7	2	4	1	8
4	8	7	3	1	6	2	9	5

HARD - 17

9	6	7	1	3	5	2	4	8
5	1	4	8	6	2	9	7	3
8	2	3	7	9	4	5	6	1
2	5	6	3	4	9	8	1	7
1	3	8	6	2	7	4	5	9
7	4	9	5	1	8	3	2	6
4	7	1	2	8	3	6	9	5
6	8	2	9	5	1	7	3	4
3	9	5	4	7	6	1	8	2

HARD - 18

7	6	4	8	9	5	3	1	2
5	3	8	6	2	1	4	7	9
9	2	1	4	7	3	5	6	8
6	4	9	5	8	7	1	2	3
3	5	7	2	1	6	8	9	4
8	1	2	3	4	9	7	5	6
1	8	5	9	3	2	6	4	7
4	9	6	7	5	8	2	3	1
2	7	3	1	6	4	9	8	5

HARD - 19

1	8	4	5	7	6	2	9	3
6	5	9	8	2	3	1	7	4
3	2	7	1	4	9	6	5	8
8	4	5	7	6	2	9	3	1
2	7	3	4	9	1	8	6	5
9	6	1	3	5	8	7	4	2
5	3	8	6	1	7	4	2	9
7	1	2	9	3	4	5	8	6
4	9	6	2	8	5	3	1	7

HARD - 20

5	6	3	2	9	7	4	1	8
9	4	1	6	8	5	7	3	2
8	2	7	1	4	3	6	5	9
7	3	6	4	2	8	1	9	5
2	9	4	7	5	1	3	8	6
1	8	5	9	3	6	2	4	7
4	5	9	3	7	2	8	6	1
6	7	8	5	1	4	9	2	3
3	1	2	8	6	9	5	7	4

HARD - 21

6	9	1	8	4	5	2	3	7
5	3	4	1	7	2	8	6	9
7	2	8	3	9	6	1	4	5
4	1	2	7	3	9	5	8	6
9	5	3	2	6	8	7	1	4
8	7	6	5	1	4	9	2	3
3	8	5	4	2	7	6	9	1
2	4	9	6	5	1	3	7	8
1	6	7	9	8	3	4	5	2

HARD - 22

9	7	2	3	1	8	6	4	5
6	8	1	5	4	9	7	3	2
3	5	4	2	7	6	1	8	9
7	3	9	8	2	4	5	6	1
8	2	6	9	5	1	4	7	3
4	1	5	6	3	7	2	9	8
2	6	3	4	8	5	9	1	7
1	4	8	7	9	2	3	5	6
5	9	7	1	6	3	8	2	4

HARD - 23

8	2	9	6	5	1	7	3	4
7	1	5	8	4	3	6	2	9
3	4	6	2	9	7	8	1	5
6	7	8	3	2	9	5	4	1
4	5	2	1	8	6	3	9	7
1	9	3	4	7	5	2	8	6
9	6	1	5	3	2	4	7	8
2	8	7	9	6	4	1	5	3
5	3	4	7	1	8	9	6	2

HARD - 24

3	5	1	9	7	2	4	6	8
7	4	2	8	1	6	9	5	3
6	8	9	3	4	5	1	7	2
9	2	8	1	3	7	6	4	5
5	3	6	2	9	4	8	1	7
4	1	7	5	6	8	3	2	9
2	7	4	6	8	9	5	3	1
1	9	5	4	2	3	7	8	6
8	6	3	7	5	1	2	9	4

HARD - 25

5	6	7	2	8	4	3	9	1
9	1	2	3	7	5	4	6	8
4	8	3	9	1	6	2	5	7
7	3	4	8	2	9	6	1	5
1	2	8	5	6	7	9	4	3
6	9	5	4	3	1	7	8	2
8	7	6	1	4	3	5	2	9
3	5	1	6	9	2	8	7	4
2	4	9	7	5	8	1	3	6

HARD - 26

2	6	8	4	5	7	9	1	3
9	3	1	2	8	6	5	4	7
7	4	5	3	9	1	6	2	8
6	5	4	9	1	3	8	7	2
8	7	9	5	4	2	3	6	1
3	1	2	6	7	8	4	5	9
5	8	6	7	2	9	1	3	4
1	2	3	8	6	4	7	9	5
4	9	7	1	3	5	2	8	6

HARD - 27

8	6	4	2	7	5	1	9	3
9	3	7	6	8	1	5	4	2
2	1	5	4	3	9	7	6	8
4	9	2	1	6	7	8	3	5
5	7	6	8	2	3	4	1	9
1	8	3	9	5	4	6	2	7
6	4	8	5	9	2	3	7	1
7	5	9	3	1	6	2	8	4
3	2	1	7	4	8	9	5	6

HARD - 28

9	8	5	7	3	6	4	1	2
1	3	6	5	4	2	9	7	8
2	4	7	9	8	1	3	5	6
3	9	1	2	6	4	5	8	7
7	5	4	8	1	9	2	6	3
6	2	8	3	7	5	1	9	4
4	7	2	1	9	8	6	3	5
8	6	9	4	5	3	7	2	1
5	1	3	6	2	7	8	4	9

HARD - 29

5	2	9	4	3	1	8	7	6
3	7	4	6	5	8	2	9	1
6	1	8	7	9	2	4	5	3
9	5	2	1	4	3	7	6	8
1	6	7	8	2	9	5	3	4
4	8	3	5	6	7	9	1	2
7	9	6	3	8	4	1	2	5
8	3	1	2	7	5	6	4	9
2	4	5	9	1	6	3	8	7

HARD - 30

1	6	9	3	7	5	4	8	2
5	3	8	2	1	4	7	9	6
7	2	4	8	9	6	1	5	3
6	8	7	9	4	2	5	3	1
2	1	5	7	8	3	9	6	4
9	4	3	6	5	1	8	2	7
3	5	1	4	2	8	6	7	9
8	7	6	1	3	9	2	4	5
4	9	2	5	6	7	3	1	8

HARD - 31

1	4	7	8	5	9	2	3	6
6	3	5	7	4	2	1	9	8
8	9	2	1	3	6	4	7	5
9	6	1	2	7	8	5	4	3
3	2	8	4	6	5	7	1	9
7	5	4	9	1	3	6	8	2
2	7	6	3	8	4	9	5	1
4	8	9	5	2	1	3	6	7
5	1	3	6	9	7	8	2	4

HARD - 32

1	9	2	5	6	3	8	7	4
6	4	7	1	2	8	3	9	5
5	3	8	4	9	7	6	2	1
8	1	4	3	7	2	9	5	6
9	5	6	8	1	4	7	3	2
7	2	3	6	5	9	4	1	8
3	7	5	2	4	6	1	8	9
4	8	1	9	3	5	2	6	7
2	6	9	7	8	1	5	4	3

HARD - 33

1	9	4	5	3	6	8	2	7
7	5	2	1	4	8	9	6	3
3	6	8	7	9	2	1	5	4
6	8	9	2	5	4	3	7	1
5	4	3	8	7	1	6	9	2
2	7	1	3	6	9	5	4	8
9	1	5	4	8	7	2	3	6
8	3	7	6	2	5	4	1	9
4	2	6	9	1	3	7	8	5

HARD - 34

3	2	1	7	9	4	8	6	5
5	9	8	6	1	3	4	7	2
4	6	7	2	5	8	1	9	3
2	3	5	4	7	1	6	8	9
9	7	4	8	2	6	5	3	1
8	1	6	9	3	5	7	2	4
6	8	3	5	4	9	2	1	7
7	4	9	1	8	2	3	5	6
1	5	2	3	6	7	9	4	8

HARD - 35

7	5	8	9	6	4	2	1	3
6	3	4	5	1	2	8	9	7
1	2	9	3	7	8	4	5	6
3	7	5	1	8	9	6	4	2
8	1	6	4	2	5	3	7	9
4	9	2	7	3	6	5	8	1
9	6	7	8	5	3	1	2	4
5	4	3	2	9	1	7	6	8
2	8	1	6	4	7	9	3	5

HARD - 36

6	2	7	8	4	5	9	1	3
8	9	3	7	6	1	4	2	5
4	5	1	9	3	2	6	7	8
5	6	4	1	9	8	2	3	7
7	1	2	3	5	6	8	9	4
9	3	8	4	2	7	5	6	1
2	7	9	5	8	3	1	4	6
3	8	6	2	1	4	7	5	9
1	4	5	6	7	9	3	8	2

HARD - 37

2	6	8	3	7	4	5	1	9
4	3	1	9	8	5	2	7	6
5	7	9	2	1	6	8	4	3
8	4	6	5	3	9	7	2	1
9	2	5	7	4	1	3	6	8
7	1	3	6	2	8	4	9	5
1	5	2	8	9	7	6	3	4
6	9	7	4	5	3	1	8	2
3	8	4	1	6	2	9	5	7

HARD - 38

6	1	4	2	9	5	7	3	8
9	2	7	1	8	3	4	6	5
8	5	3	4	6	7	1	9	2
4	6	8	3	5	1	9	2	7
7	9	2	6	4	8	5	1	3
5	3	1	7	2	9	6	8	4
1	7	9	8	3	4	2	5	6
2	8	5	9	7	6	3	4	1
3	4	6	5	1	2	8	7	9

HARD - 39

5	7	1	3	4	6	2	8	9
3	4	6	9	2	8	1	5	7
8	2	9	5	1	7	6	4	3
9	1	8	2	3	5	7	6	4
6	5	2	7	9	4	3	1	8
7	3	4	8	6	1	9	2	5
4	8	3	1	7	2	5	9	6
2	9	5	6	8	3	4	7	1
1	6	7	4	5	9	8	3	2

HARD - 40

5	6	3	2	4	9	7	1	8
9	4	1	6	7	8	2	5	3
8	7	2	3	1	5	6	4	9
1	9	4	8	2	7	5	3	6
7	3	6	1	5	4	9	8	2
2	8	5	9	3	6	4	7	1
4	1	9	7	6	3	8	2	5
6	2	7	5	8	1	3	9	4
3	5	8	4	9	2	1	6	7

HARD - 41

8	9	4	2	1	3	7	5	6
6	1	5	7	8	4	9	3	2
2	3	7	6	5	9	4	8	1
7	8	9	3	6	2	5	1	4
5	4	3	8	9	1	6	2	7
1	2	6	4	7	5	8	9	3
9	7	1	5	3	6	2	4	8
3	6	2	9	4	8	1	7	5
4	5	8	1	2	7	3	6	9

HARD - 42

3	8	1	2	6	7	4	9	5
7	6	9	4	1	5	3	2	8
5	4	2	9	8	3	1	6	7
8	7	3	1	4	6	9	5	2
2	1	4	5	9	8	7	3	6
9	5	6	3	7	2	8	1	4
6	9	5	7	3	4	2	8	1
4	3	8	6	2	1	5	7	9
1	2	7	8	5	9	6	4	3

HARD - 43

2	3	7	6	4	8	9	1	5
6	1	9	5	7	3	4	8	2
4	5	8	2	1	9	3	7	6
7	8	6	3	2	4	1	5	9
5	4	3	9	6	1	7	2	8
1	9	2	7	8	5	6	4	3
8	6	4	1	9	2	5	3	7
3	7	1	8	5	6	2	9	4
9	2	5	4	3	7	8	6	1

HARD - 44

6	2	3	5	1	9	8	4	7
9	4	5	7	2	8	3	1	6
8	7	1	4	6	3	9	5	2
7	8	6	1	4	5	2	3	9
1	3	4	6	9	2	7	8	5
2	5	9	3	8	7	4	6	1
3	6	2	8	7	1	5	9	4
4	9	8	2	5	6	1	7	3
5	1	7	9	3	4	6	2	8

HARD - 45

5	8	3	1	9	4	7	2	6
9	4	1	2	6	7	5	3	8
7	2	6	5	8	3	1	4	9
3	7	4	6	5	8	2	9	1
8	5	2	7	1	9	3	6	4
6	1	9	4	3	2	8	5	7
2	3	7	9	4	1	6	8	5
4	6	8	3	7	5	9	1	2
1	9	5	8	2	6	4	7	3

HARD - 46

3	1	9	2	4	8	5	6	7
7	8	2	9	5	6	4	3	1
5	4	6	3	7	1	8	2	9
6	9	3	5	8	4	7	1	2
1	7	5	6	9	2	3	4	8
8	2	4	1	3	7	6	9	5
4	6	8	7	1	9	2	5	3
9	3	7	4	2	5	1	8	6
2	5	1	8	6	3	9	7	4

HARD - 47

4	5	6	9	8	3	1	7	2
3	9	1	2	6	7	4	8	5
7	2	8	1	5	4	3	9	6
1	4	3	5	9	8	2	6	7
8	7	9	3	2	6	5	1	4
5	6	2	4	7	1	8	3	9
9	1	4	6	3	5	7	2	8
6	8	5	7	1	2	9	4	3
2	3	7	8	4	9	6	5	1

HARD - 48

8	7	1	9	3	4	2	5	6
2	6	9	1	7	5	4	3	8
5	3	4	8	2	6	1	7	9
3	5	6	4	9	1	8	2	7
1	4	8	2	6	7	5	9	3
9	2	7	3	5	8	6	4	1
6	8	3	5	4	9	7	1	2
4	1	2	7	8	3	9	6	5
7	9	5	6	1	2	3	8	4

HARD - 49

6	8	2	4	5	1	7	3	9
4	3	5	8	9	7	1	6	2
7	1	9	3	6	2	4	5	8
8	7	4	2	3	9	6	1	5
5	9	6	7	1	4	8	2	3
3	2	1	6	8	5	9	7	4
2	4	8	5	7	6	3	9	1
9	6	3	1	2	8	5	4	7
1	5	7	9	4	3	2	8	6

HARD - 50

1	6	2	4	9	7	5	8	3
9	5	8	2	6	3	7	1	4
7	3	4	8	1	5	9	6	2
2	8	3	9	5	1	4	7	6
6	7	9	3	2	4	1	5	8
5	4	1	7	8	6	3	2	9
4	2	6	5	7	9	8	3	1
3	1	5	6	4	8	2	9	7
8	9	7	1	3	2	6	4	5

HARD - 51

2	6	9	5	4	7	1	3	8
3	1	8	6	2	9	7	4	5
4	7	5	1	3	8	6	9	2
1	3	7	8	9	5	2	6	4
9	8	6	4	7	2	3	5	1
5	2	4	3	6	1	8	7	9
8	4	3	9	1	6	5	2	7
6	5	2	7	8	4	9	1	3
7	9	1	2	5	3	4	8	6

HARD - 52

3	1	6	8	7	4	2	5	9
7	9	2	3	5	6	8	1	4
8	5	4	9	1	2	3	6	7
4	8	5	7	3	1	6	9	2
1	6	9	2	4	5	7	8	3
2	7	3	6	8	9	5	4	1
5	3	8	1	9	7	4	2	6
6	4	1	5	2	3	9	7	8
9	2	7	4	6	8	1	3	5

HARD - 53

7	9	2	3	4	6	5	1	8
4	5	8	7	1	2	6	9	3
6	3	1	9	5	8	2	4	7
5	7	3	6	8	4	9	2	1
2	4	6	1	7	9	3	8	5
8	1	9	5	2	3	7	6	4
3	6	5	4	9	1	8	7	2
9	8	4	2	3	7	1	5	6
1	2	7	8	6	5	4	3	9

HARD - 54

7	3	5	2	8	6	1	4	9
9	1	2	3	7	4	5	8	6
4	6	8	9	5	1	2	3	7
3	5	4	7	6	9	8	1	2
8	9	1	5	2	3	7	6	4
2	7	6	1	4	8	3	9	5
6	8	7	4	3	5	9	2	1
5	4	9	8	1	2	6	7	3
1	2	3	6	9	7	4	5	8

HARD - 55

2	4	6	1	3	8	7	9	5
5	9	3	7	6	4	8	2	1
7	8	1	2	9	5	4	3	6
4	6	2	8	1	9	5	7	3
8	1	5	4	7	3	9	6	2
9	3	7	5	2	6	1	8	4
3	5	8	9	4	2	6	1	7
6	7	4	3	8	1	2	5	9
1	2	9	6	5	7	3	4	8

HARD - 56

9	4	2	1	3	5	6	8	7
1	8	5	7	2	6	9	3	4
6	3	7	4	8	9	5	1	2
5	1	9	6	7	4	3	2	8
3	7	6	2	5	8	4	9	1
4	2	8	3	9	1	7	6	5
8	6	3	5	1	7	2	4	9
7	9	4	8	6	2	1	5	3
2	5	1	9	4	3	8	7	6

HARD - 57

9	3	4	6	1	7	2	8	5
7	5	1	2	3	8	9	6	4
2	6	8	4	9	5	3	7	1
4	8	6	7	2	1	5	3	9
3	7	9	8	5	4	6	1	2
1	2	5	3	6	9	7	4	8
8	4	2	5	7	3	1	9	6
6	1	7	9	8	2	4	5	3
5	9	3	1	4	6	8	2	7

HARD - 58

6	2	8	3	1	4	5	7	9
3	9	1	5	2	7	8	4	6
7	4	5	6	9	8	2	3	1
1	6	2	4	8	5	7	9	3
5	3	7	9	6	1	4	8	2
9	8	4	2	7	3	6	1	5
2	1	9	7	4	6	3	5	8
4	5	6	8	3	9	1	2	7
8	7	3	1	5	2	9	6	4

HARD - 59

6	4	9	7	5	8	3	2	1
2	5	7	9	3	1	8	6	4
8	3	1	4	2	6	9	5	7
7	2	3	8	9	5	1	4	6
5	1	8	2	6	4	7	9	3
4	9	6	3	1	7	5	8	2
9	8	2	6	7	3	4	1	5
1	7	4	5	8	2	6	3	9
3	6	5	1	4	9	2	7	8

HARD - 60

6	2	8	7	1	5	9	4	3
4	7	9	2	8	3	5	6	1
3	5	1	6	4	9	8	7	2
2	1	5	3	7	6	4	8	9
7	4	3	5	9	8	2	1	6
9	8	6	1	2	4	3	5	7
5	6	2	4	3	7	1	9	8
1	9	4	8	6	2	7	3	5
8	3	7	9	5	1	6	2	4

HARD - 61

5	9	2	6	3	7	8	4	1
4	8	7	5	1	2	6	9	3
6	1	3	8	4	9	7	2	5
1	6	9	4	7	3	5	8	2
3	4	5	2	6	8	1	7	9
2	7	8	1	9	5	3	6	4
8	2	1	9	5	6	4	3	7
9	3	4	7	8	1	2	5	6
7	5	6	3	2	4	9	1	8

HARD - 62

9	2	5	7	8	6	1	3	4
8	6	7	4	1	3	5	2	9
3	1	4	2	5	9	8	7	6
2	4	8	6	3	7	9	1	5
6	7	9	5	4	1	2	8	3
5	3	1	9	2	8	6	4	7
1	5	6	8	7	4	3	9	2
4	9	3	1	6	2	7	5	8
7	8	2	3	9	5	4	6	1

HARD - 63

2	1	4	8	5	7	9	3	6
5	6	3	9	4	2	1	8	7
9	7	8	6	1	3	4	2	5
4	9	2	3	7	8	5	6	1
6	8	1	2	9	5	7	4	3
3	5	7	4	6	1	2	9	8
1	2	9	7	3	6	8	5	4
7	4	6	5	8	9	3	1	2
8	3	5	1	2	4	6	7	9

HARD - 64

9	5	1	7	8	4	2	3	6
6	7	3	9	2	5	8	1	4
4	8	2	6	3	1	9	7	5
1	4	7	5	9	6	3	2	8
8	6	9	3	1	2	4	5	7
3	2	5	8	4	7	1	6	9
5	1	4	2	7	8	6	9	3
2	9	6	4	5	3	7	8	1
7	3	8	1	6	9	5	4	2

HARD - 65

1	8	2	4	9	6	5	3	7
3	9	7	2	5	8	6	1	4
4	5	6	3	7	1	2	8	9
2	6	1	7	3	5	4	9	8
5	4	8	6	1	9	3	7	2
9	7	3	8	2	4	1	5	6
8	1	5	9	4	2	7	6	3
7	2	9	5	6	3	8	4	1
6	3	4	1	8	7	9	2	5

HARD - 66

3	1	6	9	5	2	4	8	7
2	5	8	1	4	7	3	6	9
9	7	4	8	6	3	5	2	1
8	9	7	2	3	1	6	4	5
6	2	1	5	8	4	7	9	3
5	4	3	7	9	6	8	1	2
4	6	2	3	7	9	1	5	8
7	8	9	6	1	5	2	3	4
1	3	5	4	2	8	9	7	6

HARD - 67

8	5	9	1	7	4	3	2	6
7	6	3	2	9	5	1	4	8
1	4	2	8	3	6	5	9	7
3	8	1	9	2	7	4	6	5
5	2	6	3	4	8	7	1	9
4	9	7	5	6	1	2	8	3
2	1	8	7	5	9	6	3	4
9	7	4	6	1	3	8	5	2
6	3	5	4	8	2	9	7	1

HARD - 68

9	5	6	2	3	7	8	1	4
1	3	2	6	4	8	5	9	7
8	4	7	5	1	9	6	2	3
4	2	8	7	9	3	1	6	5
3	1	5	4	6	2	7	8	9
7	6	9	1	8	5	4	3	2
2	8	4	3	5	1	9	7	6
5	9	3	8	7	6	2	4	1
6	7	1	9	2	4	3	5	8

HARD - 69

7	6	5	9	8	3	2	1	4
1	4	9	5	6	2	8	7	3
3	2	8	4	7	1	6	5	9
9	1	6	3	5	7	4	8	2
2	7	3	8	9	4	5	6	1
5	8	4	1	2	6	9	3	7
8	5	1	2	3	9	7	4	6
4	9	7	6	1	5	3	2	8
6	3	2	7	4	8	1	9	5

HARD - 70

1	8	3	5	7	2	9	6	4
7	2	4	3	9	6	5	8	1
5	6	9	4	8	1	7	2	3
4	1	2	6	3	5	8	9	7
3	7	5	8	2	9	4	1	6
8	9	6	1	4	7	3	5	2
9	4	1	2	5	3	6	7	8
2	3	7	9	6	8	1	4	5
6	5	8	7	1	4	2	3	9

HARD - 71

1	9	6	4	5	8	3	2	7
7	2	8	3	9	6	1	5	4
4	5	3	1	2	7	8	6	9
2	6	7	8	4	3	5	9	1
9	1	4	2	6	5	7	8	3
8	3	5	9	7	1	2	4	6
5	7	2	6	3	9	4	1	8
6	4	1	7	8	2	9	3	5
3	8	9	5	1	4	6	7	2

HARD - 72

5	2	6	9	4	1	3	7	8
4	7	8	6	3	2	9	1	5
3	1	9	8	5	7	4	2	6
8	5	7	2	9	4	1	6	3
6	3	2	1	8	5	7	9	4
9	4	1	3	7	6	5	8	2
7	9	5	4	6	8	2	3	1
2	8	4	7	1	3	6	5	9
1	6	3	5	2	9	8	4	7

HARD - 73

9	2	7	1	5	4	3	8	6
8	3	4	2	6	7	5	1	9
5	1	6	9	8	3	7	4	2
4	8	3	6	2	5	9	7	1
1	9	5	3	7	8	6	2	4
7	6	2	4	1	9	8	3	5
2	7	1	8	9	6	4	5	3
3	5	9	7	4	1	2	6	8
6	4	8	5	3	2	1	9	7

HARD - 74

7	3	9	1	6	4	2	5	8
2	4	8	7	3	5	6	1	9
6	1	5	9	8	2	7	4	3
9	6	7	2	5	8	1	3	4
1	5	3	6	4	7	9	8	2
4	8	2	3	1	9	5	6	7
8	2	4	5	7	6	3	9	1
5	9	1	4	2	3	8	7	6
3	7	6	8	9	1	4	2	5

HARD - 75

5	8	1	4	3	6	7	2	9
4	3	6	9	7	2	8	5	1
7	2	9	5	8	1	3	6	4
2	9	8	1	5	3	6	4	7
3	4	7	8	6	9	5	1	2
1	6	5	7	2	4	9	8	3
9	7	2	6	4	5	1	3	8
8	5	3	2	1	7	4	9	6
6	1	4	3	9	8	2	7	5

HARD - 76

4	5	3	9	8	7	6	2	1
7	1	9	2	3	6	4	8	5
6	2	8	4	1	5	7	9	3
1	4	5	7	6	8	9	3	2
9	8	2	1	4	3	5	6	7
3	6	7	5	2	9	8	1	4
2	9	6	3	5	4	1	7	8
5	7	1	8	9	2	3	4	6
8	3	4	6	7	1	2	5	9

HARD - 77

8	9	4	5	7	1	2	6	3
6	3	7	9	2	4	8	5	1
2	5	1	8	3	6	9	4	7
7	6	5	2	4	8	3	1	9
9	8	3	7	1	5	6	2	4
1	4	2	3	6	9	7	8	5
5	7	8	1	9	2	4	3	6
3	2	6	4	5	7	1	9	8
4	1	9	6	8	3	5	7	2

HARD - 78

6	4	2	7	8	1	9	5	3
9	7	1	3	2	5	4	6	8
3	5	8	9	4	6	7	2	1
4	6	7	5	1	3	2	8	9
5	8	3	2	7	9	1	4	6
2	1	9	4	6	8	3	7	5
1	9	4	8	5	7	6	3	2
8	2	6	1	3	4	5	9	7
7	3	5	6	9	2	8	1	4

HARD - 79

3	5	8	9	7	6	4	2	1
7	9	4	3	2	1	8	6	5
1	6	2	5	8	4	9	3	7
8	2	3	1	5	7	6	4	9
5	7	9	4	6	3	1	8	2
4	1	6	8	9	2	5	7	3
9	8	7	2	4	5	3	1	6
6	3	5	7	1	8	2	9	4
2	4	1	6	3	9	7	5	8

HARD - 80

9	4	1	3	6	8	5	2	7
5	2	7	9	1	4	8	3	6
8	3	6	7	2	5	9	4	1
7	5	2	4	3	1	6	9	8
1	9	3	2	8	6	4	7	5
6	8	4	5	9	7	2	1	3
4	1	9	6	5	3	7	8	2
2	6	8	1	7	9	3	5	4
3	7	5	8	4	2	1	6	9

HARD - 81

9	5	2	1	4	3	7	8	6
1	4	3	6	7	8	9	2	5
6	7	8	5	9	2	4	3	1
3	2	5	4	8	6	1	7	9
4	1	6	7	2	9	3	5	8
7	8	9	3	5	1	6	4	2
2	3	1	8	6	4	5	9	7
5	9	4	2	1	7	8	6	3
8	6	7	9	3	5	2	1	4

HARD - 82

7	9	3	6	1	2	8	5	4
8	5	6	4	7	3	9	1	2
1	2	4	8	5	9	7	3	6
9	6	7	2	8	1	3	4	5
3	4	8	5	6	7	1	2	9
2	1	5	3	9	4	6	8	7
4	7	2	9	3	8	5	6	1
5	8	1	7	2	6	4	9	3
6	3	9	1	4	5	2	7	8

HARD - 83

2	9	7	1	8	6	4	5	3
1	5	4	7	9	3	6	2	8
6	3	8	5	2	4	1	7	9
7	2	6	8	1	9	3	4	5
9	8	1	4	3	5	2	6	7
3	4	5	2	6	7	9	8	1
4	7	9	3	5	2	8	1	6
8	6	2	9	7	1	5	3	4
5	1	3	6	4	8	7	9	2

HARD - 84

2	5	6	4	1	9	8	7	3
3	8	4	2	7	5	6	1	9
7	1	9	3	8	6	2	4	5
4	2	5	7	3	1	9	6	8
9	7	1	6	2	8	5	3	4
6	3	8	9	5	4	1	2	7
8	9	3	1	6	7	4	5	2
5	6	7	8	4	2	3	9	1
1	4	2	5	9	3	7	8	6

HARD - 85

6	1	4	8	9	5	7	3	2
5	2	8	7	3	1	6	9	4
7	3	9	2	4	6	1	8	5
1	9	3	6	7	2	5	4	8
2	8	5	3	1	4	9	7	6
4	7	6	5	8	9	3	2	1
3	5	7	1	2	8	4	6	9
8	4	1	9	6	3	2	5	7
9	6	2	4	5	7	8	1	3

HARD - 86

6	2	5	1	8	3	7	4	9
7	1	3	5	4	9	2	8	6
8	4	9	6	7	2	3	1	5
1	6	8	7	2	5	9	3	4
9	7	4	8	3	6	1	5	2
3	5	2	9	1	4	6	7	8
2	3	1	4	6	8	5	9	7
4	9	7	2	5	1	8	6	3
5	8	6	3	9	7	4	2	1

HARD - 87

2	4	9	3	7	5	6	8	1
3	5	8	2	6	1	9	7	4
1	7	6	9	8	4	2	5	3
7	9	3	4	1	8	5	6	2
4	2	1	6	5	9	8	3	7
8	6	5	7	3	2	1	4	9
9	3	7	8	2	6	4	1	5
5	8	4	1	9	7	3	2	6
6	1	2	5	4	3	7	9	8

HARD - 88

8	3	6	4	5	9	7	1	2
7	1	2	3	6	8	9	4	5
9	4	5	7	1	2	8	6	3
3	5	8	6	7	1	4	2	9
4	7	9	5	2	3	1	8	6
2	6	1	9	8	4	3	5	7
5	8	3	1	9	6	2	7	4
1	9	7	2	4	5	6	3	8
6	2	4	8	3	7	5	9	1

HARD - 89

6	9	1	4	7	2	3	8	5
3	7	8	9	6	5	1	2	4
5	2	4	8	1	3	7	6	9
7	3	2	5	8	4	6	9	1
1	5	9	2	3	6	4	7	8
4	8	6	1	9	7	5	3	2
8	6	5	7	2	1	9	4	3
9	4	3	6	5	8	2	1	7
2	1	7	3	4	9	8	5	6

HARD - 90

2	1	5	7	4	3	6	8	9
3	8	9	1	5	6	7	4	2
6	4	7	2	9	8	1	5	3
1	7	8	9	2	5	4	3	6
9	3	2	4	6	1	8	7	5
4	5	6	3	8	7	9	2	1
7	9	4	6	3	2	5	1	8
5	6	3	8	1	4	2	9	7
8	2	1	5	7	9	3	6	4

HARD - 91

7	9	8	6	2	3	1	4	5
1	5	6	7	8	4	2	3	9
3	2	4	5	1	9	7	8	6
2	4	3	8	6	1	9	5	7
8	6	5	4	9	7	3	1	2
9	7	1	2	3	5	8	6	4
4	3	2	1	7	6	5	9	8
6	1	7	9	5	8	4	2	3
5	8	9	3	4	2	6	7	1

HARD - 92

6	8	2	5	7	3	4	9	1
4	7	1	9	2	6	5	3	8
5	9	3	8	1	4	6	7	2
1	3	4	2	6	7	8	5	9
2	6	8	4	9	5	3	1	7
9	5	7	1	3	8	2	6	4
3	4	5	7	8	9	1	2	6
8	2	9	6	5	1	7	4	3
7	1	6	3	4	2	9	8	5

HARD - 93

7	3	2	4	9	1	6	5	8
6	1	5	3	2	8	7	9	4
9	4	8	5	7	6	1	2	3
3	2	9	1	4	5	8	6	7
5	6	1	7	8	2	3	4	9
8	7	4	9	6	3	5	1	2
4	8	7	6	5	9	2	3	1
2	5	3	8	1	4	9	7	6
1	9	6	2	3	7	4	8	5

HARD - 94

3	7	1	6	4	2	5	8	9
2	6	4	5	8	9	3	7	1
5	9	8	1	3	7	6	2	4
7	5	3	8	9	1	4	6	2
6	4	9	2	5	3	8	1	7
8	1	2	4	7	6	9	5	3
4	8	7	9	1	5	2	3	6
1	2	5	3	6	4	7	9	8
9	3	6	7	2	8	1	4	5

HARD - 95

5	2	1	8	3	4	9	6	7
3	6	8	5	9	7	1	2	4
7	4	9	6	1	2	5	8	3
9	5	4	1	6	3	2	7	8
2	1	3	7	5	8	6	4	9
6	8	7	4	2	9	3	1	5
1	7	6	9	8	5	4	3	2
8	9	2	3	4	1	7	5	6
4	3	5	2	7	6	8	9	1

HARD - 96

1	2	5	7	3	9	8	6	4
4	9	3	2	8	6	1	5	7
6	8	7	1	4	5	9	2	3
8	6	9	5	2	4	3	7	1
2	3	4	6	1	7	5	8	9
5	7	1	8	9	3	2	4	6
7	4	2	3	5	1	6	9	8
9	1	8	4	6	2	7	3	5
3	5	6	9	7	8	4	1	2

HARD - 97

4	9	2	5	6	1	8	3	7
7	3	6	8	9	4	5	1	2
1	8	5	7	3	2	9	6	4
5	2	4	9	7	6	1	8	3
8	1	9	4	2	3	6	7	5
3	6	7	1	8	5	4	2	9
6	4	8	2	5	7	3	9	1
9	7	1	3	4	8	2	5	6
2	5	3	6	1	9	7	4	8

HARD - 98

4	6	2	5	9	1	3	7	8
5	8	9	7	3	2	1	4	6
1	3	7	6	4	8	2	9	5
9	5	3	2	6	7	8	1	4
8	4	1	9	5	3	6	2	7
2	7	6	8	1	4	9	5	3
7	9	5	3	2	6	4	8	1
3	2	4	1	8	5	7	6	9
6	1	8	4	7	9	5	3	2

HARD - 99

6	1	3	5	2	4	8	7	9
2	4	5	9	8	7	6	1	3
9	8	7	1	3	6	5	2	4
8	5	1	4	7	9	3	6	2
4	6	9	2	5	3	7	8	1
7	3	2	6	1	8	9	4	5
5	7	6	3	4	2	1	9	8
1	9	4	8	6	5	2	3	7
3	2	8	7	9	1	4	5	6

HARD - 100

5	8	1	3	9	4	6	7	2
3	7	6	5	2	8	4	9	1
9	4	2	6	7	1	8	3	5
1	9	5	4	8	6	3	2	7
6	3	7	1	5	2	9	8	4
4	2	8	7	3	9	5	1	6
7	6	3	9	1	5	2	4	8
8	5	9	2	4	7	1	6	3
2	1	4	8	6	3	7	5	9

HARD - 101

6	8	2	4	1	9	7	3	5
4	1	3	7	8	5	2	6	9
5	7	9	3	2	6	4	8	1
7	4	1	6	5	3	9	2	8
8	2	6	9	7	1	3	5	4
3	9	5	2	4	8	6	1	7
2	3	8	5	9	7	1	4	6
1	6	7	8	3	4	5	9	2
9	5	4	1	6	2	8	7	3

HARD - 102

1	3	9	8	2	6	5	7	4
8	6	4	3	5	7	9	2	1
2	5	7	4	9	1	6	3	8
4	2	8	7	1	5	3	9	6
9	1	3	6	8	2	7	4	5
6	7	5	9	3	4	8	1	2
5	4	6	1	7	9	2	8	3
7	8	1	2	6	3	4	5	9
3	9	2	5	4	8	1	6	7

HARD - 103

2	6	1	5	3	8	9	4	7
3	8	9	1	7	4	6	2	5
4	5	7	2	6	9	1	8	3
8	9	5	3	2	6	4	7	1
6	2	4	8	1	7	3	5	9
1	7	3	9	4	5	8	6	2
5	3	6	7	8	1	2	9	4
9	1	8	4	5	2	7	3	6
7	4	2	6	9	3	5	1	8

HARD - 104

8	1	3	9	6	4	7	2	5
9	5	6	7	1	2	3	4	8
7	2	4	3	5	8	6	9	1
5	4	2	6	8	1	9	3	7
6	7	1	2	3	9	8	5	4
3	9	8	4	7	5	1	6	2
4	3	5	1	9	7	2	8	6
1	8	9	5	2	6	4	7	3
2	6	7	8	4	3	5	1	9

HARD - 105

1	3	7	5	2	6	8	9	4
6	5	8	1	4	9	7	3	2
4	9	2	7	8	3	1	6	5
5	8	9	3	7	1	2	4	6
7	4	6	8	5	2	9	1	3
3	2	1	6	9	4	5	7	8
8	7	3	9	6	5	4	2	1
2	1	5	4	3	7	6	8	9
9	6	4	2	1	8	3	5	7

HARD - 106

2	3	4	7	6	5	9	8	1
5	8	9	2	1	4	3	6	7
1	7	6	8	3	9	2	5	4
4	5	8	6	7	3	1	9	2
3	6	7	1	9	2	8	4	5
9	1	2	4	5	8	6	7	3
7	4	3	9	2	6	5	1	8
8	9	5	3	4	1	7	2	6
6	2	1	5	8	7	4	3	9

HARD - 107

8	4	6	5	7	2	3	9	1
2	5	9	6	3	1	8	7	4
7	3	1	4	8	9	5	6	2
3	7	2	1	5	4	9	8	6
1	9	8	3	2	6	7	4	5
5	6	4	7	9	8	2	1	3
9	1	3	2	6	7	4	5	8
4	8	5	9	1	3	6	2	7
6	2	7	8	4	5	1	3	9

HARD - 108

2	7	9	8	5	3	6	4	1
4	8	6	7	2	1	5	3	9
5	1	3	4	6	9	2	7	8
8	4	1	9	3	2	7	6	5
3	6	7	5	8	4	1	9	2
9	5	2	1	7	6	3	8	4
1	2	4	6	9	7	8	5	3
6	9	5	3	1	8	4	2	7
7	3	8	2	4	5	9	1	6

HARD - 109

3	5	4	6	1	2	8	7	9
7	6	1	8	9	5	2	3	4
9	8	2	3	7	4	6	5	1
4	1	7	5	8	6	9	2	3
6	3	8	7	2	9	1	4	5
2	9	5	4	3	1	7	6	8
5	7	3	1	6	8	4	9	2
1	4	9	2	5	7	3	8	6
8	2	6	9	4	3	5	1	7

HARD - 110

8	7	3	6	9	4	1	2	5
2	6	9	8	5	1	7	4	3
4	5	1	7	2	3	9	8	6
7	9	8	3	4	5	2	6	1
1	2	5	9	8	6	4	3	7
6	3	4	2	1	7	5	9	8
3	4	7	1	6	9	8	5	2
5	1	2	4	3	8	6	7	9
9	8	6	5	7	2	3	1	4

HARD - 111

1	2	9	5	7	6	3	8	4
4	5	8	2	1	3	7	6	9
6	7	3	4	8	9	2	5	1
5	3	4	8	6	7	9	1	2
2	1	6	9	3	5	8	4	7
9	8	7	1	4	2	5	3	6
3	6	5	7	2	1	4	9	8
8	9	2	6	5	4	1	7	3
7	4	1	3	9	8	6	2	5

HARD - 112

6	4	2	1	3	5	9	8	7
1	3	5	7	8	9	2	6	4
7	8	9	4	2	6	3	5	1
4	5	8	9	1	2	7	3	6
3	9	6	8	7	4	1	2	5
2	7	1	5	6	3	8	4	9
9	6	3	2	5	1	4	7	8
8	2	4	6	9	7	5	1	3
5	1	7	3	4	8	6	9	2

HARD - 113

2	1	3	4	6	5	9	7	8
8	9	6	2	7	3	1	5	4
4	7	5	1	8	9	2	3	6
5	4	9	3	1	2	6	8	7
1	3	7	8	4	6	5	9	2
6	2	8	9	5	7	4	1	3
3	6	2	5	9	8	7	4	1
7	5	4	6	3	1	8	2	9
9	8	1	7	2	4	3	6	5

HARD - 114

3	4	9	5	2	6	1	8	7
5	8	7	9	3	1	4	6	2
2	6	1	4	8	7	9	5	3
6	9	2	1	7	8	5	3	4
8	7	4	3	5	9	2	1	6
1	5	3	2	6	4	7	9	8
4	1	6	7	9	3	8	2	5
7	3	5	8	1	2	6	4	9
9	2	8	6	4	5	3	7	1

HARD - 115

9	3	6	8	5	1	7	2	4
4	7	5	3	9	2	8	1	6
8	1	2	6	7	4	9	5	3
7	4	1	9	6	8	2	3	5
6	8	3	4	2	5	1	7	9
5	2	9	1	3	7	4	6	8
1	9	7	5	8	3	6	4	2
3	6	4	2	1	9	5	8	7
2	5	8	7	4	6	3	9	1

HARD - 116

6	2	3	7	8	4	9	1	5
7	4	1	5	3	9	2	6	8
9	8	5	2	1	6	4	3	7
1	5	8	4	6	3	7	2	9
3	6	2	8	9	7	1	5	4
4	7	9	1	5	2	6	8	3
2	3	6	9	7	8	5	4	1
8	1	7	6	4	5	3	9	2
5	9	4	3	2	1	8	7	6

HARD - 117

7	9	2	8	5	6	1	3	4
1	5	8	7	4	3	6	9	2
6	4	3	1	2	9	5	8	7
2	1	6	5	3	8	7	4	9
9	8	7	4	6	2	3	1	5
4	3	5	9	1	7	8	2	6
8	7	1	6	9	4	2	5	3
5	2	4	3	7	1	9	6	8
3	6	9	2	8	5	4	7	1

HARD - 118

3	6	7	5	1	2	8	9	4
5	8	2	4	7	9	3	6	1
1	4	9	8	3	6	7	2	5
9	5	1	6	4	3	2	8	7
2	3	6	7	8	5	4	1	9
8	7	4	9	2	1	6	5	3
4	1	8	2	5	7	9	3	6
6	2	5	3	9	4	1	7	8
7	9	3	1	6	8	5	4	2

HARD - 119

1	9	2	4	8	5	7	6	3
4	7	6	9	2	3	8	5	1
5	8	3	1	6	7	4	2	9
9	5	4	8	7	2	3	1	6
7	6	1	3	9	4	2	8	5
3	2	8	5	1	6	9	7	4
2	1	9	6	4	8	5	3	7
6	3	7	2	5	9	1	4	8
8	4	5	7	3	1	6	9	2

HARD - 120

3	7	1	6	4	2	5	8	9
2	5	6	7	8	9	1	4	3
8	9	4	5	1	3	6	7	2
6	1	8	2	5	7	9	3	4
7	2	9	4	3	1	8	6	5
5	4	3	8	9	6	2	1	7
4	3	2	9	6	8	7	5	1
1	6	7	3	2	5	4	9	8
9	8	5	1	7	4	3	2	6

HARD - 121

6	2	4	7	8	3	5	9	1
1	3	8	5	2	9	4	6	7
5	7	9	6	1	4	3	8	2
2	1	3	4	6	5	9	7	8
8	9	6	2	3	7	1	4	5
7	4	5	8	9	1	2	3	6
4	6	2	9	5	8	7	1	3
3	5	7	1	4	6	8	2	9
9	8	1	3	7	2	6	5	4

HARD - 122

2	5	6	4	8	1	7	3	9
3	1	7	2	5	9	4	6	8
8	9	4	7	3	6	5	2	1
7	8	2	1	9	5	3	4	6
1	3	9	6	2	4	8	5	7
6	4	5	3	7	8	9	1	2
9	6	1	5	4	7	2	8	3
5	7	3	8	6	2	1	9	4
4	2	8	9	1	3	6	7	5

HARD - 123

8	2	3	5	9	1	7	4	6
9	1	4	7	6	3	8	2	5
5	6	7	2	4	8	1	9	3
6	8	2	1	3	9	5	7	4
1	3	5	8	7	4	2	6	9
4	7	9	6	5	2	3	8	1
2	9	6	3	1	7	4	5	8
3	4	8	9	2	5	6	1	7
7	5	1	4	8	6	9	3	2

HARD - 124

6	3	1	8	9	5	4	7	2
7	8	4	6	3	2	1	5	9
9	5	2	7	4	1	8	3	6
5	6	9	2	7	8	3	1	4
4	1	3	5	6	9	7	2	8
2	7	8	3	1	4	9	6	5
1	4	5	9	2	3	6	8	7
8	9	7	1	5	6	2	4	3
3	2	6	4	8	7	5	9	1

HARD - 125

6	1	3	2	5	9	8	4	7
2	7	8	6	4	1	3	9	5
4	5	9	8	3	7	1	6	2
3	6	5	7	9	4	2	8	1
7	4	1	3	8	2	9	5	6
8	9	2	1	6	5	7	3	4
1	3	6	4	2	8	5	7	9
5	8	7	9	1	6	4	2	3
9	2	4	5	7	3	6	1	8

HARD - 126

3	9	8	7	5	6	2	4	1
7	1	6	2	9	4	3	8	5
5	2	4	8	1	3	7	6	9
9	5	2	3	4	8	6	1	7
8	4	1	6	7	5	9	2	3
6	7	3	9	2	1	8	5	4
1	3	9	4	8	2	5	7	6
4	8	7	5	6	9	1	3	2
2	6	5	1	3	7	4	9	8

HARD - 127

7	3	9	1	6	4	2	5	8
2	4	8	7	3	5	6	1	9
6	1	5	9	8	2	7	3	4
9	7	4	6	1	8	5	2	3
5	6	1	2	9	3	8	4	7
8	2	3	4	5	7	1	9	6
4	5	7	3	2	6	9	8	1
1	8	6	5	4	9	3	7	2
3	9	2	8	7	1	4	6	5

HARD - 128

1	2	4	8	3	7	9	5	6
6	7	5	9	2	4	1	8	3
8	9	3	1	6	5	4	7	2
5	1	2	7	4	8	6	3	9
4	6	9	2	5	3	7	1	8
7	3	8	6	1	9	2	4	5
3	4	6	5	7	2	8	9	1
2	8	7	3	9	1	5	6	4
9	5	1	4	8	6	3	2	7

HARD - 129

9	1	4	6	8	7	2	5	3
8	2	7	1	3	5	4	6	9
6	5	3	9	2	4	7	8	1
4	7	5	2	1	6	3	9	8
1	9	2	3	5	8	6	4	7
3	8	6	7	4	9	1	2	5
7	6	8	4	9	1	5	3	2
2	4	9	5	7	3	8	1	6
5	3	1	8	6	2	9	7	4

HARD - 130

3	8	9	2	7	6	1	5	4
1	6	2	3	5	4	8	9	7
5	7	4	9	8	1	3	2	6
7	4	3	1	6	2	9	8	5
8	9	1	7	3	5	6	4	2
6	2	5	8	4	9	7	1	3
4	5	8	6	1	3	2	7	9
2	3	7	4	9	8	5	6	1
9	1	6	5	2	7	4	3	8

HARD - 131

7	8	2	1	3	9	4	5	6
9	1	5	4	2	6	3	7	8
6	4	3	8	5	7	2	1	9
5	6	4	2	8	3	7	9	1
8	2	7	9	6	1	5	3	4
3	9	1	7	4	5	6	8	2
2	5	8	3	1	4	9	6	7
1	7	6	5	9	2	8	4	3
4	3	9	6	7	8	1	2	5

HARD - 132

5	8	4	1	3	7	2	6	9
7	9	3	5	6	2	1	4	8
1	6	2	8	9	4	5	7	3
8	1	6	9	2	5	4	3	7
9	4	5	3	7	8	6	1	2
3	2	7	6	4	1	8	9	5
6	7	1	2	5	3	9	8	4
4	5	8	7	1	9	3	2	6
2	3	9	4	8	6	7	5	1

HARD - 133

4	6	8	7	3	9	1	2	5
7	1	2	6	8	5	3	4	9
3	5	9	4	2	1	8	6	7
8	9	4	3	6	2	5	7	1
5	3	7	9	1	4	6	8	2
6	2	1	8	5	7	4	9	3
2	4	3	5	7	8	9	1	6
9	7	5	1	4	6	2	3	8
1	8	6	2	9	3	7	5	4

HARD - 134

2	3	6	7	1	4	8	9	5
9	7	1	3	5	8	4	6	2
8	5	4	9	2	6	1	3	7
6	2	8	5	3	9	7	1	4
4	9	7	2	8	1	6	5	3
3	1	5	6	4	7	2	8	9
1	4	2	8	9	3	5	7	6
7	8	9	4	6	5	3	2	1
5	6	3	1	7	2	9	4	8

HARD - 135

7	3	5	2	6	1	9	8	4
6	1	8	4	3	9	7	2	5
2	4	9	7	8	5	1	6	3
5	9	3	8	7	2	4	1	6
1	7	4	6	5	3	2	9	8
8	2	6	9	1	4	3	5	7
9	5	7	1	4	8	6	3	2
4	8	1	3	2	6	5	7	9
3	6	2	5	9	7	8	4	1

HARD - 136

4	8	2	9	7	3	5	1	6
7	3	5	4	6	1	8	2	9
1	9	6	2	8	5	3	4	7
2	5	7	8	3	6	4	9	1
3	1	8	7	9	4	6	5	2
6	4	9	1	5	2	7	3	8
5	2	1	6	4	7	9	8	3
9	6	3	5	2	8	1	7	4
8	7	4	3	1	9	2	6	5

HARD - 137

8	6	2	7	3	9	1	4	5
1	5	9	8	2	4	7	3	6
7	3	4	1	6	5	2	8	9
2	8	1	9	4	3	5	6	7
9	7	5	2	8	6	3	1	4
6	4	3	5	7	1	9	2	8
3	9	7	4	1	8	6	5	2
4	2	6	3	5	7	8	9	1
5	1	8	6	9	2	4	7	3

HARD - 138

8	7	3	6	9	4	1	2	5
1	6	4	7	2	5	3	9	8
2	5	9	1	8	3	7	6	4
5	1	2	3	6	9	8	4	7
7	4	6	2	1	8	5	3	9
9	3	8	4	5	7	6	1	2
3	9	7	5	4	1	2	8	6
6	8	5	9	3	2	4	7	1
4	2	1	8	7	6	9	5	3

HARD - 139

8	3	2	6	1	5	9	7	4
7	1	5	8	4	9	3	2	6
9	4	6	7	3	2	1	8	5
1	8	9	5	2	3	6	4	7
3	5	4	1	7	6	2	9	8
6	2	7	9	8	4	5	3	1
4	9	8	3	6	1	7	5	2
2	6	3	4	5	7	8	1	9
5	7	1	2	9	8	4	6	3

HARD - 140

1	3	5	2	9	4	6	8	7
2	4	6	3	7	8	9	5	1
9	7	8	1	5	6	3	2	4
5	1	4	8	3	7	2	9	6
8	2	3	6	1	9	7	4	5
6	9	7	5	4	2	1	3	8
4	6	1	9	2	5	8	7	3
7	8	2	4	6	3	5	1	9
3	5	9	7	8	1	4	6	2

HARD - 141

9	5	2	3	7	4	6	1	8
3	6	1	8	5	9	2	7	4
4	7	8	6	2	1	5	9	3
1	3	7	9	8	2	4	5	6
8	9	6	5	4	3	1	2	7
5	2	4	1	6	7	8	3	9
7	4	5	2	9	6	3	8	1
2	1	9	4	3	8	7	6	5
6	8	3	7	1	5	9	4	2

HARD - 142

7	4	2	6	8	3	1	5	9
5	1	6	4	2	9	8	7	3
8	9	3	5	7	1	6	4	2
1	6	4	9	5	2	3	8	7
2	5	8	7	3	4	9	1	6
3	7	9	1	6	8	5	2	4
4	2	1	3	9	5	7	6	8
6	3	5	8	4	7	2	9	1
9	8	7	2	1	6	4	3	5

HARD - 143

1	2	9	6	5	4	3	8	7
5	3	8	9	7	1	2	4	6
6	4	7	2	8	3	5	9	1
7	8	6	4	9	5	1	3	2
9	5	4	3	1	2	6	7	8
2	1	3	7	6	8	9	5	4
4	7	1	5	3	6	8	2	9
3	6	2	8	4	9	7	1	5
8	9	5	1	2	7	4	6	3

HARD - 144

5	4	3	2	8	6	9	7	1
2	7	9	1	5	4	3	6	8
6	8	1	7	3	9	5	4	2
7	2	5	3	4	1	8	9	6
8	3	6	5	9	7	2	1	4
1	9	4	6	2	8	7	3	5
4	6	8	9	7	2	1	5	3
9	5	2	4	1	3	6	8	7
3	1	7	8	6	5	4	2	9

HARD - 145

4	7	8	9	1	6	5	3	2
3	1	5	4	8	2	6	9	7
2	6	9	3	5	7	8	1	4
5	9	4	2	3	8	7	6	1
1	2	6	5	7	9	4	8	3
7	8	3	1	6	4	9	2	5
9	5	1	8	4	3	2	7	6
6	4	2	7	9	1	3	5	8
8	3	7	6	2	5	1	4	9

HARD - 146

5	9	3	8	6	4	7	1	2
6	7	1	3	5	2	4	9	8
8	4	2	7	9	1	3	6	5
7	2	9	1	3	5	8	4	6
3	5	4	6	7	8	1	2	9
1	6	8	4	2	9	5	7	3
9	1	5	2	8	7	6	3	4
4	8	6	9	1	3	2	5	7
2	3	7	5	4	6	9	8	1

HARD - 147

8	5	6	1	2	9	4	3	7
3	4	9	5	8	7	6	2	1
7	2	1	3	4	6	9	8	5
9	6	4	7	1	2	3	5	8
5	7	2	6	3	8	1	9	4
1	8	3	9	5	4	2	7	6
6	3	8	4	9	5	7	1	2
4	1	5	2	7	3	8	6	9
2	9	7	8	6	1	5	4	3

HARD - 148

9	4	2	3	7	6	8	1	5
8	6	7	2	1	5	9	3	4
1	5	3	9	4	8	7	6	2
7	8	9	4	2	1	3	5	6
5	2	6	7	8	3	1	4	9
3	1	4	5	6	9	2	7	8
4	7	1	6	9	2	5	8	3
6	9	5	8	3	7	4	2	1
2	3	8	1	5	4	6	9	7

HARD - 149

8	1	4	5	3	9	6	2	7
6	3	5	8	7	2	9	4	1
7	9	2	1	6	4	8	5	3
5	8	3	6	9	7	2	1	4
9	7	6	4	2	1	5	3	8
4	2	1	3	5	8	7	6	9
3	6	8	7	1	5	4	9	2
1	4	9	2	8	6	3	7	5
2	5	7	9	4	3	1	8	6

HARD - 150

9	5	7	6	3	8	4	1	2
6	2	8	9	4	1	3	5	7
1	3	4	2	5	7	9	6	8
2	4	5	3	8	9	1	7	6
3	8	9	1	7	6	5	2	4
7	6	1	5	2	4	8	9	3
8	9	2	4	6	5	7	3	1
5	7	3	8	1	2	6	4	9
4	1	6	7	9	3	2	8	5

HARD - 151

8	1	4	6	7	9	5	2	3
6	9	5	3	4	2	7	8	1
3	2	7	5	8	1	9	4	6
1	5	6	9	3	4	2	7	8
2	7	8	1	6	5	3	9	4
9	4	3	7	2	8	6	1	5
7	3	2	8	1	6	4	5	9
5	6	1	4	9	7	8	3	2
4	8	9	2	5	3	1	6	7

HARD - 152

3	4	1	9	7	2	8	5	6
2	5	7	4	8	6	9	3	1
9	6	8	3	5	1	2	7	4
8	9	4	6	3	5	1	2	7
1	2	5	8	4	7	3	6	9
7	3	6	2	1	9	4	8	5
5	8	9	1	6	3	7	4	2
4	7	2	5	9	8	6	1	3
6	1	3	7	2	4	5	9	8

HARD - 153

5	6	8	4	9	2	3	1	7
9	3	7	8	6	1	5	4	2
1	4	2	3	7	5	6	9	8
3	7	6	1	5	9	2	8	4
2	9	5	6	4	8	1	7	3
4	8	1	2	3	7	9	5	6
7	1	3	5	2	4	8	6	9
8	2	4	9	1	6	7	3	5
6	5	9	7	8	3	4	2	1

HARD - 154

8	3	9	2	4	1	5	6	7
6	7	1	9	3	5	8	2	4
5	4	2	6	8	7	3	9	1
3	5	4	1	7	6	2	8	9
2	6	8	3	9	4	7	1	5
9	1	7	5	2	8	4	3	6
7	8	3	4	6	9	1	5	2
4	9	5	8	1	2	6	7	3
1	2	6	7	5	3	9	4	8

HARD - 155

6	4	9	2	3	7	8	5	1
2	3	1	6	8	5	4	7	9
7	5	8	1	9	4	2	6	3
4	8	5	7	1	3	6	9	2
1	6	7	4	2	9	5	3	8
9	2	3	5	6	8	7	1	4
5	1	6	9	4	2	3	8	7
3	7	4	8	5	1	9	2	6
8	9	2	3	7	6	1	4	5

HARD - 156

7	1	4	2	8	3	9	5	6
3	2	8	9	6	5	4	7	1
5	6	9	1	7	4	2	3	8
8	7	2	4	3	6	5	1	9
1	5	6	7	9	8	3	4	2
9	4	3	5	1	2	6	8	7
4	3	1	6	2	7	8	9	5
6	8	7	3	5	9	1	2	4
2	9	5	8	4	1	7	6	3

HARD - 157

2	8	9	4	1	3	7	5	6
4	6	1	9	5	7	8	3	2
7	3	5	8	6	2	1	4	9
8	9	3	5	7	4	6	2	1
1	5	2	3	8	6	4	9	7
6	4	7	1	2	9	5	8	3
9	1	4	6	3	8	2	7	5
3	7	6	2	4	5	9	1	8
5	2	8	7	9	1	3	6	4

HARD - 158

2	4	3	5	1	8	9	7	6
9	5	7	4	2	6	3	1	8
6	1	8	9	3	7	5	4	2
8	6	1	7	5	3	4	2	9
4	3	5	8	9	2	1	6	7
7	2	9	1	6	4	8	3	5
1	9	2	6	4	5	7	8	3
5	7	6	3	8	1	2	9	4
3	8	4	2	7	9	6	5	1

HARD - 159

9	7	3	8	5	4	1	2	6
2	6	1	3	9	7	5	8	4
5	8	4	1	6	2	3	7	9
7	4	8	9	3	1	2	6	5
1	2	5	7	8	6	4	9	3
6	3	9	2	4	5	8	1	7
8	9	6	5	2	3	7	4	1
3	1	2	4	7	9	6	5	8
4	5	7	6	1	8	9	3	2

HARD - 160

4	7	8	9	2	1	6	3	5
5	6	3	4	8	7	9	2	1
2	9	1	6	5	3	8	4	7
1	4	6	8	9	5	3	7	2
7	8	2	3	1	6	5	9	4
9	3	5	2	7	4	1	8	6
3	1	9	5	4	2	7	6	8
8	2	7	1	6	9	4	5	3
6	5	4	7	3	8	2	1	9

HARD - 161

2	6	1	3	9	8	7	5	4
5	7	8	4	6	2	3	1	9
9	3	4	7	5	1	2	8	6
3	2	5	9	8	7	4	6	1
4	8	7	5	1	6	9	2	3
6	1	9	2	3	4	5	7	8
1	9	2	8	4	5	6	3	7
7	4	6	1	2	3	8	9	5
8	5	3	6	7	9	1	4	2

HARD - 162

4	1	5	2	3	9	7	6	8
2	7	3	4	8	6	1	5	9
8	9	6	5	7	1	2	3	4
9	6	7	1	2	5	4	8	3
3	4	1	8	6	7	9	2	5
5	8	2	3	9	4	6	7	1
6	2	4	9	5	3	8	1	7
7	5	9	6	1	8	3	4	2
1	3	8	7	4	2	5	9	6

HARD - 163

4	5	3	8	9	6	2	7	1
2	7	8	1	5	4	3	6	9
1	9	6	3	2	7	8	5	4
5	2	4	7	8	1	9	3	6
3	8	7	5	6	9	4	1	2
6	1	9	2	4	3	5	8	7
8	6	2	9	1	5	7	4	3
9	3	1	4	7	8	6	2	5
7	4	5	6	3	2	1	9	8

HARD - 164

2	8	5	3	7	4	1	9	6
9	3	4	5	6	1	8	7	2
1	7	6	2	9	8	3	5	4
7	4	8	6	2	3	9	1	5
3	6	1	9	8	5	2	4	7
5	9	2	4	1	7	6	8	3
8	1	3	7	5	6	4	2	9
6	5	9	1	4	2	7	3	8
4	2	7	8	3	9	5	6	1

HARD - 165

4	6	9	3	1	7	2	8	5
7	5	1	4	2	8	3	6	9
8	3	2	6	5	9	1	4	7
6	4	3	9	8	5	7	1	2
9	7	8	2	4	1	6	5	3
2	1	5	7	3	6	4	9	8
5	8	4	1	7	3	9	2	6
1	9	7	8	6	2	5	3	4
3	2	6	5	9	4	8	7	1

HARD - 166

7	3	9	2	8	4	1	6	5
6	1	4	5	9	7	8	2	3
8	5	2	1	6	3	4	7	9
1	4	6	8	3	9	7	5	2
5	7	8	6	1	2	3	9	4
9	2	3	7	4	5	6	1	8
2	6	1	3	5	8	9	4	7
4	8	7	9	2	6	5	3	1
3	9	5	4	7	1	2	8	6

HARD - 167

8	5	2	3	7	9	1	4	6
1	7	9	2	6	4	8	3	5
3	4	6	8	1	5	9	2	7
6	3	8	5	4	7	2	9	1
4	2	5	1	9	8	6	7	3
9	1	7	6	3	2	4	5	8
5	8	1	9	2	3	7	6	4
2	6	4	7	5	1	3	8	9
7	9	3	4	8	6	5	1	2

HARD - 168

9	1	5	6	4	8	7	3	2
8	4	3	2	9	7	1	6	5
7	6	2	5	1	3	4	9	8
3	7	1	9	2	5	8	4	6
5	9	8	3	6	4	2	1	7
4	2	6	8	7	1	3	5	9
2	8	4	1	5	6	9	7	3
1	5	9	7	3	2	6	8	4
6	3	7	4	8	9	5	2	1

HARD - 169

5	2	4	7	1	6	8	3	9
7	3	9	8	5	2	1	6	4
1	6	8	3	9	4	5	7	2
6	9	3	1	4	5	2	8	7
2	1	5	6	8	7	4	9	3
4	8	7	9	2	3	6	1	5
3	5	6	4	7	1	9	2	8
9	4	1	2	3	8	7	5	6
8	7	2	5	6	9	3	4	1

HARD - 170

1	7	8	2	4	6	3	5	9
5	9	6	7	1	3	8	2	4
2	3	4	9	5	8	1	6	7
8	6	1	4	7	5	2	9	3
7	4	3	8	9	2	6	1	5
9	2	5	3	6	1	4	7	8
6	1	9	5	8	4	7	3	2
4	5	2	6	3	7	9	8	1
3	8	7	1	2	9	5	4	6

HARD - 171

4	8	5	9	6	1	7	3	2
6	7	9	5	3	2	4	1	8
2	3	1	4	7	8	9	6	5
8	2	4	7	1	6	5	9	3
5	6	3	8	4	9	1	2	7
9	1	7	2	5	3	6	8	4
7	9	6	3	8	5	2	4	1
1	5	8	6	2	4	3	7	9
3	4	2	1	9	7	8	5	6

HARD - 172

4	8	3	1	6	2	5	9	7
7	9	5	4	8	3	1	6	2
1	6	2	7	9	5	3	8	4
9	1	4	6	7	8	2	5	3
8	3	7	2	5	4	6	1	9
2	5	6	9	3	1	4	7	8
3	7	9	5	2	6	8	4	1
5	2	1	8	4	7	9	3	6
6	4	8	3	1	9	7	2	5

HARD - 173

9	8	4	3	5	1	2	7	6
3	2	6	8	7	4	5	1	9
1	5	7	9	2	6	4	8	3
2	6	5	4	8	7	3	9	1
7	3	8	2	1	9	6	4	5
4	9	1	6	3	5	7	2	8
8	4	2	1	6	3	9	5	7
6	7	9	5	4	8	1	3	2
5	1	3	7	9	2	8	6	4

HARD - 174

6	2	3	5	4	1	8	7	9
7	4	5	8	2	9	1	6	3
8	9	1	3	6	7	5	2	4
2	3	4	1	9	8	7	5	6
9	1	6	2	7	5	4	3	8
5	7	8	6	3	4	9	1	2
1	6	7	9	8	2	3	4	5
4	8	2	7	5	3	6	9	1
3	5	9	4	1	6	2	8	7

HARD - 175

9	4	6	5	8	2	7	1	3
1	7	3	4	9	6	5	8	2
8	2	5	3	7	1	9	4	6
4	1	2	7	5	9	6	3	8
3	6	7	8	1	4	2	9	5
5	9	8	2	6	3	4	7	1
2	8	4	6	3	7	1	5	9
7	5	1	9	2	8	3	6	4
6	3	9	1	4	5	8	2	7

HARD - 176

3	9	2	5	7	1	4	8	6
8	7	1	9	4	6	3	2	5
5	4	6	2	8	3	9	1	7
7	3	8	1	2	9	6	5	4
9	1	4	6	5	8	7	3	2
2	6	5	4	3	7	1	9	8
4	2	7	3	9	5	8	6	1
1	5	9	8	6	4	2	7	3
6	8	3	7	1	2	5	4	9

HARD - 177

8	3	7	2	6	5	1	4	9
1	2	6	8	4	9	5	7	3
5	9	4	7	3	1	8	6	2
4	6	3	5	8	2	7	9	1
2	5	9	6	1	7	4	3	8
7	1	8	4	9	3	2	5	6
9	8	2	3	5	4	6	1	7
3	7	5	1	2	6	9	8	4
6	4	1	9	7	8	3	2	5

HARD - 178

6	9	8	1	7	5	3	2	4
7	4	5	9	3	2	8	1	6
1	3	2	6	8	4	7	9	5
2	7	3	5	1	8	4	6	9
4	8	6	7	2	9	1	5	3
5	1	9	4	6	3	2	8	7
9	5	1	2	4	7	6	3	8
8	2	4	3	9	6	5	7	1
3	6	7	8	5	1	9	4	2

HARD - 179

9	6	8	1	3	5	2	4	7
1	5	7	2	8	4	9	3	6
3	4	2	9	6	7	8	1	5
6	7	4	3	1	2	5	9	8
2	9	1	6	5	8	4	7	3
5	8	3	4	7	9	6	2	1
8	2	6	7	4	3	1	5	9
7	1	9	5	2	6	3	8	4
4	3	5	8	9	1	7	6	2

HARD - 180

3	4	1	9	7	8	2	5	6
7	2	8	4	5	6	1	9	3
9	5	6	1	2	3	4	7	8
2	6	9	7	4	5	3	8	1
8	3	4	6	1	9	5	2	7
1	7	5	3	8	2	9	6	4
5	9	3	8	6	4	7	1	2
6	1	2	5	3	7	8	4	9
4	8	7	2	9	1	6	3	5

HARD - 181

1	6	9	8	5	3	2	7	4
7	8	2	6	1	4	9	3	5
3	5	4	9	7	2	6	1	8
2	7	6	1	3	5	8	4	9
9	4	1	2	6	8	3	5	7
8	3	5	4	9	7	1	2	6
4	1	3	7	8	6	5	9	2
5	2	8	3	4	9	7	6	1
6	9	7	5	2	1	4	8	3

HARD - 182

5	6	1	2	9	4	7	8	3
7	3	8	1	5	6	4	9	2
9	2	4	8	7	3	1	6	5
2	8	9	4	3	7	5	1	6
1	7	6	9	8	5	2	3	4
4	5	3	6	2	1	8	7	9
6	1	5	7	4	9	3	2	8
3	9	2	5	1	8	6	4	7
8	4	7	3	6	2	9	5	1

HARD - 183

3	6	7	2	5	8	9	1	4
9	1	4	7	6	3	8	2	5
8	2	5	1	9	4	3	7	6
4	9	2	6	1	5	7	8	3
1	7	8	3	4	2	5	6	9
6	5	3	8	7	9	1	4	2
5	8	6	4	3	1	2	9	7
7	3	1	9	2	6	4	5	8
2	4	9	5	8	7	6	3	1

HARD - 184

6	7	4	3	5	1	8	9	2
2	3	5	9	7	8	1	6	4
8	1	9	4	2	6	5	3	7
1	2	8	6	3	9	4	7	5
3	4	6	7	8	5	9	2	1
9	5	7	2	1	4	6	8	3
4	6	2	1	9	7	3	5	8
7	8	1	5	6	3	2	4	9
5	9	3	8	4	2	7	1	6

HARD - 185

5	8	6	4	9	1	7	2	3
9	1	4	2	7	3	8	6	5
3	2	7	6	5	8	4	1	9
6	3	1	9	4	2	5	7	8
2	7	8	1	3	5	9	4	6
4	5	9	8	6	7	2	3	1
7	4	5	3	8	6	1	9	2
1	9	3	5	2	4	6	8	7
8	6	2	7	1	9	3	5	4

HARD - 186

3	8	1	9	2	6	4	5	7
7	5	2	3	4	8	9	1	6
9	4	6	7	1	5	8	2	3
5	9	8	4	3	7	2	6	1
6	7	4	2	5	1	3	8	9
1	2	3	8	6	9	7	4	5
8	3	5	6	7	2	1	9	4
4	6	9	1	8	3	5	7	2
2	1	7	5	9	4	6	3	8

HARD - 187

7	4	5	1	6	8	9	2	3
9	2	3	5	7	4	8	1	6
1	8	6	3	2	9	4	5	7
5	9	7	8	3	6	1	4	2
2	3	4	9	5	1	6	7	8
6	1	8	2	4	7	5	3	9
8	5	1	7	9	3	2	6	4
3	6	2	4	8	5	7	9	1
4	7	9	6	1	2	3	8	5

HARD - 188

7	5	3	8	9	2	4	1	6
4	9	8	6	7	1	2	5	3
1	6	2	4	5	3	8	7	9
2	7	6	5	3	8	9	4	1
5	4	9	7	1	6	3	8	2
3	8	1	2	4	9	7	6	5
6	2	4	9	8	5	1	3	7
8	1	5	3	2	7	6	9	4
9	3	7	1	6	4	5	2	8

HARD - 189

9	4	8	3	2	1	6	7	5
5	3	1	7	8	6	9	2	4
2	6	7	5	9	4	8	3	1
1	8	5	6	7	9	3	4	2
4	2	3	8	1	5	7	9	6
6	7	9	2	4	3	1	5	8
3	5	2	1	6	7	4	8	9
7	9	6	4	5	8	2	1	3
8	1	4	9	3	2	5	6	7

HARD - 190

7	6	9	3	8	1	4	5	2
2	8	4	5	7	6	9	1	3
3	1	5	2	9	4	8	6	7
1	9	6	4	5	3	2	7	8
5	3	8	7	1	2	6	4	9
4	7	2	9	6	8	1	3	5
9	5	1	8	4	7	3	2	6
8	4	3	6	2	5	7	9	1
6	2	7	1	3	9	5	8	4

HARD - 191

1	7	2	5	4	3	6	9	8
3	6	8	2	7	9	5	1	4
4	9	5	6	8	1	7	3	2
5	3	6	7	2	8	9	4	1
9	4	7	3	1	5	8	2	6
8	2	1	4	9	6	3	7	5
2	1	9	8	5	7	4	6	3
6	5	4	9	3	2	1	8	7
7	8	3	1	6	4	2	5	9

HARD - 192

4	8	2	7	5	3	6	9	1
7	5	6	9	2	1	3	8	4
3	9	1	8	6	4	2	5	7
5	6	9	1	4	7	8	2	3
1	2	3	5	9	8	7	4	6
8	4	7	6	3	2	9	1	5
6	3	4	2	1	9	5	7	8
2	7	5	4	8	6	1	3	9
9	1	8	3	7	5	4	6	2

HARD - 193

8	9	3	7	1	6	4	2	5
7	5	6	2	9	4	1	3	8
4	2	1	8	3	5	6	9	7
1	8	5	4	2	3	7	6	9
9	7	4	6	8	1	3	5	2
6	3	2	5	7	9	8	1	4
3	1	7	9	4	2	5	8	6
5	4	9	1	6	8	2	7	3
2	6	8	3	5	7	9	4	1

HARD - 194

4	8	7	5	9	6	1	3	2
2	6	1	3	4	7	5	9	8
9	5	3	2	1	8	7	4	6
5	9	6	7	2	1	4	8	3
8	7	4	9	5	3	2	6	1
3	1	2	6	8	4	9	5	7
1	4	5	8	6	2	3	7	9
7	2	8	4	3	9	6	1	5
6	3	9	1	7	5	8	2	4

HARD - 195

8	5	6	9	1	7	2	3	4
9	7	2	3	5	4	6	1	8
4	1	3	6	8	2	9	5	7
1	8	9	5	2	3	4	7	6
5	2	4	8	7	6	1	9	3
6	3	7	1	4	9	5	8	2
3	6	8	4	9	5	7	2	1
2	4	5	7	3	1	8	6	9
7	9	1	2	6	8	3	4	5

HARD - 196

7	5	4	3	6	1	9	2	8
8	9	2	7	4	5	1	3	6
1	3	6	9	2	8	4	7	5
2	8	9	1	3	7	6	5	4
5	4	1	6	8	2	7	9	3
6	7	3	5	9	4	2	8	1
4	1	5	2	7	3	8	6	9
9	2	8	4	5	6	3	1	7
3	6	7	8	1	9	5	4	2

HARD - 197

5	1	2	3	7	4	8	9	6
4	6	9	8	5	1	7	2	3
7	3	8	2	9	6	5	1	4
1	4	6	5	8	9	3	7	2
8	2	3	1	6	7	4	5	9
9	5	7	4	3	2	1	6	8
6	8	5	9	1	3	2	4	7
2	9	1	7	4	8	6	3	5
3	7	4	6	2	5	9	8	1

HARD - 198

1	9	8	7	4	6	3	5	2
7	3	5	2	1	9	8	4	6
2	6	4	3	8	5	9	1	7
8	5	2	6	3	4	1	7	9
3	1	6	9	7	8	5	2	4
9	4	7	5	2	1	6	3	8
4	8	9	1	5	2	7	6	3
6	7	1	4	9	3	2	8	5
5	2	3	8	6	7	4	9	1

HARD - 199

9	3	5	8	2	1	6	7	4
7	2	6	5	4	9	1	8	3
8	4	1	3	7	6	9	5	2
6	7	2	1	9	4	5	3	8
1	9	3	7	8	5	2	4	6
5	8	4	2	6	3	7	1	9
3	1	8	9	5	2	4	6	7
2	6	7	4	1	8	3	9	5
4	5	9	6	3	7	8	2	1

HARD - 200

6	2	8	9	4	1	3	7	5
4	9	3	7	6	5	2	8	1
1	5	7	3	2	8	9	4	6
5	7	6	8	3	9	1	2	4
2	4	1	5	7	6	8	3	9
8	3	9	2	1	4	5	6	7
9	8	2	4	5	7	6	1	3
7	6	5	1	8	3	4	9	2
3	1	4	6	9	2	7	5	8

HARD - 201

2	1	6	9	3	7	4	5	8
5	3	9	2	4	8	1	6	7
7	8	4	6	5	1	2	9	3
9	4	7	3	6	2	8	1	5
1	2	5	4	8	9	3	7	6
3	6	8	1	7	5	9	4	2
4	9	3	7	2	6	5	8	1
8	7	2	5	1	4	6	3	9
6	5	1	8	9	3	7	2	4

HARD - 202

4	9	7	6	5	8	3	2	1
2	6	5	1	4	3	9	8	7
1	3	8	9	7	2	4	6	5
5	4	1	3	8	7	2	9	6
6	8	2	4	1	9	7	5	3
3	7	9	5	2	6	8	1	4
9	5	6	8	3	4	1	7	2
8	2	3	7	6	1	5	4	9
7	1	4	2	9	5	6	3	8

HARD - 203

6	7	1	3	5	4	9	8	2
4	8	9	6	1	2	3	5	7
2	3	5	9	8	7	6	4	1
1	2	7	4	9	6	8	3	5
5	4	8	7	3	1	2	9	6
3	9	6	5	2	8	1	7	4
7	6	3	2	4	9	5	1	8
9	1	2	8	7	5	4	6	3
8	5	4	1	6	3	7	2	9

HARD - 204

2	4	9	5	6	3	1	7	8
7	5	1	4	2	8	3	9	6
6	3	8	1	9	7	4	5	2
3	7	4	8	1	6	9	2	5
1	2	5	9	7	4	8	6	3
8	9	6	2	3	5	7	4	1
9	1	3	7	5	2	6	8	4
4	6	2	3	8	9	5	1	7
5	8	7	6	4	1	2	3	9

VERY HARD - 1

8	7	5	9	1	6	4	2	3
6	4	9	7	3	2	1	5	8
1	2	3	4	8	5	9	7	6
2	3	7	6	5	4	8	1	9
9	8	4	3	2	1	5	6	7
5	6	1	8	9	7	3	4	2
3	1	2	5	7	9	6	8	4
7	9	6	1	4	8	2	3	5
4	5	8	2	6	3	7	9	1

VERY HARD - 2

1	9	3	2	6	7	8	5	4
5	8	7	4	3	1	9	2	6
2	6	4	8	9	5	3	1	7
3	2	8	9	1	4	6	7	5
4	7	5	3	2	6	1	8	9
6	1	9	7	5	8	4	3	2
7	3	6	5	8	9	2	4	1
8	5	1	6	4	2	7	9	3
9	4	2	1	7	3	5	6	8

VERY HARD - 3

1	4	2	7	5	3	6	9	8
9	3	8	2	1	6	7	4	5
5	6	7	4	9	8	1	2	3
7	5	1	6	8	4	2	3	9
2	9	3	5	7	1	4	8	6
6	8	4	9	3	2	5	1	7
8	1	5	3	2	7	9	6	4
3	7	6	1	4	9	8	5	2
4	2	9	8	6	5	3	7	1

VERY HARD - 4

6	9	3	8	1	4	2	7	5
5	8	1	2	6	7	4	9	3
7	2	4	5	9	3	1	6	8
4	3	6	7	2	1	5	8	9
8	7	9	6	4	5	3	1	2
2	1	5	9	3	8	6	4	7
9	4	8	3	5	6	7	2	1
3	6	7	1	8	2	9	5	4
1	5	2	4	7	9	8	3	6

VERY HARD - 5

4	2	7	8	1	6	5	3	9
1	9	6	5	2	3	8	4	7
8	3	5	9	4	7	2	6	1
7	8	3	4	9	5	6	1	2
2	6	4	7	8	1	3	9	5
5	1	9	6	3	2	7	8	4
3	5	1	2	6	4	9	7	8
9	4	2	3	7	8	1	5	6
6	7	8	1	5	9	4	2	3

VERY HARD - 6

6	1	3	4	9	2	5	8	7
4	2	7	8	1	5	9	6	3
9	8	5	3	7	6	1	2	4
1	9	8	6	4	7	2	3	5
7	3	2	9	5	8	4	1	6
5	6	4	1	2	3	8	7	9
3	7	9	5	8	1	6	4	2
8	5	6	2	3	4	7	9	1
2	4	1	7	6	9	3	5	8

VERY HARD - 7

1	7	4	8	5	9	6	2	3
5	3	8	4	6	2	9	1	7
6	9	2	3	1	7	4	5	8
9	8	6	1	7	5	3	4	2
4	1	5	9	2	3	8	7	6
7	2	3	6	4	8	5	9	1
2	5	9	7	8	6	1	3	4
3	6	1	2	9	4	7	8	5
8	4	7	5	3	1	2	6	9

VERY HARD - 8

8	9	1	7	6	3	5	4	2
3	2	4	8	1	5	6	7	9
7	6	5	9	4	2	3	8	1
9	4	2	6	5	7	1	3	8
1	3	6	4	9	8	7	2	5
5	8	7	3	2	1	4	9	6
4	1	9	2	3	6	8	5	7
2	5	8	1	7	4	9	6	3
6	7	3	5	8	9	2	1	4

VERY HARD - 9

7	3	6	1	4	2	5	8	9
2	8	4	5	9	3	1	7	6
5	1	9	6	7	8	4	2	3
6	7	3	4	5	9	2	1	8
8	4	5	2	3	1	6	9	7
9	2	1	8	6	7	3	4	5
3	5	7	9	2	4	8	6	1
1	9	2	3	8	6	7	5	4
4	6	8	7	1	5	9	3	2

VERY HARD - 10

1	2	4	8	9	5	7	3	6
8	9	7	3	6	4	2	1	5
3	6	5	2	1	7	4	8	9
9	1	8	7	4	3	5	6	2
2	5	3	6	8	9	1	4	7
7	4	6	5	2	1	3	9	8
4	8	1	9	5	2	6	7	3
6	3	2	1	7	8	9	5	4
5	7	9	4	3	6	8	2	1

VERY HARD - 11

5	4	2	8	9	1	7	6	3
6	9	1	3	4	7	8	2	5
8	3	7	5	2	6	4	9	1
4	2	8	1	6	5	3	7	9
9	1	6	2	7	3	5	8	4
7	5	3	9	8	4	6	1	2
3	7	9	4	1	8	2	5	6
1	6	4	7	5	2	9	3	8
2	8	5	6	3	9	1	4	7

VERY HARD - 12

4	5	3	8	1	9	6	2	7
8	6	9	5	7	2	1	3	4
7	2	1	6	3	4	8	5	9
1	4	6	2	5	8	9	7	3
2	8	7	3	9	6	5	4	1
3	9	5	1	4	7	2	6	8
6	1	8	7	2	3	4	9	5
5	3	4	9	6	1	7	8	2
9	7	2	4	8	5	3	1	6

VERY HARD - 13

3	5	6	2	4	7	8	1	9
8	4	9	1	3	6	7	2	5
1	2	7	9	8	5	6	3	4
9	7	2	3	5	4	1	6	8
5	6	3	7	1	8	4	9	2
4	1	8	6	9	2	3	5	7
2	3	4	8	6	9	5	7	1
6	9	5	4	7	1	2	8	3
7	8	1	5	2	3	9	4	6

VERY HARD - 14

3	1	2	5	9	8	7	4	6
9	8	4	6	7	2	3	5	1
7	5	6	1	3	4	8	2	9
4	2	7	9	5	6	1	8	3
1	6	9	4	8	3	5	7	2
8	3	5	7	2	1	9	6	4
6	7	8	3	4	9	2	1	5
5	9	1	2	6	7	4	3	8
2	4	3	8	1	5	6	9	7

VERY HARD - 15

5	6	4	8	1	9	3	7	2
8	1	3	2	6	7	9	4	5
2	9	7	3	4	5	6	8	1
3	7	9	5	8	1	2	6	4
4	5	2	9	7	6	8	1	3
1	8	6	4	3	2	5	9	7
6	2	8	1	5	4	7	3	9
7	4	5	6	9	3	1	2	8
9	3	1	7	2	8	4	5	6

VERY HARD - 16

6	1	5	7	8	3	9	2	4
3	9	2	6	4	5	8	1	7
8	7	4	2	9	1	6	5	3
1	8	6	9	2	4	7	3	5
7	4	3	1	5	6	2	8	9
5	2	9	8	3	7	1	4	6
4	6	1	3	7	2	5	9	8
9	3	7	5	1	8	4	6	2
2	5	8	4	6	9	3	7	1

VERY HARD - 17

4	7	6	8	5	3	1	9	2
5	2	3	7	1	9	6	8	4
8	1	9	4	2	6	3	5	7
9	8	1	2	4	7	5	6	3
2	5	4	3	6	8	7	1	9
3	6	7	1	9	5	4	2	8
1	4	8	6	7	2	9	3	5
6	3	5	9	8	4	2	7	1
7	9	2	5	3	1	8	4	6

VERY HARD - 18

5	1	8	6	9	3	2	4	7
6	2	3	5	7	4	9	1	8
4	7	9	8	1	2	5	6	3
1	3	6	2	5	9	7	8	4
8	5	4	1	3	7	6	2	9
2	9	7	4	8	6	1	3	5
9	6	2	3	4	5	8	7	1
7	4	1	9	2	8	3	5	6
3	8	5	7	6	1	4	9	2

VERY HARD - 19

3	9	6	7	1	2	8	4	5
1	8	7	9	4	5	3	2	6
2	4	5	6	8	3	9	7	1
9	1	2	8	5	7	4	6	3
4	7	3	1	6	9	2	5	8
6	5	8	2	3	4	7	1	9
5	3	1	4	7	8	6	9	2
7	6	9	3	2	1	5	8	4
8	2	4	5	9	6	1	3	7

VERY HARD - 20

8	2	4	3	1	5	6	7	9
3	7	9	8	6	4	5	2	1
1	5	6	2	9	7	4	3	8
2	3	1	5	8	6	9	4	7
6	8	7	1	4	9	3	5	2
9	4	5	7	2	3	8	1	6
5	1	2	9	3	8	7	6	4
7	6	8	4	5	2	1	9	3
4	9	3	6	7	1	2	8	5

VERY HARD - 21

4	5	3	8	9	7	6	2	1
6	9	2	5	4	1	3	8	7
8	7	1	2	3	6	4	9	5
5	4	9	3	1	2	8	7	6
7	1	6	9	5	8	2	3	4
2	3	8	6	7	4	1	5	9
3	6	4	7	8	9	5	1	2
9	2	5	1	6	3	7	4	8
1	8	7	4	2	5	9	6	3

VERY HARD - 22

2	6	1	5	8	7	9	4	3
4	3	9	1	6	2	7	5	8
7	8	5	3	9	4	1	6	2
6	5	2	8	4	9	3	1	7
8	9	7	6	1	3	4	2	5
1	4	3	2	7	5	6	8	9
3	1	8	9	5	6	2	7	4
5	2	4	7	3	1	8	9	6
9	7	6	4	2	8	5	3	1

VERY HARD - 23

4	6	5	8	2	9	7	1	3
2	3	7	6	4	1	5	9	8
1	9	8	7	3	5	2	4	6
3	7	1	9	8	2	6	5	4
5	8	9	3	6	4	1	2	7
6	4	2	5	1	7	8	3	9
7	2	4	1	9	6	3	8	5
8	1	6	4	5	3	9	7	2
9	5	3	2	7	8	4	6	1

VERY HARD - 24

5	4	7	8	6	9	3	2	1
2	9	6	5	3	1	7	4	8
1	8	3	4	2	7	5	6	9
9	5	2	3	8	6	1	7	4
4	6	1	7	5	2	9	8	3
7	3	8	9	1	4	2	5	6
3	7	5	6	9	8	4	1	2
8	1	9	2	4	5	6	3	7
6	2	4	1	7	3	8	9	5

VERY HARD - 25

7	1	5	9	3	6	4	2	8
8	4	2	7	5	1	3	6	9
6	9	3	4	2	8	5	7	1
3	2	4	1	6	9	7	8	5
1	7	8	2	4	5	6	9	3
5	6	9	8	7	3	2	1	4
4	5	1	6	9	2	8	3	7
2	8	7	3	1	4	9	5	6
9	3	6	5	8	7	1	4	2

VERY HARD - 26

8	3	6	5	2	7	9	1	4
1	7	9	3	6	4	5	8	2
4	5	2	8	9	1	3	6	7
6	8	4	7	3	2	1	5	9
9	1	3	4	5	8	2	7	6
7	2	5	9	1	6	8	4	3
2	4	7	1	8	3	6	9	5
5	6	1	2	4	9	7	3	8
3	9	8	6	7	5	4	2	1

VERY HARD - 27

6	1	8	7	2	4	3	5	9
4	2	3	6	9	5	1	7	8
5	7	9	1	3	8	2	4	6
2	4	5	9	6	1	7	8	3
3	9	1	2	8	7	5	6	4
8	6	7	5	4	3	9	1	2
9	5	6	4	1	2	8	3	7
7	3	4	8	5	9	6	2	1
1	8	2	3	7	6	4	9	5

VERY HARD - 28

8	3	2	7	4	5	9	1	6
4	7	6	2	1	9	8	3	5
1	5	9	6	8	3	4	2	7
9	4	7	3	6	2	1	5	8
3	1	5	9	7	8	2	6	4
6	2	8	1	5	4	7	9	3
2	6	3	8	9	7	5	4	1
7	9	4	5	3	1	6	8	2
5	8	1	4	2	6	3	7	9

VERY HARD - 29

1	7	4	5	8	9	6	2	3
6	3	8	4	7	2	1	9	5
9	2	5	6	3	1	7	8	4
4	9	2	8	6	3	5	7	1
5	6	1	9	4	7	8	3	2
3	8	7	1	2	5	9	4	6
2	4	6	7	1	8	3	5	9
8	1	9	3	5	4	2	6	7
7	5	3	2	9	6	4	1	8

VERY HARD - 30

3	7	2	4	6	5	9	8	1
5	6	1	8	2	9	7	3	4
8	9	4	1	7	3	2	6	5
9	8	5	7	1	4	3	2	6
6	2	3	5	9	8	4	1	7
4	1	7	6	3	2	5	9	8
2	5	8	3	4	1	6	7	9
7	4	9	2	8	6	1	5	3
1	3	6	9	5	7	8	4	2

VERY HARD - 31

6	9	1	4	7	5	3	8	2
4	8	2	6	1	3	7	5	9
7	5	3	8	9	2	6	4	1
9	4	5	2	3	8	1	7	6
3	2	6	1	4	7	5	9	8
8	1	7	5	6	9	4	2	3
1	7	8	9	5	6	2	3	4
5	6	9	3	2	4	8	1	7
2	3	4	7	8	1	9	6	5

VERY HARD - 32

6	4	9	1	3	8	2	5	7
2	8	1	5	7	4	9	3	6
3	7	5	2	6	9	8	4	1
4	2	7	8	9	1	3	6	5
5	6	3	4	2	7	1	9	8
1	9	8	6	5	3	7	2	4
8	3	4	9	1	6	5	7	2
7	1	2	3	4	5	6	8	9
9	5	6	7	8	2	4	1	3

VERY HARD - 33

5	2	7	3	6	4	9	8	1
3	4	6	8	9	1	2	5	7
1	8	9	2	5	7	6	4	3
7	9	3	5	1	2	4	6	8
4	5	8	7	3	6	1	9	2
6	1	2	4	8	9	3	7	5
8	6	4	1	7	3	5	2	9
9	3	5	6	2	8	7	1	4
2	7	1	9	4	5	8	3	6

VERY HARD - 34

7	6	4	5	1	3	8	9	2
8	5	3	2	7	9	4	1	6
1	9	2	6	4	8	5	7	3
3	4	8	9	2	7	1	6	5
2	7	6	4	5	1	9	3	8
9	1	5	3	8	6	2	4	7
5	8	1	7	3	4	6	2	9
4	3	9	8	6	2	7	5	1
6	2	7	1	9	5	3	8	4

VERY HARD - 35

7	6	4	9	5	3	1	2	8
1	5	3	4	2	8	9	7	6
8	9	2	6	7	1	5	4	3
6	1	8	7	9	5	2	3	4
2	7	9	3	4	6	8	1	5
3	4	5	8	1	2	6	9	7
4	2	6	1	8	7	3	5	9
5	3	7	2	6	9	4	8	1
9	8	1	5	3	4	7	6	2

VERY HARD - 36

7	9	6	2	1	5	4	8	3
1	4	3	9	6	8	2	7	5
5	2	8	3	7	4	6	9	1
4	8	2	1	3	7	5	6	9
3	7	1	6	5	9	8	2	4
6	5	9	8	4	2	3	1	7
9	6	5	4	2	1	7	3	8
2	1	7	5	8	3	9	4	6
8	3	4	7	9	6	1	5	2

VERY HARD - 37

8	1	4	7	9	2	3	6	5
5	6	3	4	1	8	7	9	2
7	2	9	3	5	6	8	4	1
3	7	2	8	6	9	1	5	4
9	4	1	2	3	5	6	8	7
6	5	8	1	7	4	9	2	3
2	8	7	9	4	3	5	1	6
1	9	5	6	2	7	4	3	8
4	3	6	5	8	1	2	7	9

VERY HARD - 38

3	6	5	9	4	8	7	2	1
7	8	4	1	6	2	5	9	3
2	9	1	5	7	3	8	6	4
5	1	2	4	9	7	6	3	8
8	3	9	2	1	6	4	5	7
4	7	6	8	3	5	2	1	9
1	5	8	3	2	4	9	7	6
6	2	3	7	8	9	1	4	5
9	4	7	6	5	1	3	8	2

VERY HARD - 39

5	8	7	2	6	1	9	4	3
4	9	2	3	8	7	6	1	5
6	1	3	5	9	4	2	7	8
2	6	9	1	4	3	5	8	7
7	3	4	9	5	8	1	6	2
1	5	8	7	2	6	3	9	4
3	2	1	8	7	9	4	5	6
8	4	5	6	1	2	7	3	9
9	7	6	4	3	5	8	2	1

VERY HARD - 40

6	5	7	4	3	1	8	2	9
8	1	4	2	7	9	3	6	5
9	2	3	5	6	8	4	7	1
4	7	5	1	2	6	9	8	3
2	6	1	8	9	3	7	5	4
3	9	8	7	4	5	2	1	6
5	3	9	6	8	2	1	4	7
1	4	2	9	5	7	6	3	8
7	8	6	3	1	4	5	9	2

VERY HARD - 41

8	6	2	7	4	3	9	1	5
4	1	5	6	9	2	7	3	8
7	3	9	1	8	5	2	4	6
3	5	1	4	2	9	6	8	7
9	4	7	5	6	8	3	2	1
6	2	8	3	7	1	5	9	4
5	9	3	8	1	7	4	6	2
1	7	4	2	3	6	8	5	9
2	8	6	9	5	4	1	7	3

VERY HARD - 42

6	3	7	5	8	9	2	1	4
4	9	5	6	2	1	3	7	8
2	1	8	7	3	4	6	5	9
3	6	2	8	9	5	7	4	1
1	5	9	3	4	7	8	2	6
8	7	4	2	1	6	5	9	3
9	8	3	4	7	2	1	6	5
5	2	1	9	6	3	4	8	7
7	4	6	1	5	8	9	3	2

VERY HARD - 43

4	9	2	5	6	1	7	3	8
3	6	1	2	8	7	9	4	5
7	8	5	9	3	4	1	6	2
8	1	9	4	5	3	2	7	6
5	2	3	8	7	6	4	9	1
6	7	4	1	2	9	8	5	3
9	5	8	6	4	2	3	1	7
1	3	6	7	9	8	5	2	4
2	4	7	3	1	5	6	8	9

VERY HARD - 44

5	6	4	1	9	8	7	3	2
2	1	8	3	4	7	5	9	6
7	9	3	2	5	6	1	8	4
8	3	2	4	6	1	9	7	5
4	5	9	7	3	2	6	1	8
1	7	6	5	8	9	4	2	3
3	8	7	6	1	5	2	4	9
6	4	1	9	2	3	8	5	7
9	2	5	8	7	4	3	6	1

VERY HARD - 45

1	3	4	9	2	6	7	8	5
2	5	7	8	3	1	6	9	4
8	6	9	4	5	7	2	3	1
4	8	6	1	9	5	3	2	7
7	2	5	6	4	3	8	1	9
3	9	1	7	8	2	4	5	6
6	4	3	5	1	8	9	7	2
5	7	2	3	6	9	1	4	8
9	1	8	2	7	4	5	6	3

VERY HARD - 46

5	9	4	2	7	6	3	8	1
2	6	8	9	3	1	7	5	4
7	1	3	5	4	8	6	9	2
3	2	5	7	6	4	8	1	9
8	4	6	1	9	3	2	7	5
1	7	9	8	2	5	4	3	6
6	5	2	3	8	9	1	4	7
4	3	1	6	5	7	9	2	8
9	8	7	4	1	2	5	6	3

VERY HARD - 47

1	4	9	7	5	8	6	3	2
7	8	2	4	6	3	1	5	9
5	6	3	9	2	1	4	7	8
9	1	7	6	3	5	8	2	4
8	2	4	1	9	7	5	6	3
3	5	6	2	8	4	9	1	7
2	7	8	5	1	9	3	4	6
4	3	1	8	7	6	2	9	5
6	9	5	3	4	2	7	8	1

VERY HARD - 48

5	7	8	9	3	2	1	6	4
4	6	3	8	1	5	9	7	2
1	9	2	4	7	6	5	3	8
3	2	1	6	8	7	4	5	9
7	8	9	1	5	4	3	2	6
6	4	5	3	2	9	7	8	1
2	5	6	7	4	1	8	9	3
8	1	7	2	9	3	6	4	5
9	3	4	5	6	8	2	1	7

VERY HARD - 49

2	6	9	8	5	3	1	7	4
1	8	4	7	9	2	6	3	5
5	3	7	4	1	6	2	9	8
8	9	6	2	4	7	3	5	1
4	2	5	6	3	1	9	8	7
3	7	1	5	8	9	4	6	2
6	5	3	1	7	4	8	2	9
9	4	8	3	2	5	7	1	6
7	1	2	9	6	8	5	4	3

VERY HARD - 50

5	8	4	7	9	6	3	2	1
1	2	9	8	5	3	6	4	7
6	7	3	1	4	2	5	9	8
2	6	7	4	1	8	9	3	5
9	5	8	2	3	7	1	6	4
3	4	1	9	6	5	8	7	2
8	1	2	6	7	9	4	5	3
4	3	6	5	2	1	7	8	9
7	9	5	3	8	4	2	1	6

VERY HARD - 51

1	4	5	3	8	9	6	2	7
9	2	6	4	7	5	3	1	8
3	8	7	6	1	2	9	5	4
8	3	4	1	6	7	2	9	5
6	9	1	2	5	8	7	4	3
5	7	2	9	3	4	1	8	6
7	1	8	5	2	6	4	3	9
2	5	9	7	4	3	8	6	1
4	6	3	8	9	1	5	7	2

VERY HARD - 52

4	2	7	1	6	3	9	5	8
8	3	6	2	9	5	7	1	4
1	5	9	8	4	7	3	6	2
6	4	2	7	8	1	5	9	3
9	7	3	6	5	2	8	4	1
5	8	1	9	3	4	6	2	7
2	1	8	5	7	6	4	3	9
7	6	4	3	2	9	1	8	5
3	9	5	4	1	8	2	7	6

VERY HARD - 53

5	1	8	7	9	4	2	3	6
7	2	3	6	8	1	4	9	5
9	4	6	3	5	2	1	7	8
2	7	1	8	4	3	5	6	9
6	8	9	5	2	7	3	1	4
3	5	4	9	1	6	7	8	2
8	3	7	2	6	5	9	4	1
1	9	2	4	7	8	6	5	3
4	6	5	1	3	9	8	2	7

VERY HARD - 54

5	6	9	1	2	3	8	7	4
1	3	7	5	4	8	2	6	9
2	8	4	7	6	9	1	5	3
7	9	2	6	5	4	3	8	1
3	4	8	9	1	7	5	2	6
6	5	1	3	8	2	4	9	7
8	2	3	4	9	6	7	1	5
9	7	5	2	3	1	6	4	8
4	1	6	8	7	5	9	3	2

VERY HARD - 55

1	6	7	4	8	9	5	3	2
3	5	2	6	7	1	8	4	9
9	4	8	3	2	5	7	1	6
2	7	1	9	6	4	3	8	5
8	9	5	2	3	7	4	6	1
4	3	6	1	5	8	2	9	7
6	8	3	5	1	2	9	7	4
5	1	4	7	9	3	6	2	8
7	2	9	8	4	6	1	5	3

VERY HARD - 56

6	1	3	8	7	2	9	4	5
9	4	8	6	1	5	7	3	2
2	5	7	9	3	4	8	6	1
7	8	1	5	9	6	4	2	3
3	2	6	7	4	1	5	8	9
5	9	4	3	2	8	1	7	6
4	7	9	2	5	3	6	1	8
1	6	2	4	8	9	3	5	7
8	3	5	1	6	7	2	9	4

VERY HARD - 57

8	9	4	5	3	7	6	2	1
7	2	3	9	1	6	4	5	8
6	1	5	4	8	2	7	9	3
9	5	8	3	6	4	2	1	7
3	4	7	2	5	1	9	8	6
1	6	2	8	7	9	5	3	4
4	7	9	1	2	3	8	6	5
5	3	6	7	9	8	1	4	2
2	8	1	6	4	5	3	7	9

VERY HARD - 58

6	1	5	7	4	3	8	9	2
2	8	7	6	1	9	4	3	5
4	9	3	8	2	5	1	7	6
1	5	4	3	8	2	7	6	9
7	2	8	9	5	6	3	4	1
9	3	6	4	7	1	2	5	8
8	4	9	1	6	7	5	2	3
5	6	1	2	3	4	9	8	7
3	7	2	5	9	8	6	1	4

VERY HARD - 59

8	7	5	2	3	4	1	6	9
3	2	4	1	6	9	7	5	8
9	1	6	8	7	5	4	3	2
2	6	1	5	4	3	8	9	7
7	5	3	9	1	8	2	4	6
4	8	9	6	2	7	5	1	3
1	4	7	3	9	2	6	8	5
6	3	8	7	5	1	9	2	4
5	9	2	4	8	6	3	7	1

VERY HARD - 60

2	7	6	9	8	3	1	5	4
5	8	1	7	4	2	3	9	6
3	4	9	5	1	6	2	7	8
8	1	3	2	5	7	6	4	9
6	5	4	1	3	9	8	2	7
9	2	7	8	6	4	5	3	1
1	6	2	4	7	5	9	8	3
7	3	5	6	9	8	4	1	2
4	9	8	3	2	1	7	6	5

VERY HARD - 61

1	3	9	5	6	4	8	2	7
8	7	4	2	1	9	6	5	3
6	2	5	3	7	8	1	4	9
9	8	2	4	3	5	7	1	6
3	1	7	8	2	6	5	9	4
5	4	6	7	9	1	2	3	8
7	9	1	6	4	2	3	8	5
2	6	8	9	5	3	4	7	1
4	5	3	1	8	7	9	6	2

VERY HARD - 62

5	6	8	4	7	2	9	1	3
4	7	9	8	1	3	6	5	2
2	1	3	6	5	9	4	7	8
8	9	6	5	2	7	1	3	4
3	4	5	9	8	1	7	2	6
7	2	1	3	4	6	8	9	5
9	8	7	2	6	5	3	4	1
1	5	4	7	3	8	2	6	9
6	3	2	1	9	4	5	8	7

VERY HARD - 63

3	2	6	1	5	8	7	9	4
7	1	5	6	9	4	2	3	8
9	8	4	3	7	2	5	1	6
6	4	2	9	1	5	8	7	3
5	3	9	8	4	7	1	6	2
8	7	1	2	6	3	9	4	5
2	9	3	4	8	1	6	5	7
4	6	7	5	2	9	3	8	1
1	5	8	7	3	6	4	2	9

VERY HARD - 64

5	4	7	9	2	1	6	8	3
9	3	6	5	7	8	4	1	2
2	1	8	6	3	4	5	9	7
8	2	3	7	9	6	1	5	4
1	7	9	3	4	5	2	6	8
4	6	5	1	8	2	7	3	9
6	9	2	8	1	7	3	4	5
3	5	4	2	6	9	8	7	1
7	8	1	4	5	3	9	2	6

VERY HARD - 65

8	3	4	2	1	9	7	5	6
1	9	2	6	5	7	3	4	8
6	5	7	3	8	4	1	9	2
2	6	9	1	4	5	8	7	3
5	8	1	7	9	3	2	6	4
7	4	3	8	2	6	5	1	9
4	2	5	9	7	8	6	3	1
3	7	8	4	6	1	9	2	5
9	1	6	5	3	2	4	8	7

VERY HARD - 66

2	8	3	1	4	5	9	6	7
4	5	9	2	6	7	3	1	8
1	7	6	8	3	9	5	4	2
3	4	7	5	1	6	8	2	9
8	6	2	3	9	4	7	5	1
9	1	5	7	8	2	4	3	6
5	2	8	6	7	3	1	9	4
6	9	1	4	5	8	2	7	3
7	3	4	9	2	1	6	8	5

VERY HARD - 67

6	3	4	7	5	8	2	9	1
5	2	8	3	1	9	6	4	7
7	1	9	2	4	6	3	8	5
2	4	6	1	7	3	9	5	8
9	8	7	6	2	5	4	1	3
3	5	1	9	8	4	7	6	2
8	7	3	4	6	1	5	2	9
4	9	5	8	3	2	1	7	6
1	6	2	5	9	7	8	3	4

VERY HARD - 68

7	6	3	8	9	1	4	2	5
1	8	5	6	4	2	7	3	9
4	9	2	7	5	3	8	6	1
8	2	9	1	3	4	5	7	6
3	4	6	5	7	8	1	9	2
5	1	7	2	6	9	3	8	4
9	5	4	3	2	7	6	1	8
6	7	8	9	1	5	2	4	3
2	3	1	4	8	6	9	5	7

VERY HARD - 69

9	3	8	2	4	1	5	6	7
1	7	4	3	6	5	9	8	2
6	5	2	7	9	8	3	1	4
8	2	6	1	7	9	4	5	3
7	4	5	8	2	3	6	9	1
3	9	1	4	5	6	2	7	8
2	6	3	5	8	7	1	4	9
5	1	7	9	3	4	8	2	6
4	8	9	6	1	2	7	3	5

VERY HARD - 70

3	4	2	1	7	9	5	8	6
7	5	9	4	8	6	1	3	2
8	1	6	5	3	2	9	4	7
2	3	8	7	4	5	6	1	9
4	7	5	6	9	1	3	2	8
9	6	1	8	2	3	7	5	4
5	9	3	2	6	4	8	7	1
1	8	4	9	5	7	2	6	3
6	2	7	3	1	8	4	9	5

VERY HARD - 71

2	9	1	5	8	3	7	6	4
4	5	3	9	6	7	8	2	1
8	7	6	2	4	1	9	5	3
5	2	8	3	1	4	6	7	9
9	1	4	6	7	5	3	8	2
3	6	7	8	2	9	4	1	5
7	4	5	1	9	8	2	3	6
1	8	2	4	3	6	5	9	7
6	3	9	7	5	2	1	4	8

VERY HARD - 72

2	5	3	9	8	4	1	6	7
4	8	9	1	7	6	5	3	2
6	7	1	2	5	3	9	8	4
5	4	7	8	6	2	3	9	1
9	2	6	5	3	1	7	4	8
1	3	8	4	9	7	6	2	5
3	1	5	6	2	8	4	7	9
7	9	2	3	4	5	8	1	6
8	6	4	7	1	9	2	5	3

VERY HARD - 73

5	9	2	4	3	6	7	8	1
3	4	8	7	9	1	5	6	2
7	1	6	8	5	2	4	9	3
6	2	7	1	8	3	9	4	5
4	3	5	9	2	7	8	1	6
9	8	1	5	6	4	2	3	7
2	5	9	6	1	8	3	7	4
1	7	3	2	4	9	6	5	8
8	6	4	3	7	5	1	2	9

VERY HARD - 74

2	5	6	8	3	7	4	9	1
8	3	4	5	9	1	6	2	7
1	9	7	2	6	4	5	3	8
5	7	9	4	2	6	8	1	3
6	1	2	9	8	3	7	5	4
4	8	3	7	1	5	9	6	2
7	2	8	1	5	9	3	4	6
9	6	1	3	4	8	2	7	5
3	4	5	6	7	2	1	8	9

VERY HARD - 75

4	1	8	3	9	2	6	7	5
9	7	3	4	5	6	1	8	2
6	5	2	8	7	1	3	4	9
5	8	1	9	3	4	2	6	7
2	3	4	7	6	5	9	1	8
7	6	9	1	2	8	4	5	3
1	4	5	2	8	9	7	3	6
8	9	7	6	1	3	5	2	4
3	2	6	5	4	7	8	9	1

VERY HARD - 76

5	2	4	7	9	3	1	8	6
3	1	9	6	2	8	5	4	7
6	8	7	5	1	4	2	3	9
9	4	3	1	5	2	7	6	8
8	5	6	9	4	7	3	1	2
2	7	1	3	8	6	4	9	5
1	6	2	4	7	9	8	5	3
7	9	5	8	3	1	6	2	4
4	3	8	2	6	5	9	7	1

VERY HARD - 77

8	1	2	4	7	5	9	3	6
4	9	7	8	6	3	1	5	2
5	3	6	2	1	9	7	4	8
9	2	1	7	4	8	5	6	3
6	7	4	3	5	1	8	2	9
3	8	5	6	9	2	4	7	1
7	6	9	1	3	4	2	8	5
2	5	3	9	8	7	6	1	4
1	4	8	5	2	6	3	9	7

VERY HARD - 78

4	8	6	5	2	1	9	7	3
2	3	1	7	8	9	4	5	6
5	7	9	4	6	3	1	2	8
3	6	2	9	7	8	5	4	1
9	4	8	3	1	5	2	6	7
1	5	7	2	4	6	3	8	9
8	9	4	6	3	2	7	1	5
6	2	5	1	9	7	8	3	4
7	1	3	8	5	4	6	9	2

VERY HARD - 79

4	6	7	9	3	8	5	1	2
3	8	1	4	5	2	7	9	6
2	9	5	6	7	1	8	3	4
5	2	4	3	9	7	6	8	1
9	7	3	1	8	6	4	2	5
8	1	6	2	4	5	3	7	9
6	4	8	7	2	9	1	5	3
7	3	9	5	1	4	2	6	8
1	5	2	8	6	3	9	4	7

VERY HARD - 80

2	4	6	9	3	1	7	8	5
7	8	1	4	5	2	6	3	9
3	5	9	6	7	8	4	2	1
8	2	3	1	6	7	9	5	4
9	7	4	8	2	5	3	1	6
6	1	5	3	4	9	2	7	8
4	6	2	5	8	3	1	9	7
1	3	8	7	9	4	5	6	2
5	9	7	2	1	6	8	4	3

VERY HARD - 81

2	4	1	8	5	9	6	7	3
9	3	6	1	7	4	5	2	8
8	5	7	2	6	3	1	4	9
3	1	2	6	9	8	7	5	4
7	8	5	3	4	2	9	1	6
6	9	4	5	1	7	8	3	2
4	6	9	7	2	5	3	8	1
1	7	3	4	8	6	2	9	5
5	2	8	9	3	1	4	6	7

VERY HARD - 82

9	3	4	1	2	7	6	8	5
2	5	6	3	8	9	4	7	1
8	1	7	5	6	4	9	2	3
4	8	9	7	5	2	1	3	6
3	6	5	8	4	1	2	9	7
1	7	2	6	9	3	8	5	4
6	4	8	9	7	5	3	1	2
5	2	3	4	1	8	7	6	9
7	9	1	2	3	6	5	4	8

VERY HARD - 83

5	4	6	9	3	1	2	7	8
2	3	1	7	4	8	9	6	5
9	7	8	6	2	5	3	1	4
3	5	4	2	6	9	7	8	1
8	2	7	1	5	3	4	9	6
6	1	9	8	7	4	5	2	3
4	9	2	3	8	6	1	5	7
1	6	3	5	9	7	8	4	2
7	8	5	4	1	2	6	3	9

VERY HARD - 84

5	6	9	1	7	8	4	3	2
4	8	7	2	9	3	1	5	6
1	2	3	5	4	6	7	9	8
8	3	6	7	5	1	9	2	4
2	4	1	8	3	9	6	7	5
7	9	5	6	2	4	8	1	3
3	1	4	9	6	2	5	8	7
9	5	2	4	8	7	3	6	1
6	7	8	3	1	5	2	4	9

VERY HARD - 85

6	5	7	4	3	9	2	1	8
4	8	9	6	1	2	3	7	5
1	2	3	5	7	8	6	4	9
5	7	2	1	6	3	9	8	4
3	6	8	9	5	4	7	2	1
9	4	1	8	2	7	5	3	6
2	1	5	7	4	6	8	9	3
8	3	6	2	9	1	4	5	7
7	9	4	3	8	5	1	6	2

VERY HARD - 86

5	8	3	7	1	4	2	9	6
6	9	1	8	5	2	3	4	7
4	7	2	9	3	6	1	5	8
2	3	4	1	7	9	8	6	5
8	5	9	4	6	3	7	2	1
7	1	6	2	8	5	9	3	4
9	2	8	5	4	7	6	1	3
1	6	5	3	9	8	4	7	2
3	4	7	6	2	1	5	8	9

VERY HARD - 87

8	4	1	2	9	6	7	3	5
9	2	6	3	5	7	1	8	4
7	3	5	4	1	8	2	9	6
5	1	7	8	6	4	3	2	9
2	9	4	1	3	5	8	6	7
6	8	3	9	7	2	4	5	1
3	5	9	7	2	1	6	4	8
4	7	2	6	8	9	5	1	3
1	6	8	5	4	3	9	7	2

VERY HARD - 88

7	4	5	9	6	8	3	2	1
9	2	8	3	1	7	5	6	4
6	1	3	5	2	4	7	9	8
3	8	1	4	5	6	2	7	9
2	9	6	8	7	3	4	1	5
4	5	7	2	9	1	6	8	3
5	3	2	7	8	9	1	4	6
8	6	4	1	3	2	9	5	7
1	7	9	6	4	5	8	3	2

VERY HARD - 89

4	9	2	1	7	5	6	3	8
6	1	5	9	8	3	2	7	4
7	8	3	2	6	4	5	1	9
8	7	1	5	3	9	4	6	2
5	3	6	4	2	7	9	8	1
2	4	9	8	1	6	3	5	7
9	5	8	6	4	1	7	2	3
1	6	7	3	9	2	8	4	5
3	2	4	7	5	8	1	9	6

VERY HARD - 90

6	4	5	8	2	7	9	3	1
8	3	2	1	9	5	4	6	7
7	9	1	4	6	3	5	2	8
9	2	4	3	8	1	6	7	5
1	7	6	9	5	2	8	4	3
3	5	8	7	4	6	1	9	2
2	8	3	6	1	9	7	5	4
5	1	9	2	7	4	3	8	6
4	6	7	5	3	8	2	1	9

VERY HARD - 91

5	1	3	7	4	2	9	8	6
6	7	8	9	5	1	2	3	4
9	2	4	8	3	6	1	7	5
2	6	5	1	8	3	7	4	9
7	8	9	5	6	4	3	2	1
4	3	1	2	7	9	6	5	8
1	9	7	4	2	5	8	6	3
3	5	2	6	9	8	4	1	7
8	4	6	3	1	7	5	9	2

VERY HARD - 92

4	1	5	3	2	7	6	8	9
9	7	3	5	8	6	1	4	2
2	6	8	4	1	9	5	7	3
6	8	9	7	3	4	2	1	5
1	5	4	9	6	2	7	3	8
7	3	2	8	5	1	4	9	6
3	2	6	1	4	8	9	5	7
8	9	1	2	7	5	3	6	4
5	4	7	6	9	3	8	2	1

VERY HARD - 93

3	4	8	9	6	5	7	2	1
1	6	7	2	3	8	4	5	9
5	2	9	1	4	7	6	8	3
7	5	1	6	9	3	2	4	8
4	8	2	5	7	1	9	3	6
6	9	3	8	2	4	5	1	7
9	3	5	4	8	6	1	7	2
8	1	6	7	5	2	3	9	4
2	7	4	3	1	9	8	6	5

VERY HARD - 94

8	6	1	5	2	3	7	9	4
7	2	5	4	6	9	3	1	8
4	3	9	1	7	8	5	2	6
6	1	7	9	3	2	8	4	5
2	4	8	6	5	7	1	3	9
5	9	3	8	1	4	6	7	2
3	7	6	2	4	5	9	8	1
9	5	4	7	8	1	2	6	3
1	8	2	3	9	6	4	5	7

VERY HARD - 95

2	4	6	9	1	8	3	7	5
9	3	7	5	2	4	8	1	6
1	5	8	7	6	3	2	4	9
7	6	3	8	9	1	4	5	2
8	2	1	6	4	5	9	3	7
5	9	4	3	7	2	6	8	1
4	7	9	1	8	6	5	2	3
6	8	5	2	3	7	1	9	4
3	1	2	4	5	9	7	6	8

VERY HARD - 96

9	2	8	4	3	1	7	6	5
3	6	1	9	7	5	4	2	8
7	4	5	2	8	6	3	1	9
8	1	4	6	9	7	5	3	2
5	3	7	1	2	8	6	9	4
6	9	2	3	5	4	1	8	7
2	8	6	7	4	3	9	5	1
1	7	9	5	6	2	8	4	3
4	5	3	8	1	9	2	7	6

VERY HARD - 97

1	9	3	2	6	7	8	4	5
8	7	4	3	5	1	9	6	2
2	5	6	8	4	9	3	1	7
9	4	8	1	2	5	6	7	3
5	1	7	4	3	6	2	9	8
6	3	2	9	7	8	1	5	4
4	8	9	5	1	3	7	2	6
7	2	1	6	8	4	5	3	9
3	6	5	7	9	2	4	8	1

VERY HARD - 98

2	8	7	5	9	1	6	4	3
4	9	6	7	3	2	5	8	1
1	5	3	4	8	6	9	7	2
7	3	5	1	6	9	8	2	4
9	1	2	8	7	4	3	6	5
8	6	4	3	2	5	7	1	9
6	2	1	9	5	7	4	3	8
3	7	9	2	4	8	1	5	6
5	4	8	6	1	3	2	9	7

VERY HARD - 99

8	1	4	7	9	5	3	6	2
3	9	2	6	4	1	7	8	5
6	7	5	8	3	2	4	1	9
1	4	3	9	8	6	5	2	7
7	2	8	1	5	4	9	3	6
9	5	6	2	7	3	1	4	8
5	6	7	4	1	8	2	9	3
4	8	9	3	2	7	6	5	1
2	3	1	5	6	9	8	7	4

VERY HARD - 100

1	5	8	6	4	2	9	3	7
3	4	6	7	9	5	2	1	8
7	2	9	3	1	8	6	5	4
2	7	4	8	5	9	3	6	1
8	9	1	4	3	6	5	7	2
6	3	5	2	7	1	4	8	9
4	8	3	5	2	7	1	9	6
5	1	7	9	6	4	8	2	3
9	6	2	1	8	3	7	4	5

VERY HARD - 101

3	6	4	2	5	9	8	7	1
5	8	1	7	3	6	2	4	9
2	7	9	1	4	8	5	6	3
6	3	5	8	2	1	4	9	7
1	4	7	9	6	5	3	8	2
9	2	8	3	7	4	6	1	5
7	9	6	5	8	2	1	3	4
4	5	3	6	1	7	9	2	8
8	1	2	4	9	3	7	5	6

VERY HARD - 102

7	9	1	2	5	8	4	3	6
8	5	2	3	4	6	9	7	1
3	4	6	7	1	9	2	8	5
5	8	7	1	2	3	6	4	9
4	2	9	5	6	7	3	1	8
6	1	3	9	8	4	5	2	7
9	6	8	4	3	1	7	5	2
2	7	4	8	9	5	1	6	3
1	3	5	6	7	2	8	9	4

VERY HARD - 103

3	1	5	4	8	7	9	2	6
9	7	6	5	2	1	4	3	8
4	8	2	6	3	9	7	1	5
7	2	3	9	6	8	5	4	1
5	6	4	1	7	2	8	9	3
8	9	1	3	4	5	6	7	2
2	3	7	8	5	4	1	6	9
6	5	9	7	1	3	2	8	4
1	4	8	2	9	6	3	5	7

VERY HARD - 104

9	6	5	2	4	7	8	3	1
7	2	8	1	3	6	9	4	5
1	4	3	8	9	5	6	2	7
5	8	7	3	2	1	4	9	6
3	1	6	9	5	4	7	8	2
2	9	4	6	7	8	5	1	3
8	3	9	7	6	2	1	5	4
6	5	1	4	8	3	2	7	9
4	7	2	5	1	9	3	6	8

VERY HARD - 105

9	3	2	8	6	4	5	1	7
1	5	4	7	9	2	6	8	3
6	8	7	1	5	3	9	4	2
8	2	3	5	7	9	4	6	1
5	4	6	2	1	8	3	7	9
7	1	9	4	3	6	8	2	5
3	9	8	6	2	7	1	5	4
4	7	1	3	8	5	2	9	6
2	6	5	9	4	1	7	3	8

VERY HARD - 106

6	1	9	4	8	3	2	7	5
8	2	3	5	9	7	4	6	1
4	5	7	2	1	6	9	3	8
2	4	5	3	7	8	1	9	6
7	8	6	9	2	1	5	4	3
3	9	1	6	5	4	7	8	2
1	7	2	8	3	9	6	5	4
9	3	4	1	6	5	8	2	7
5	6	8	7	4	2	3	1	9

VERY HARD - 107

7	3	8	5	9	4	2	1	6
1	5	9	2	6	8	4	7	3
4	6	2	3	1	7	8	5	9
6	2	4	9	7	1	3	8	5
5	1	3	8	2	6	9	4	7
8	9	7	4	5	3	6	2	1
9	4	5	1	3	2	7	6	8
2	7	1	6	8	9	5	3	4
3	8	6	7	4	5	1	9	2

VERY HARD - 108

8	7	9	6	1	2	5	4	3
4	2	6	7	5	3	9	1	8
1	3	5	8	9	4	2	7	6
2	4	3	5	8	1	7	6	9
6	1	7	2	3	9	8	5	4
9	5	8	4	7	6	1	3	2
3	9	4	1	2	7	6	8	5
5	6	1	9	4	8	3	2	7
7	8	2	3	6	5	4	9	1

VERY HARD - 109

7	6	1	4	2	8	5	9	3
3	2	4	7	5	9	6	8	1
9	8	5	3	1	6	7	4	2
8	9	3	2	6	5	1	7	4
1	5	6	8	4	7	3	2	9
4	7	2	9	3	1	8	5	6
2	1	7	5	9	3	4	6	8
6	4	8	1	7	2	9	3	5
5	3	9	6	8	4	2	1	7

VERY HARD - 110

2	4	6	3	1	7	9	5	8
5	3	8	6	4	9	2	1	7
7	1	9	5	2	8	3	6	4
3	9	4	7	8	6	1	2	5
6	2	7	9	5	1	4	8	3
8	5	1	2	3	4	7	9	6
9	6	5	4	7	2	8	3	1
1	7	3	8	9	5	6	4	2
4	8	2	1	6	3	5	7	9

VERY HARD - 111

8	1	2	4	6	9	5	7	3
4	5	3	7	8	1	6	9	2
7	9	6	3	5	2	4	1	8
1	3	4	5	7	6	8	2	9
2	7	8	1	9	4	3	5	6
9	6	5	2	3	8	7	4	1
5	4	9	8	1	3	2	6	7
6	8	7	9	2	5	1	3	4
3	2	1	6	4	7	9	8	5

VERY HARD - 112

4	3	1	5	2	6	8	7	9
2	8	9	1	3	7	6	4	5
7	6	5	9	8	4	2	1	3
6	2	8	7	9	5	4	3	1
9	7	4	3	1	2	5	6	8
5	1	3	4	6	8	7	9	2
3	5	2	6	4	9	1	8	7
8	9	6	2	7	1	3	5	4
1	4	7	8	5	3	9	2	6

VERY HARD - 113

6	5	8	9	7	3	4	2	1
1	7	2	5	8	4	9	3	6
3	4	9	1	2	6	8	7	5
5	2	3	4	6	7	1	9	8
8	6	4	3	1	9	7	5	2
7	9	1	8	5	2	6	4	3
4	3	6	2	9	1	5	8	7
9	8	7	6	3	5	2	1	4
2	1	5	7	4	8	3	6	9

VERY HARD - 114

2	1	4	3	7	8	9	6	5
9	8	6	5	4	2	1	3	7
3	7	5	9	1	6	8	4	2
1	3	8	4	6	5	2	7	9
7	6	2	8	3	9	4	5	1
5	4	9	1	2	7	3	8	6
6	9	1	7	8	4	5	2	3
4	5	7	2	9	3	6	1	8
8	2	3	6	5	1	7	9	4

VERY HARD - 115

6	5	7	1	4	2	3	8	9
8	4	1	9	3	5	6	2	7
3	2	9	6	8	7	1	4	5
7	1	5	3	9	4	8	6	2
9	8	6	5	2	1	7	3	4
4	3	2	8	7	6	5	9	1
1	9	4	7	6	3	2	5	8
5	6	8	2	1	9	4	7	3
2	7	3	4	5	8	9	1	6

VERY HARD - 116

1	6	8	7	9	4	2	5	3
4	7	2	8	3	5	1	6	9
9	5	3	1	2	6	4	8	7
2	9	1	5	4	3	8	7	6
8	3	7	6	1	2	5	9	4
5	4	6	9	7	8	3	2	1
7	2	5	3	6	1	9	4	8
3	8	9	4	5	7	6	1	2
6	1	4	2	8	9	7	3	5

VERY HARD - 117

1	5	6	8	7	9	4	3	2
9	7	8	2	3	4	6	5	1
3	4	2	6	1	5	8	7	9
6	9	1	4	8	7	5	2	3
7	8	4	5	2	3	1	9	6
2	3	5	1	9	6	7	4	8
8	6	3	7	5	2	9	1	4
5	1	9	3	4	8	2	6	7
4	2	7	9	6	1	3	8	5

VERY HARD - 118

3	1	9	2	6	7	8	5	4
5	8	7	1	3	4	9	2	6
4	2	6	8	9	5	1	3	7
6	5	8	7	4	9	2	1	3
2	4	3	6	1	8	5	7	9
7	9	1	3	5	2	6	4	8
8	7	4	5	2	6	3	9	1
1	6	2	9	7	3	4	8	5
9	3	5	4	8	1	7	6	2

VERY HARD - 119

4	5	7	2	1	6	3	8	9
3	2	1	4	9	8	6	7	5
6	8	9	7	5	3	2	1	4
5	9	8	1	4	2	7	6	3
1	4	3	5	6	7	8	9	2
7	6	2	8	3	9	4	5	1
2	7	5	3	8	1	9	4	6
9	3	4	6	7	5	1	2	8
8	1	6	9	2	4	5	3	7

VERY HARD - 120

6	4	5	9	8	7	1	3	2
8	3	7	6	2	1	4	5	9
9	2	1	3	4	5	6	7	8
4	9	3	2	5	6	7	8	1
5	7	8	1	9	4	2	6	3
1	6	2	8	7	3	5	9	4
7	5	9	4	3	2	8	1	6
3	1	4	5	6	8	9	2	7
2	8	6	7	1	9	3	4	5

VERY HARD - 121

9	8	7	6	5	2	4	3	1
2	1	3	9	8	4	6	5	7
5	4	6	3	7	1	2	9	8
7	3	2	1	6	5	9	8	4
1	6	5	4	9	8	3	7	2
4	9	8	2	3	7	1	6	5
3	5	4	7	2	6	8	1	9
8	2	9	5	1	3	7	4	6
6	7	1	8	4	9	5	2	3

VERY HARD - 122

4	1	9	7	2	3	8	6	5
6	2	7	1	5	8	9	4	3
5	3	8	6	9	4	1	7	2
9	8	2	3	7	1	6	5	4
7	4	5	2	6	9	3	8	1
3	6	1	8	4	5	2	9	7
2	5	6	9	3	7	4	1	8
8	9	4	5	1	2	7	3	6
1	7	3	4	8	6	5	2	9

VERY HARD - 123

7	4	5	9	1	8	3	6	2
3	2	6	4	7	5	9	8	1
1	8	9	3	6	2	7	5	4
8	6	1	5	2	9	4	3	7
9	7	3	1	4	6	5	2	8
2	5	4	8	3	7	6	1	9
4	9	2	6	5	1	8	7	3
6	1	8	7	9	3	2	4	5
5	3	7	2	8	4	1	9	6

VERY HARD - 124

4	3	7	8	1	9	2	6	5
5	6	2	3	7	4	9	1	8
9	1	8	5	6	2	4	7	3
6	2	3	1	8	7	5	4	9
7	4	1	2	9	5	3	8	6
8	5	9	6	4	3	7	2	1
3	7	6	4	5	1	8	9	2
2	8	4	9	3	6	1	5	7
1	9	5	7	2	8	6	3	4

VERY HARD - 125

7	8	6	3	4	5	9	1	2
1	4	5	8	9	2	6	7	3
3	9	2	6	7	1	8	5	4
9	6	8	5	1	4	3	2	7
4	2	3	9	8	7	1	6	5
5	7	1	2	3	6	4	8	9
2	5	4	1	6	3	7	9	8
6	3	9	7	5	8	2	4	1
8	1	7	4	2	9	5	3	6

VERY HARD - 126

9	1	3	5	2	4	7	6	8
6	2	5	9	7	8	4	1	3
7	8	4	3	6	1	9	5	2
8	5	2	7	1	9	3	4	6
3	4	7	6	5	2	1	8	9
1	9	6	8	4	3	5	2	7
2	6	9	1	3	5	8	7	4
4	3	1	2	8	7	6	9	5
5	7	8	4	9	6	2	3	1

VERY HARD - 127

5	6	8	7	3	9	4	1	2
2	4	9	8	5	1	3	6	7
1	3	7	4	2	6	9	8	5
7	5	2	3	4	8	1	9	6
3	9	1	6	7	5	8	2	4
4	8	6	1	9	2	7	5	3
8	7	5	2	1	4	6	3	9
9	1	3	5	6	7	2	4	8
6	2	4	9	8	3	5	7	1

VERY HARD - 128

3	5	8	2	6	9	4	1	7
4	9	1	7	8	5	2	6	3
6	2	7	3	1	4	8	5	9
2	7	4	5	9	1	3	8	6
5	3	9	8	7	6	1	2	4
8	1	6	4	2	3	7	9	5
9	4	5	1	3	2	6	7	8
7	6	2	9	4	8	5	3	1
1	8	3	6	5	7	9	4	2

VERY HARD - 129

7	1	3	6	9	4	8	5	2
5	9	4	3	2	8	1	6	7
8	2	6	5	7	1	3	4	9
4	7	9	1	6	5	2	3	8
1	6	8	4	3	2	7	9	5
3	5	2	7	8	9	4	1	6
9	3	5	2	1	7	6	8	4
6	4	7	8	5	3	9	2	1
2	8	1	9	4	6	5	7	3

VERY HARD - 130

7	4	2	6	8	3	1	9	5
3	6	5	9	4	1	2	8	7
9	1	8	7	2	5	4	6	3
6	2	1	3	9	4	5	7	8
8	3	4	1	5	7	9	2	6
5	7	9	8	6	2	3	1	4
1	8	6	5	3	9	7	4	2
4	5	7	2	1	6	8	3	9
2	9	3	4	7	8	6	5	1

VERY HARD - 131

3	4	2	7	8	6	5	9	1
1	7	5	3	9	4	8	2	6
8	9	6	2	5	1	3	7	4
5	6	8	1	7	2	4	3	9
7	2	1	9	4	3	6	8	5
9	3	4	5	6	8	2	1	7
4	5	7	8	3	9	1	6	2
2	8	9	6	1	5	7	4	3
6	1	3	4	2	7	9	5	8

VERY HARD - 132

1	2	6	7	8	5	9	3	4
9	7	4	6	1	3	8	2	5
3	5	8	9	2	4	1	7	6
5	3	1	2	9	6	4	8	7
4	6	9	3	7	8	5	1	2
2	8	7	5	4	1	3	6	9
6	9	5	1	3	7	2	4	8
8	1	2	4	6	9	7	5	3
7	4	3	8	5	2	6	9	1

VERY HARD - 133

1	8	2	4	9	6	7	3	5
9	6	3	5	7	1	8	2	4
7	4	5	2	3	8	1	6	9
2	9	1	3	6	4	5	8	7
3	7	4	1	8	5	2	9	6
6	5	8	7	2	9	4	1	3
8	3	7	6	5	2	9	4	1
5	1	9	8	4	3	6	7	2
4	2	6	9	1	7	3	5	8

VERY HARD - 134

5	6	9	4	7	3	1	8	2
1	3	8	6	5	2	9	7	4
4	7	2	8	9	1	6	3	5
2	9	7	1	3	5	8	4	6
8	1	3	7	4	6	5	2	9
6	5	4	2	8	9	7	1	3
7	2	1	5	6	4	3	9	8
9	8	6	3	2	7	4	5	1
3	4	5	9	1	8	2	6	7

VERY HARD - 135

4	1	9	8	6	7	3	5	2
8	2	3	9	4	5	7	1	6
6	7	5	3	1	2	9	4	8
7	6	8	2	9	4	1	3	5
9	3	2	5	7	1	8	6	4
5	4	1	6	8	3	2	7	9
1	5	6	7	2	9	4	8	3
2	8	4	1	3	6	5	9	7
3	9	7	4	5	8	6	2	1

VERY HARD - 136

2	4	9	3	6	5	7	8	1
1	5	6	4	7	8	3	9	2
3	7	8	2	9	1	4	6	5
7	6	3	9	4	2	5	1	8
9	1	4	8	5	6	2	7	3
5	8	2	7	1	3	6	4	9
6	3	7	1	2	9	8	5	4
8	9	5	6	3	4	1	2	7
4	2	1	5	8	7	9	3	6

VERY HARD - 137

1	2	6	7	5	3	8	9	4
9	4	5	2	8	1	7	6	3
7	3	8	4	6	9	5	1	2
4	1	2	9	3	5	6	8	7
3	6	7	8	1	2	9	4	5
8	5	9	6	7	4	3	2	1
2	8	3	5	4	6	1	7	9
5	7	4	1	9	8	2	3	6
6	9	1	3	2	7	4	5	8

VERY HARD - 138

1	4	8	2	5	6	7	3	9
5	3	9	8	1	7	2	4	6
7	6	2	9	3	4	5	8	1
2	1	5	7	8	9	3	6	4
8	9	3	4	6	5	1	2	7
6	7	4	1	2	3	8	9	5
3	2	7	6	9	1	4	5	8
9	5	1	3	4	8	6	7	2
4	8	6	5	7	2	9	1	3

VERY HARD - 139

6	8	3	2	7	4	9	1	5
7	2	5	1	3	9	6	4	8
9	4	1	6	8	5	3	2	7
5	7	2	8	1	3	4	9	6
8	9	6	5	4	2	1	7	3
3	1	4	7	9	6	8	5	2
4	6	9	3	5	7	2	8	1
1	3	7	9	2	8	5	6	4
2	5	8	4	6	1	7	3	9

VERY HARD - 140

5	7	9	3	6	8	4	1	2
6	8	3	4	1	2	5	7	9
2	1	4	5	7	9	3	6	8
8	2	6	9	4	5	1	3	7
9	3	7	1	2	6	8	4	5
1	4	5	7	8	3	9	2	6
3	9	2	6	5	4	7	8	1
4	6	1	8	9	7	2	5	3
7	5	8	2	3	1	6	9	4

VERY HARD - 141

3	6	7	5	1	4	9	8	2
8	9	2	7	3	6	5	1	4
1	4	5	8	9	2	6	7	3
5	1	4	2	8	9	3	6	7
6	2	3	4	5	7	8	9	1
9	7	8	3	6	1	4	2	5
7	5	1	6	4	8	2	3	9
2	3	6	9	7	5	1	4	8
4	8	9	1	2	3	7	5	6

VERY HARD - 142

1	4	9	3	7	2	6	5	8
8	2	5	9	1	6	4	7	3
3	6	7	5	4	8	9	1	2
7	3	6	4	5	9	8	2	1
9	8	2	6	3	1	7	4	5
4	5	1	2	8	7	3	6	9
6	9	4	8	2	5	1	3	7
5	1	8	7	6	3	2	9	4
2	7	3	1	9	4	5	8	6

VERY HARD - 143

1	9	2	5	8	4	3	7	6
7	8	5	6	1	3	9	4	2
4	6	3	7	9	2	8	5	1
2	1	8	3	6	5	7	9	4
9	5	6	4	7	1	2	8	3
3	7	4	8	2	9	6	1	5
6	3	9	1	5	8	4	2	7
5	2	7	9	4	6	1	3	8
8	4	1	2	3	7	5	6	9

VERY HARD - 144

8	3	4	2	6	7	5	1	9
7	5	1	3	8	9	4	6	2
9	6	2	5	1	4	3	7	8
4	9	6	1	5	3	2	8	7
1	8	7	4	2	6	9	5	3
3	2	5	7	9	8	1	4	6
6	7	3	9	4	5	8	2	1
2	4	9	8	7	1	6	3	5
5	1	8	6	3	2	7	9	4

VERY HARD - 145

4	8	3	9	1	2	6	7	5
9	6	2	5	8	7	1	4	3
5	7	1	3	4	6	9	2	8
8	3	9	6	7	4	5	1	2
1	5	6	2	3	9	7	8	4
7	2	4	1	5	8	3	9	6
3	9	8	7	2	5	4	6	1
6	4	5	8	9	1	2	3	7
2	1	7	4	6	3	8	5	9

VERY HARD - 146

6	9	1	8	2	3	4	7	5
3	8	7	6	5	4	9	1	2
5	2	4	7	1	9	6	3	8
8	5	2	9	3	1	7	6	4
1	3	6	4	7	8	2	5	9
4	7	9	5	6	2	3	8	1
9	1	3	2	8	7	5	4	6
2	6	8	3	4	5	1	9	7
7	4	5	1	9	6	8	2	3

VERY HARD - 147

1	7	4	5	8	9	6	2	3
6	3	8	4	7	2	1	9	5
2	9	5	6	3	1	4	7	8
9	5	6	8	4	7	3	1	2
4	1	7	9	2	3	8	5	6
8	2	3	1	5	6	7	4	9
5	4	9	7	6	8	2	3	1
7	8	2	3	1	5	9	6	4
3	6	1	2	9	4	5	8	7

VERY HARD - 148

1	6	7	3	8	9	5	2	4
9	2	5	6	4	7	1	8	3
4	8	3	2	5	1	9	7	6
7	9	6	8	3	4	2	5	1
5	4	8	9	1	2	6	3	7
2	3	1	7	6	5	8	4	9
6	5	4	1	7	8	3	9	2
8	1	2	4	9	3	7	6	5
3	7	9	5	2	6	4	1	8

VERY HARD - 149

8	3	6	9	2	5	1	4	7
4	9	1	8	6	7	2	3	5
5	7	2	3	4	1	8	6	9
7	2	5	6	8	3	4	9	1
9	6	8	2	1	4	7	5	3
1	4	3	5	7	9	6	2	8
2	1	9	7	5	6	3	8	4
6	5	4	1	3	8	9	7	2
3	8	7	4	9	2	5	1	6

VERY HARD - 150

4	6	2	5	9	1	3	7	8
5	8	3	2	7	6	1	4	9
9	7	1	3	8	4	5	6	2
2	3	8	4	1	5	6	9	7
7	5	9	6	2	8	4	3	1
1	4	6	7	3	9	8	2	5
3	1	5	9	4	2	7	8	6
6	2	7	8	5	3	9	1	4
8	9	4	1	6	7	2	5	3

VERY HARD - 151

6	2	8	7	4	3	5	1	9
5	1	4	6	2	9	8	7	3
7	9	3	1	5	8	6	4	2
8	6	1	4	3	7	2	9	5
4	5	7	9	6	2	1	3	8
2	3	9	8	1	5	4	6	7
3	4	5	2	9	1	7	8	6
1	7	2	3	8	6	9	5	4
9	8	6	5	7	4	3	2	1

VERY HARD - 152

6	7	1	9	8	4	3	2	5
8	2	3	7	1	5	9	6	4
9	4	5	6	3	2	8	7	1
5	6	9	2	4	7	1	3	8
7	1	8	5	9	3	6	4	2
4	3	2	1	6	8	5	9	7
2	9	4	8	5	6	7	1	3
1	5	7	3	2	9	4	8	6
3	8	6	4	7	1	2	5	9

VERY HARD - 153

7	3	2	5	4	1	6	9	8
6	4	5	8	9	7	1	2	3
1	8	9	2	6	3	4	7	5
4	1	6	7	3	9	5	8	2
3	2	7	6	8	5	9	1	4
9	5	8	1	2	4	3	6	7
5	9	1	3	7	8	2	4	6
8	6	3	4	1	2	7	5	9
2	7	4	9	5	6	8	3	1

VERY HARD - 154

8	9	7	3	4	1	2	6	5
5	2	6	9	8	7	1	4	3
4	1	3	2	6	5	7	8	9
2	5	1	4	3	8	6	9	7
7	4	9	1	2	6	3	5	8
6	3	8	7	5	9	4	1	2
3	6	4	8	9	2	5	7	1
1	8	5	6	7	3	9	2	4
9	7	2	5	1	4	8	3	6

VERY HARD - 155

4	8	1	9	3	7	2	6	5
6	9	2	4	1	5	7	8	3
5	3	7	6	2	8	9	4	1
7	2	4	3	6	9	5	1	8
9	1	5	8	7	4	3	2	6
8	6	3	2	5	1	4	7	9
1	7	6	5	9	2	8	3	4
3	4	9	7	8	6	1	5	2
2	5	8	1	4	3	6	9	7

VERY HARD - 156

2	8	3	4	5	9	7	6	1
5	4	1	7	8	6	3	9	2
9	7	6	2	1	3	4	5	8
1	5	7	8	4	2	9	3	6
4	2	9	6	3	7	1	8	5
6	3	8	1	9	5	2	4	7
3	6	4	5	2	1	8	7	9
7	9	2	3	6	8	5	1	4
8	1	5	9	7	4	6	2	3

VERY HARD - 157

7	6	3	8	5	9	4	2	1
8	1	2	4	3	7	9	6	5
5	4	9	6	1	2	3	7	8
2	3	7	9	4	1	8	5	6
4	5	6	7	2	8	1	9	3
9	8	1	3	6	5	2	4	7
3	7	8	5	9	4	6	1	2
1	9	5	2	8	6	7	3	4
6	2	4	1	7	3	5	8	9

VERY HARD - 158

5	2	6	1	4	8	3	7	9
7	3	4	6	2	9	8	1	5
8	9	1	5	7	3	6	2	4
9	5	3	7	1	6	4	8	2
2	4	7	9	8	5	1	3	6
1	6	8	4	3	2	5	9	7
4	1	9	8	6	7	2	5	3
3	8	5	2	9	4	7	6	1
6	7	2	3	5	1	9	4	8

VERY HARD - 159

2	9	5	7	6	3	1	4	8
6	7	8	1	9	4	2	5	3
4	1	3	5	2	8	6	9	7
5	8	7	9	3	2	4	6	1
3	4	6	8	5	1	7	2	9
9	2	1	4	7	6	3	8	5
8	6	4	3	1	9	5	7	2
1	5	9	2	4	7	8	3	6
7	3	2	6	8	5	9	1	4

VERY HARD - 160

3	2	9	5	1	8	4	7	6
4	7	8	9	3	6	5	1	2
5	1	6	2	4	7	3	8	9
9	5	1	4	2	3	7	6	8
2	3	7	8	6	5	1	9	4
8	6	4	1	7	9	2	3	5
6	4	3	7	9	2	8	5	1
1	9	5	3	8	4	6	2	7
7	8	2	6	5	1	9	4	3

VERY HARD - 161

8	1	3	6	7	9	4	5	2
4	2	5	1	8	3	7	9	6
9	7	6	2	4	5	1	8	3
3	4	2	8	9	6	5	7	1
1	5	7	4	3	2	9	6	8
6	8	9	5	1	7	3	2	4
5	3	4	9	6	8	2	1	7
7	9	8	3	2	1	6	4	5
2	6	1	7	5	4	8	3	9

VERY HARD - 162

9	1	5	6	7	4	8	2	3
8	6	3	5	1	2	9	7	4
7	4	2	3	8	9	6	5	1
4	5	1	2	3	6	7	8	9
2	8	7	1	9	5	3	4	6
6	3	9	8	4	7	2	1	5
1	9	4	7	6	8	5	3	2
3	2	8	9	5	1	4	6	7
5	7	6	4	2	3	1	9	8

VERY HARD - 163

3	9	2	6	4	1	5	8	7
4	5	7	9	3	8	6	2	1
6	8	1	5	7	2	4	3	9
5	4	3	8	6	9	7	1	2
9	2	6	4	1	7	3	5	8
1	7	8	2	5	3	9	6	4
2	6	4	7	8	5	1	9	3
7	3	9	1	2	6	8	4	5
8	1	5	3	9	4	2	7	6

VERY HARD - 164

2	7	6	1	8	9	3	4	5
3	8	9	5	7	4	2	6	1
1	4	5	6	2	3	8	9	7
5	1	7	2	9	8	4	3	6
8	2	4	7	3	6	5	1	9
6	9	3	4	1	5	7	2	8
9	6	8	3	4	7	1	5	2
7	3	1	9	5	2	6	8	4
4	5	2	8	6	1	9	7	3

VERY HARD - 165

5	7	2	9	8	6	4	3	1
9	3	4	2	7	1	8	6	5
6	8	1	4	3	5	9	7	2
4	6	7	8	9	2	1	5	3
3	2	5	6	1	4	7	8	9
1	9	8	7	5	3	6	2	4
8	1	3	5	6	9	2	4	7
2	5	6	1	4	7	3	9	8
7	4	9	3	2	8	5	1	6

VERY HARD - 166

2	8	6	4	9	3	7	5	1
1	5	3	7	8	2	9	4	6
7	4	9	1	5	6	2	8	3
4	2	7	5	6	1	8	3	9
3	1	5	8	4	9	6	7	2
6	9	8	2	3	7	4	1	5
8	7	1	9	2	5	3	6	4
9	3	4	6	1	8	5	2	7
5	6	2	3	7	4	1	9	8

VERY HARD - 167

6	2	5	7	1	4	8	9	3
1	4	9	8	6	3	5	7	2
3	7	8	2	5	9	6	1	4
8	1	7	4	9	2	3	5	6
2	3	6	5	7	1	9	4	8
9	5	4	3	8	6	7	2	1
4	6	3	9	2	7	1	8	5
5	9	2	1	3	8	4	6	7
7	8	1	6	4	5	2	3	9

VERY HARD - 168

1	4	3	6	7	5	9	8	2
2	7	5	1	9	8	4	6	3
9	6	8	4	3	2	1	7	5
8	2	7	3	5	4	6	9	1
4	5	1	2	6	9	7	3	8
6	3	9	7	8	1	2	5	4
3	8	4	9	2	7	5	1	6
5	9	2	8	1	6	3	4	7
7	1	6	5	4	3	8	2	9

VERY HARD - 169

5	1	3	6	7	2	4	8	9
7	9	2	4	5	8	3	6	1
4	8	6	1	9	3	7	2	5
1	5	7	9	6	4	8	3	2
8	3	4	7	2	5	1	9	6
2	6	9	8	3	1	5	4	7
3	4	5	2	1	6	9	7	8
9	2	1	3	8	7	6	5	4
6	7	8	5	4	9	2	1	3

VERY HARD - 170

5	6	9	8	2	3	1	7	4
1	7	2	6	9	4	8	3	5
8	4	3	5	1	7	2	9	6
6	8	7	3	4	5	9	1	2
2	1	4	9	6	8	7	5	3
9	3	5	2	7	1	4	6	8
4	9	8	1	5	6	3	2	7
7	2	6	4	3	9	5	8	1
3	5	1	7	8	2	6	4	9

VERY HARD - 171

9	3	4	5	2	1	6	7	8
8	5	6	9	3	7	1	4	2
7	1	2	8	6	4	5	3	9
2	6	1	3	9	5	4	8	7
4	8	3	1	7	6	2	9	5
5	7	9	2	4	8	3	1	6
1	9	7	6	5	3	8	2	4
3	2	5	4	8	9	7	6	1
6	4	8	7	1	2	9	5	3

VERY HARD - 172

4	8	9	7	3	2	1	5	6
3	2	7	1	6	5	4	9	8
1	6	5	4	8	9	2	7	3
5	7	3	6	1	8	9	4	2
9	4	6	3	2	7	8	1	5
2	1	8	5	9	4	3	6	7
8	3	1	9	7	6	5	2	4
6	5	2	8	4	1	7	3	9
7	9	4	2	5	3	6	8	1

VERY HARD - 173

2	4	7	9	5	8	3	6	1
3	8	9	7	1	6	4	5	2
1	6	5	2	3	4	7	8	9
5	9	3	1	4	7	8	2	6
7	2	4	6	8	5	9	1	3
8	1	6	3	9	2	5	4	7
4	3	2	5	6	9	1	7	8
6	5	1	8	7	3	2	9	4
9	7	8	4	2	1	6	3	5

VERY HARD - 174

5	3	8	1	2	6	4	7	9
1	9	4	7	5	3	8	2	6
7	2	6	4	9	8	5	1	3
3	4	1	2	7	9	6	5	8
6	8	7	3	4	5	1	9	2
2	5	9	6	8	1	7	3	4
8	7	2	5	3	4	9	6	1
4	6	3	9	1	7	2	8	5
9	1	5	8	6	2	3	4	7

VERY HARD - 175

3	7	9	6	5	4	8	2	1
4	1	2	7	3	8	5	6	9
5	6	8	1	2	9	3	7	4
6	9	1	4	7	5	2	8	3
8	4	3	9	6	2	7	1	5
2	5	7	8	1	3	4	9	6
9	8	5	2	4	6	1	3	7
7	2	4	3	9	1	6	5	8
1	3	6	5	8	7	9	4	2

VERY HARD - 176

3	8	4	7	5	1	6	9	2
2	1	6	8	3	9	5	4	7
9	5	7	6	2	4	8	3	1
5	6	2	1	4	7	3	8	9
1	4	3	5	9	8	7	2	6
8	7	9	3	6	2	1	5	4
6	9	5	4	7	3	2	1	8
4	3	8	2	1	6	9	7	5
7	2	1	9	8	5	4	6	3

VERY HARD - 177

2	4	8	7	3	9	1	6	5
7	9	3	6	1	5	4	8	2
1	5	6	4	2	8	7	9	3
8	6	2	5	4	1	3	7	9
5	1	4	9	7	3	8	2	6
9	3	7	8	6	2	5	1	4
4	8	1	3	9	6	2	5	7
3	2	9	1	5	7	6	4	8
6	7	5	2	8	4	9	3	1

VERY HARD - 178

1	5	6	8	7	9	4	2	3
3	4	9	1	6	2	7	5	8
7	2	8	5	3	4	6	1	9
2	6	4	9	1	8	5	3	7
9	1	7	4	5	3	2	8	6
8	3	5	7	2	6	1	9	4
5	7	3	6	8	1	9	4	2
6	9	2	3	4	5	8	7	1
4	8	1	2	9	7	3	6	5

VERY HARD - 179

8	5	6	2	9	1	4	7	3
4	2	9	3	6	7	5	1	8
7	3	1	8	5	4	9	6	2
6	1	7	9	8	5	3	2	4
3	8	5	6	4	2	1	9	7
9	4	2	1	7	3	8	5	6
2	6	3	5	1	8	7	4	9
5	9	4	7	3	6	2	8	1
1	7	8	4	2	9	6	3	5

VERY HARD - 180

1	5	2	3	8	6	4	9	7
9	8	4	1	7	5	3	2	6
3	7	6	9	4	2	8	1	5
6	9	3	7	1	8	2	5	4
5	1	8	2	9	4	7	6	3
2	4	7	6	5	3	9	8	1
4	2	9	5	3	1	6	7	8
7	3	1	8	6	9	5	4	2
8	6	5	4	2	7	1	3	9

VERY HARD - 181

7	3	6	8	2	5	9	1	4
2	5	1	9	6	4	3	8	7
9	8	4	1	7	3	6	5	2
5	2	3	6	1	9	7	4	8
1	9	8	4	3	7	5	2	6
4	6	7	5	8	2	1	9	3
3	7	5	2	9	8	4	6	1
6	4	2	3	5	1	8	7	9
8	1	9	7	4	6	2	3	5

VERY HARD - 182

9	4	5	3	2	1	7	6	8
1	7	8	4	9	6	2	3	5
2	3	6	5	7	8	9	1	4
3	2	4	7	6	5	8	9	1
5	6	7	1	8	9	3	4	2
8	9	1	2	4	3	5	7	6
6	1	2	8	3	7	4	5	9
7	8	9	6	5	4	1	2	3
4	5	3	9	1	2	6	8	7

VERY HARD - 183

5	3	1	9	4	2	8	7	6
8	7	9	3	6	1	2	4	5
4	6	2	5	7	8	3	9	1
6	9	5	8	3	4	1	2	7
1	2	8	6	9	7	4	5	3
3	4	7	1	2	5	6	8	9
7	8	3	4	1	9	5	6	2
2	1	4	7	5	6	9	3	8
9	5	6	2	8	3	7	1	4

VERY HARD - 184

2	1	5	6	7	8	4	9	3
3	4	9	5	1	2	6	7	8
6	8	7	9	3	4	1	5	2
7	2	4	8	9	1	3	6	5
5	6	3	2	4	7	8	1	9
8	9	1	3	5	6	7	2	4
9	5	8	1	6	3	2	4	7
1	7	2	4	8	5	9	3	6
4	3	6	7	2	9	5	8	1

VERY HARD - 185

9	4	2	6	3	1	7	8	5
1	6	3	5	7	8	9	2	4
5	8	7	9	4	2	3	1	6
7	5	1	2	8	9	4	6	3
4	2	9	3	6	7	8	5	1
8	3	6	4	1	5	2	9	7
3	7	5	8	2	6	1	4	9
6	1	8	7	9	4	5	3	2
2	9	4	1	5	3	6	7	8

VERY HARD - 186

9	3	2	7	1	4	8	6	5
6	4	5	9	2	8	7	3	1
1	7	8	3	5	6	2	9	4
2	5	7	1	3	9	4	8	6
8	9	1	4	6	2	3	5	7
3	6	4	8	7	5	1	2	9
7	2	3	5	9	1	6	4	8
5	8	6	2	4	7	9	1	3
4	1	9	6	8	3	5	7	2

VERY HARD - 187

7	1	8	6	2	3	9	5	4
6	2	5	1	9	4	3	7	8
3	4	9	7	5	8	6	2	1
8	5	6	4	1	9	2	3	7
4	7	1	5	3	2	8	9	6
2	9	3	8	6	7	4	1	5
9	8	7	2	4	1	5	6	3
1	6	2	3	8	5	7	4	9
5	3	4	9	7	6	1	8	2

VERY HARD - 188

6	1	3	2	5	9	8	4	7
2	7	8	6	4	1	3	9	5
4	5	9	8	3	7	1	2	6
5	4	7	1	6	8	9	3	2
9	3	6	5	2	4	7	1	8
8	2	1	9	7	3	5	6	4
1	8	4	7	9	2	6	5	3
7	6	2	3	1	5	4	8	9
3	9	5	4	8	6	2	7	1

VERY HARD - 189

7	6	8	5	4	1	3	2	9
5	1	9	8	2	3	6	4	7
3	2	4	6	7	9	5	8	1
8	5	3	2	6	7	9	1	4
4	9	1	3	8	5	2	7	6
2	7	6	1	9	4	8	3	5
9	4	5	7	3	8	1	6	2
1	3	2	4	5	6	7	9	8
6	8	7	9	1	2	4	5	3

VERY HARD - 190

1	3	7	6	2	8	4	5	9
2	8	4	7	9	5	1	3	6
6	5	9	3	4	1	7	2	8
8	9	1	2	5	6	3	7	4
7	4	6	1	3	9	2	8	5
3	2	5	8	7	4	9	6	1
9	7	8	5	1	2	6	4	3
5	1	3	4	6	7	8	9	2
4	6	2	9	8	3	5	1	7

VERY HARD - 191

5	8	2	6	3	1	4	7	9
9	4	7	5	8	2	1	6	3
3	6	1	4	7	9	8	5	2
4	3	8	9	1	6	5	2	7
7	2	6	3	5	4	9	8	1
1	5	9	8	2	7	3	4	6
6	9	5	2	4	3	7	1	8
8	7	3	1	6	5	2	9	4
2	1	4	7	9	8	6	3	5

VERY HARD - 192

7	2	8	1	5	6	4	3	9
3	9	4	7	8	2	5	1	6
1	5	6	4	9	3	7	8	2
5	4	9	2	1	7	3	6	8
2	6	1	8	3	4	9	7	5
8	7	3	9	6	5	1	2	4
6	8	5	3	4	1	2	9	7
9	1	2	5	7	8	6	4	3
4	3	7	6	2	9	8	5	1

VERY HARD - 193

5	6	3	7	8	9	2	4	1
9	7	2	1	3	4	8	6	5
1	4	8	5	6	2	9	3	7
8	1	4	9	2	6	7	5	3
6	5	9	8	7	3	4	1	2
2	3	7	4	1	5	6	9	8
3	2	1	6	9	7	5	8	4
4	8	6	2	5	1	3	7	9
7	9	5	3	4	8	1	2	6

VERY HARD - 194

5	4	6	2	8	3	9	1	7
2	7	1	5	4	9	8	6	3
8	3	9	6	1	7	5	2	4
7	6	2	8	3	4	1	9	5
9	1	4	7	6	5	2	3	8
3	5	8	9	2	1	7	4	6
1	2	5	4	7	6	3	8	9
4	9	3	1	5	8	6	7	2
6	8	7	3	9	2	4	5	1

VERY HARD - 195

1	4	6	5	3	9	7	8	2
5	9	3	7	8	2	6	1	4
2	8	7	1	4	6	3	9	5
6	5	9	4	7	8	2	3	1
3	2	4	9	1	5	8	7	6
7	1	8	2	6	3	5	4	9
8	6	1	3	2	4	9	5	7
4	3	5	6	9	7	1	2	8
9	7	2	8	5	1	4	6	3

VERY HARD - 196

9	3	6	7	4	8	5	1	2
5	7	8	3	1	2	9	6	4
2	1	4	5	9	6	8	3	7
4	9	1	8	3	5	2	7	6
3	8	7	2	6	1	4	5	9
6	5	2	9	7	4	1	8	3
8	4	5	6	2	7	3	9	1
7	2	3	1	5	9	6	4	8
1	6	9	4	8	3	7	2	5

VERY HARD - 197

9	3	8	5	6	2	7	1	4
2	4	7	9	8	1	5	6	3
5	6	1	3	7	4	9	2	8
1	7	6	4	2	3	8	9	5
8	5	9	7	1	6	4	3	2
3	2	4	8	5	9	1	7	6
6	8	5	2	9	7	3	4	1
4	9	2	1	3	5	6	8	7
7	1	3	6	4	8	2	5	9

VERY HARD - 198

9	3	8	5	7	4	2	6	1
6	2	7	1	3	9	8	4	5
1	5	4	8	6	2	9	3	7
3	9	2	4	8	5	7	1	6
4	1	5	6	9	7	3	2	8
7	8	6	2	1	3	5	9	4
8	6	9	3	5	1	4	7	2
2	7	1	9	4	8	6	5	3
5	4	3	7	2	6	1	8	9

VERY HARD - 199

9	8	6	7	3	5	4	1	2
7	1	3	9	4	2	5	6	8
2	4	5	1	6	8	9	7	3
6	5	9	8	7	1	2	3	4
1	3	8	6	2	4	7	5	9
4	2	7	3	5	9	1	8	6
3	6	4	5	9	7	8	2	1
8	7	2	4	1	6	3	9	5
5	9	1	2	8	3	6	4	7

VERY HARD - 200

5	4	2	8	9	1	7	6	3
6	9	1	3	4	7	5	2	8
3	7	8	6	2	5	9	4	1
2	1	4	5	7	8	3	9	6
8	5	6	9	3	2	4	1	7
7	3	9	4	1	6	8	5	2
4	8	3	1	6	9	2	7	5
1	2	5	7	8	4	6	3	9
9	6	7	2	5	3	1	8	4

VERY HARD - 201

5	1	2	4	9	3	7	6	8
3	4	8	2	6	7	5	1	9
9	6	7	5	1	8	2	3	4
8	2	9	7	3	4	1	5	6
6	5	1	8	2	9	3	4	7
7	3	4	6	5	1	9	8	2
2	7	5	1	4	6	8	9	3
1	9	6	3	8	2	4	7	5
4	8	3	9	7	5	6	2	1

VERY HARD - 202

8	2	3	7	9	4	1	6	5
1	7	5	3	6	2	4	9	8
4	9	6	1	5	8	2	7	3
2	6	8	9	1	3	7	5	4
7	1	9	6	4	5	3	8	2
3	5	4	8	2	7	6	1	9
9	3	2	5	7	1	8	4	6
6	4	1	2	8	9	5	3	7
5	8	7	4	3	6	9	2	1

VERY HARD - 203

1	2	8	7	9	5	6	3	4
7	6	3	1	4	8	2	9	5
9	4	5	3	2	6	7	1	8
6	3	4	5	1	2	8	7	9
8	7	2	9	6	4	1	5	3
5	9	1	8	7	3	4	2	6
2	1	6	4	5	9	3	8	7
3	5	7	6	8	1	9	4	2
4	8	9	2	3	7	5	6	1

VERY HARD - 204

9	1	3	2	7	5	8	6	4
5	4	6	1	3	8	9	7	2
7	2	8	6	9	4	5	1	3
3	8	9	4	6	7	1	2	5
1	5	4	8	2	3	6	9	7
2	6	7	5	1	9	3	4	8
6	7	5	9	8	2	4	3	1
4	9	2	3	5	1	7	8	6
8	3	1	7	4	6	2	5	9

INSANE - 1

9	8	6	7	1	4	3	2	5
7	4	5	8	3	2	1	9	6
3	2	1	9	6	5	8	7	4
6	3	4	2	5	1	9	8	7
2	5	9	3	7	8	6	4	1
8	1	7	6	4	9	2	5	3
5	7	8	1	9	6	4	3	2
1	9	3	4	2	7	5	6	8
4	6	2	5	8	3	7	1	9

INSANE - 2

8	4	5	7	3	1	2	9	6
6	1	2	5	9	4	7	8	3
3	9	7	2	8	6	4	5	1
4	3	1	6	2	5	9	7	8
2	8	6	3	7	9	1	4	5
5	7	9	1	4	8	3	6	2
9	5	8	4	1	3	6	2	7
7	6	3	9	5	2	8	1	4
1	2	4	8	6	7	5	3	9

INSANE - 3

4	6	3	9	5	7	8	2	1
5	8	7	2	3	1	6	9	4
2	9	1	4	6	8	7	5	3
3	1	6	5	2	4	9	8	7
7	2	5	8	1	9	4	3	6
8	4	9	3	7	6	2	1	5
6	3	2	7	9	5	1	4	8
1	5	4	6	8	2	3	7	9
9	7	8	1	4	3	5	6	2

INSANE - 4

3	4	5	2	1	8	6	9	7
8	2	9	5	6	7	1	3	4
6	1	7	4	9	3	8	5	2
5	3	1	9	2	4	7	8	6
9	6	2	8	7	5	4	1	3
7	8	4	1	3	6	5	2	9
1	9	8	7	4	2	3	6	5
2	7	6	3	5	1	9	4	8
4	5	3	6	8	9	2	7	1

INSANE - 5

1	5	8	6	4	2	9	3	7
3	4	6	7	9	5	2	1	8
7	2	9	3	1	8	6	5	4
2	7	4	8	5	9	3	6	1
8	9	1	4	3	6	5	7	2
5	6	3	2	7	1	4	8	9
9	1	7	5	6	4	8	2	3
4	8	5	1	2	3	7	9	6
6	3	2	9	8	7	1	4	5

INSANE - 6

6	7	2	9	8	1	5	4	3
5	1	4	7	3	2	6	8	9
8	3	9	5	6	4	2	7	1
4	5	7	2	1	6	3	9	8
3	6	1	8	7	9	4	5	2
2	9	8	4	5	3	7	1	6
9	4	6	1	2	5	8	3	7
1	8	3	6	4	7	9	2	5
7	2	5	3	9	8	1	6	4

INSANE - 7

6	7	8	1	9	5	2	4	3
4	5	2	3	8	6	1	9	7
3	9	1	4	2	7	5	6	8
9	3	4	8	5	2	6	7	1
8	2	7	6	1	9	3	5	4
5	1	6	7	3	4	9	8	2
2	6	3	5	4	8	7	1	9
1	8	5	9	7	3	4	2	6
7	4	9	2	6	1	8	3	5

INSANE - 8

2	4	8	6	5	3	1	7	9
1	9	7	2	8	4	5	6	3
3	6	5	9	1	7	8	2	4
7	1	2	3	4	8	6	9	5
4	5	3	1	9	6	7	8	2
6	8	9	5	7	2	4	3	1
8	7	1	4	2	9	3	5	6
9	3	4	7	6	5	2	1	8
5	2	6	8	3	1	9	4	7

INSANE - 9

2	6	1	5	3	8	9	4	7
3	4	8	9	1	7	2	6	5
9	7	5	4	2	6	8	1	3
1	5	3	6	7	9	4	8	2
7	2	4	3	8	1	6	5	9
6	8	9	2	5	4	7	3	1
8	9	2	1	4	5	3	7	6
4	1	6	7	9	3	5	2	8
5	3	7	8	6	2	1	9	4

INSANE - 10

8	3	4	7	6	1	9	5	2
5	7	9	8	4	2	6	1	3
2	1	6	3	9	5	8	7	4
4	9	5	6	1	8	3	2	7
6	8	1	2	3	7	4	9	5
3	2	7	9	5	4	1	6	8
1	6	8	5	2	3	7	4	9
7	4	2	1	8	9	5	3	6
9	5	3	4	7	6	2	8	1

INSANE - 11

8	5	6	7	1	9	2	3	4
1	3	4	6	8	2	5	7	9
2	7	9	5	3	4	1	8	6
4	6	2	3	5	7	9	1	8
9	8	5	2	4	1	7	6	3
3	1	7	9	6	8	4	5	2
6	4	8	1	9	5	3	2	7
7	9	1	8	2	3	6	4	5
5	2	3	4	7	6	8	9	1

INSANE - 12

1	4	5	6	7	3	2	9	8
9	8	3	2	4	1	7	6	5
6	7	2	9	8	5	4	1	3
3	5	4	1	9	8	6	2	7
2	6	9	3	5	7	1	8	4
8	1	7	4	6	2	5	3	9
4	3	1	7	2	9	8	5	6
5	9	6	8	1	4	3	7	2
7	2	8	5	3	6	9	4	1

INSANE - 13

5	7	9	8	2	3	4	6	1
8	6	1	9	5	4	3	2	7
3	4	2	6	1	7	8	5	9
1	2	4	7	8	9	5	3	6
9	8	7	5	3	6	1	4	2
6	5	3	2	4	1	9	7	8
4	9	6	1	7	5	2	8	3
7	3	8	4	9	2	6	1	5
2	1	5	3	6	8	7	9	4

INSANE - 14

9	4	7	5	3	8	2	1	6
6	2	3	1	4	9	8	5	7
5	8	1	6	7	2	3	9	4
2	9	4	3	8	6	5	7	1
7	6	8	4	5	1	9	3	2
3	1	5	9	2	7	6	4	8
4	7	6	2	9	5	1	8	3
8	5	2	7	1	3	4	6	9
1	3	9	8	6	4	7	2	5

INSANE - 15

6	1	4	3	9	5	7	8	2
2	7	5	1	6	8	4	3	9
3	9	8	7	4	2	6	5	1
8	6	1	5	2	9	3	4	7
9	2	3	4	8	7	5	1	6
4	5	7	6	1	3	2	9	8
1	8	6	2	3	4	9	7	5
7	3	2	9	5	1	8	6	4
5	4	9	8	7	6	1	2	3

INSANE - 16

5	7	8	9	3	6	2	4	1
2	9	6	5	4	1	3	7	8
3	1	4	2	7	8	9	5	6
7	5	2	1	6	3	8	9	4
6	3	1	4	8	9	5	2	7
4	8	9	7	5	2	1	6	3
9	2	7	8	1	4	6	3	5
1	6	5	3	2	7	4	8	9
8	4	3	6	9	5	7	1	2

INSANE - 17

3	8	7	2	6	1	9	4	5
9	1	4	3	7	5	8	6	2
2	5	6	4	8	9	3	1	7
8	7	9	1	4	2	6	5	3
1	3	5	7	9	6	2	8	4
6	4	2	8	5	3	1	7	9
7	9	1	5	2	8	4	3	6
5	6	8	9	3	4	7	2	1
4	2	3	6	1	7	5	9	8

INSANE - 18

7	5	3	6	1	9	8	2	4
8	1	2	7	3	4	5	9	6
4	9	6	8	5	2	7	1	3
9	4	1	2	8	3	6	7	5
2	7	5	9	4	6	3	8	1
6	3	8	1	7	5	9	4	2
3	2	4	5	9	7	1	6	8
1	6	9	3	2	8	4	5	7
5	8	7	4	6	1	2	3	9

INSANE - 19

7	6	3	1	5	4	9	8	2
5	2	1	3	8	9	6	4	7
8	4	9	7	2	6	5	3	1
2	7	8	4	6	3	1	9	5
4	1	5	8	9	2	7	6	3
3	9	6	5	7	1	4	2	8
6	5	2	9	3	7	8	1	4
1	3	7	6	4	8	2	5	9
9	8	4	2	1	5	3	7	6

INSANE - 20

5	7	1	4	6	2	3	8	9
3	4	6	5	8	9	1	2	7
9	8	2	1	3	7	5	6	4
2	1	3	9	4	6	8	7	5
4	6	7	2	5	8	9	3	1
8	5	9	7	1	3	2	4	6
1	3	5	8	7	4	6	9	2
7	9	8	6	2	1	4	5	3
6	2	4	3	9	5	7	1	8

INSANE - 21

5	7	4	6	1	9	2	3	8
3	8	2	4	7	5	6	1	9
6	1	9	2	8	3	5	4	7
1	5	7	3	4	6	9	8	2
8	2	6	1	9	7	3	5	4
4	9	3	8	5	2	1	7	6
7	3	8	9	6	1	4	2	5
9	4	1	5	2	8	7	6	3
2	6	5	7	3	4	8	9	1

INSANE - 22

1	2	8	6	4	9	3	7	5
6	9	5	8	3	7	4	2	1
4	7	3	1	5	2	6	9	8
2	8	9	4	1	3	5	6	7
5	6	7	2	9	8	1	4	3
3	4	1	5	7	6	9	8	2
9	5	6	7	2	1	8	3	4
8	1	2	3	6	4	7	5	9
7	3	4	9	8	5	2	1	6

INSANE - 23

5	9	3	7	4	8	6	1	2
8	7	1	2	9	6	5	4	3
4	6	2	1	3	5	8	7	9
7	2	8	3	1	9	4	5	6
9	4	6	5	7	2	3	8	1
3	1	5	6	8	4	2	9	7
2	5	7	4	6	1	9	3	8
1	8	4	9	2	3	7	6	5
6	3	9	8	5	7	1	2	4

INSANE - 24

8	6	2	5	4	1	7	9	3
4	9	1	3	8	7	6	2	5
7	3	5	9	6	2	4	8	1
1	4	7	8	9	6	5	3	2
6	2	9	1	3	5	8	7	4
5	8	3	7	2	4	9	1	6
2	5	8	4	7	3	1	6	9
9	1	6	2	5	8	3	4	7
3	7	4	6	1	9	2	5	8

INSANE - 25

3	7	4	8	5	6	9	2	1
9	5	6	2	4	1	8	7	3
2	8	1	9	3	7	4	6	5
8	4	9	1	6	2	5	3	7
5	2	7	4	9	3	1	8	6
6	1	3	7	8	5	2	9	4
4	6	2	5	7	8	3	1	9
1	3	5	6	2	9	7	4	8
7	9	8	3	1	4	6	5	2

INSANE - 26

1	9	2	5	8	4	3	6	7
6	7	5	3	1	2	8	9	4
4	3	8	9	7	6	2	1	5
8	2	6	1	4	9	5	7	3
9	1	3	8	5	7	6	4	2
5	4	7	2	6	3	9	8	1
3	6	9	7	2	1	4	5	8
7	5	4	6	3	8	1	2	9
2	8	1	4	9	5	7	3	6

INSANE - 27

6	8	3	5	1	9	4	2	7
7	1	5	6	4	2	8	9	3
2	9	4	7	8	3	5	1	6
5	2	7	8	3	4	9	6	1
3	6	9	1	7	5	2	8	4
8	4	1	2	9	6	7	3	5
1	3	2	4	5	8	6	7	9
9	5	8	3	6	7	1	4	2
4	7	6	9	2	1	3	5	8

INSANE - 28

8	1	4	3	5	2	7	6	9
2	6	5	8	7	9	3	4	1
7	3	9	4	6	1	2	8	5
5	9	6	7	2	8	1	3	4
4	7	2	9	1	3	6	5	8
1	8	3	5	4	6	9	2	7
6	2	7	1	8	5	4	9	3
3	5	1	6	9	4	8	7	2
9	4	8	2	3	7	5	1	6

INSANE - 29

2	1	8	3	4	5	9	6	7
4	5	3	9	6	7	2	8	1
9	6	7	8	2	1	4	5	3
8	7	4	1	3	9	5	2	6
3	9	1	6	5	2	7	4	8
6	2	5	7	8	4	3	1	9
7	3	2	5	1	8	6	9	4
5	8	9	4	7	6	1	3	2
1	4	6	2	9	3	8	7	5

INSANE - 30

8	3	6	5	2	7	9	4	1
7	5	1	4	9	6	2	8	3
4	9	2	1	3	8	6	7	5
5	4	8	3	7	9	1	2	6
6	1	9	2	8	4	3	5	7
2	7	3	6	5	1	8	9	4
1	2	5	8	4	3	7	6	9
3	8	7	9	6	5	4	1	2
9	6	4	7	1	2	5	3	8

INSANE - 31

3	5	9	7	2	8	4	1	6
7	8	4	5	1	6	3	9	2
2	6	1	4	9	3	5	8	7
1	9	5	3	6	2	7	4	8
8	7	3	1	5	4	6	2	9
6	4	2	8	7	9	1	5	3
9	3	7	2	4	1	8	6	5
4	2	8	6	3	5	9	7	1
5	1	6	9	8	7	2	3	4

INSANE - 32

1	2	4	6	8	3	9	5	7
8	7	5	4	2	9	6	3	1
9	6	3	5	7	1	8	4	2
6	5	8	3	9	2	1	7	4
7	1	9	8	5	4	2	6	3
3	4	2	1	6	7	5	8	9
5	9	1	7	4	6	3	2	8
2	8	7	9	3	5	4	1	6
4	3	6	2	1	8	7	9	5

INSANE - 33

2	3	1	5	7	4	6	9	8
9	4	6	8	3	1	7	2	5
5	7	8	6	9	2	3	4	1
6	1	9	4	8	5	2	7	3
4	8	2	3	1	7	9	5	6
7	5	3	9	2	6	1	8	4
1	6	4	2	5	9	8	3	7
8	2	5	7	6	3	4	1	9
3	9	7	1	4	8	5	6	2

INSANE - 34

9	8	7	3	6	2	4	1	5
2	4	5	8	7	1	3	9	6
6	1	3	4	9	5	7	2	8
1	2	4	6	3	8	9	5	7
8	7	6	2	5	9	1	4	3
3	5	9	7	1	4	6	8	2
7	9	2	5	4	3	8	6	1
4	3	8	1	2	6	5	7	9
5	6	1	9	8	7	2	3	4

INSANE - 35

9	7	4	5	1	2	3	6	8
5	6	3	7	9	8	1	4	2
8	2	1	3	4	6	9	5	7
3	8	6	9	7	5	2	1	4
1	4	2	8	6	3	5	7	9
7	5	9	1	2	4	8	3	6
4	9	8	6	3	1	7	2	5
6	1	5	2	8	7	4	9	3
2	3	7	4	5	9	6	8	1

INSANE - 36

2	8	1	5	3	4	9	7	6
3	5	4	9	6	7	8	1	2
9	6	7	1	8	2	4	5	3
8	3	2	4	9	1	7	6	5
5	1	9	7	2	6	3	8	4
7	4	6	8	5	3	1	2	9
1	2	5	3	4	8	6	9	7
4	9	8	6	7	5	2	3	1
6	7	3	2	1	9	5	4	8

INSANE - 37

7	3	9	2	8	4	1	5	6
1	4	8	9	6	5	7	3	2
5	2	6	7	1	3	8	9	4
3	8	5	1	2	9	6	4	7
6	1	7	4	3	8	9	2	5
2	9	4	6	5	7	3	8	1
4	7	1	3	9	2	5	6	8
8	6	3	5	4	1	2	7	9
9	5	2	8	7	6	4	1	3

INSANE - 38

9	3	7	4	6	2	8	5	1
8	5	4	3	1	9	2	7	6
2	6	1	5	7	8	9	3	4
3	8	5	2	9	4	6	1	7
6	7	9	1	5	3	4	2	8
1	4	2	7	8	6	5	9	3
5	9	3	8	4	7	1	6	2
4	2	6	9	3	1	7	8	5
7	1	8	6	2	5	3	4	9

INSANE - 39

6	3	1	8	9	5	2	4	7
5	9	8	7	4	2	1	6	3
4	7	2	1	3	6	5	9	8
8	5	7	3	2	9	6	1	4
1	6	9	4	8	7	3	2	5
2	4	3	5	6	1	7	8	9
3	1	6	9	5	8	4	7	2
9	2	5	6	7	4	8	3	1
7	8	4	2	1	3	9	5	6

INSANE - 40

8	4	5	7	6	3	1	9	2
3	9	6	8	1	2	5	4	7
2	1	7	5	4	9	3	6	8
7	6	2	4	8	5	9	1	3
9	8	4	6	3	1	7	2	5
1	5	3	9	2	7	4	8	6
5	2	8	3	9	4	6	7	1
4	3	1	2	7	6	8	5	9
6	7	9	1	5	8	2	3	4

INSANE - 41

1	3	4	6	2	5	9	7	8
8	7	5	4	9	3	6	1	2
6	9	2	1	8	7	5	3	4
2	8	6	9	3	4	1	5	7
7	1	3	5	6	2	4	8	9
5	4	9	8	7	1	2	6	3
9	5	7	2	1	8	3	4	6
4	6	8	3	5	9	7	2	1
3	2	1	7	4	6	8	9	5

INSANE - 42

5	3	8	4	9	7	2	1	6
2	7	4	6	1	5	9	3	8
9	6	1	3	8	2	7	5	4
7	5	6	1	4	8	3	9	2
4	9	2	7	6	3	1	8	5
1	8	3	2	5	9	6	4	7
8	1	5	9	7	6	4	2	3
6	2	9	8	3	4	5	7	1
3	4	7	5	2	1	8	6	9

INSANE - 43

5	7	9	3	8	4	6	1	2
2	4	8	7	1	6	5	3	9
6	3	1	5	9	2	8	4	7
9	1	2	4	7	5	3	8	6
4	8	7	1	6	3	2	9	5
3	6	5	8	2	9	1	7	4
8	2	4	6	3	7	9	5	1
1	5	6	9	4	8	7	2	3
7	9	3	2	5	1	4	6	8

INSANE - 44

3	8	9	2	7	1	5	4	6
1	7	2	5	6	4	8	9	3
6	5	4	8	3	9	7	1	2
8	6	1	9	2	7	3	5	4
9	4	5	3	8	6	2	7	1
2	3	7	4	1	5	9	6	8
4	1	8	7	5	3	6	2	9
7	2	6	1	9	8	4	3	5
5	9	3	6	4	2	1	8	7

INSANE - 45

8	2	4	5	7	1	6	9	3
6	1	3	4	9	2	5	8	7
7	9	5	8	6	3	1	4	2
3	7	6	9	5	8	4	2	1
5	8	2	3	1	4	9	7	6
9	4	1	7	2	6	8	3	5
1	6	8	2	3	9	7	5	4
4	3	7	6	8	5	2	1	9
2	5	9	1	4	7	3	6	8

INSANE - 46

8	5	6	2	7	1	9	4	3
3	7	2	9	5	4	8	6	1
1	4	9	6	3	8	2	5	7
4	6	3	5	1	9	7	2	8
2	1	5	7	8	3	4	9	6
7	9	8	4	2	6	3	1	5
9	2	7	8	6	5	1	3	4
6	3	4	1	9	7	5	8	2
5	8	1	3	4	2	6	7	9

INSANE - 47

3	5	4	6	2	9	1	8	7
9	6	1	5	7	8	4	2	3
2	7	8	4	3	1	6	5	9
7	2	6	1	9	5	8	3	4
5	8	9	7	4	3	2	1	6
1	4	3	2	8	6	7	9	5
6	9	2	8	5	7	3	4	1
8	3	7	9	1	4	5	6	2
4	1	5	3	6	2	9	7	8

INSANE - 48

7	2	1	6	3	5	9	4	8
5	9	6	1	8	4	2	7	3
3	8	4	9	2	7	6	5	1
4	6	7	5	1	2	8	3	9
9	3	2	8	4	6	5	1	7
1	5	8	7	9	3	4	6	2
6	7	9	3	5	8	1	2	4
8	4	5	2	7	1	3	9	6
2	1	3	4	6	9	7	8	5

INSANE - 49

1	6	2	4	7	8	3	5	9
3	5	7	2	6	9	4	1	8
4	9	8	5	3	1	6	2	7
2	3	5	9	4	7	8	6	1
8	7	1	3	5	6	9	4	2
6	4	9	8	1	2	5	7	3
5	8	6	1	2	3	7	9	4
9	1	4	7	8	5	2	3	6
7	2	3	6	9	4	1	8	5

INSANE - 50

7	6	2	4	3	8	9	5	1
8	5	4	6	9	1	2	3	7
9	3	1	7	2	5	4	6	8
6	8	5	2	7	3	1	4	9
3	2	9	8	1	4	6	7	5
1	4	7	5	6	9	8	2	3
2	1	6	3	8	7	5	9	4
5	7	8	9	4	2	3	1	6
4	9	3	1	5	6	7	8	2

INSANE - 51

8	1	2	6	7	4	9	5	3
9	3	6	1	5	2	8	4	7
4	5	7	3	8	9	1	2	6
5	2	9	4	1	3	7	6	8
7	4	8	5	9	6	3	1	2
1	6	3	7	2	8	5	9	4
3	8	5	2	6	1	4	7	9
6	7	4	9	3	5	2	8	1
2	9	1	8	4	7	6	3	5

INSANE - 52

4	6	9	2	7	1	5	8	3
5	1	2	9	3	8	6	4	7
8	3	7	4	5	6	2	1	9
6	8	1	3	9	7	4	2	5
3	9	4	5	8	2	7	6	1
2	7	5	1	6	4	3	9	8
7	2	8	6	1	3	9	5	4
1	5	6	7	4	9	8	3	2
9	4	3	8	2	5	1	7	6

INSANE - 53

7	9	2	3	4	6	5	1	8
4	5	8	7	1	2	3	6	9
3	1	6	5	9	8	4	7	2
2	7	1	6	5	4	8	9	3
5	8	3	1	7	9	6	2	4
6	4	9	8	2	3	7	5	1
8	2	5	9	3	7	1	4	6
9	3	7	4	6	1	2	8	5
1	6	4	2	8	5	9	3	7

INSANE - 54

6	1	8	4	9	5	7	3	2
5	2	4	7	8	3	6	9	1
9	3	7	1	2	6	4	5	8
2	8	9	3	1	4	5	6	7
1	5	6	9	7	2	3	8	4
7	4	3	6	5	8	1	2	9
4	7	2	5	3	9	8	1	6
3	9	1	8	6	7	2	4	5
8	6	5	2	4	1	9	7	3

INSANE - 55

5	6	7	1	9	8	2	3	4
1	2	9	4	7	3	8	6	5
3	4	8	6	2	5	9	7	1
8	7	5	2	4	1	6	9	3
2	9	3	7	5	6	4	1	8
4	1	6	8	3	9	5	2	7
7	5	4	9	1	2	3	8	6
9	8	1	3	6	4	7	5	2
6	3	2	5	8	7	1	4	9

INSANE - 56

7	6	2	4	8	5	3	1	9
4	9	1	7	2	3	8	5	6
5	8	3	1	9	6	4	2	7
6	7	4	3	1	9	5	8	2
3	5	9	8	6	2	1	7	4
2	1	8	5	4	7	6	9	3
8	4	6	2	7	1	9	3	5
9	2	5	6	3	8	7	4	1
1	3	7	9	5	4	2	6	8

INSANE - 57

1	5	9	4	2	6	3	7	8
2	8	4	9	3	7	6	5	1
3	7	6	5	8	1	4	9	2
7	4	1	6	9	3	8	2	5
6	2	8	7	4	5	9	1	3
5	9	3	2	1	8	7	4	6
4	6	7	8	5	2	1	3	9
9	3	2	1	6	4	5	8	7
8	1	5	3	7	9	2	6	4

INSANE - 58

6	8	7	5	4	2	9	3	1
9	2	5	8	3	1	4	6	7
1	3	4	9	6	7	2	5	8
8	7	9	3	2	6	5	1	4
4	1	6	7	9	5	8	2	3
2	5	3	1	8	4	7	9	6
7	4	1	2	5	3	6	8	9
5	6	8	4	1	9	3	7	2
3	9	2	6	7	8	1	4	5

INSANE - 59

9	6	2	5	3	8	1	4	7
7	1	3	9	4	2	6	8	5
4	8	5	7	1	6	2	3	9
5	3	7	1	9	4	8	2	6
1	2	8	3	6	5	9	7	4
6	4	9	2	8	7	3	5	1
3	5	6	4	2	1	7	9	8
8	9	4	6	7	3	5	1	2
2	7	1	8	5	9	4	6	3

INSANE - 60

7	3	6	4	5	9	8	2	1
9	2	5	8	3	1	4	7	6
8	1	4	2	7	6	5	3	9
6	9	7	3	4	8	2	1	5
4	8	1	9	2	5	7	6	3
3	5	2	1	6	7	9	8	4
5	6	3	7	9	2	1	4	8
2	4	8	5	1	3	6	9	7
1	7	9	6	8	4	3	5	2

INSANE - 61

8	3	2	7	4	6	9	1	5
4	5	6	3	1	9	2	8	7
9	7	1	5	8	2	6	4	3
7	4	9	2	6	8	5	3	1
2	8	5	1	9	3	4	7	6
6	1	3	4	7	5	8	2	9
1	2	4	9	5	7	3	6	8
5	6	7	8	3	4	1	9	2
3	9	8	6	2	1	7	5	4

INSANE - 62

1	7	6	5	4	3	9	2	8
5	9	8	1	7	2	4	6	3
3	4	2	6	8	9	7	1	5
7	3	5	8	9	6	1	4	2
4	8	9	7	2	1	5	3	6
2	6	1	4	3	5	8	7	9
9	2	4	3	1	8	6	5	7
8	5	7	2	6	4	3	9	1
6	1	3	9	5	7	2	8	4

INSANE - 63

2	4	6	1	9	8	5	7	3
7	3	8	6	2	5	1	9	4
5	1	9	7	4	3	2	6	8
3	8	4	2	6	1	9	5	7
1	7	2	5	8	9	3	4	6
6	9	5	4	3	7	8	1	2
9	2	7	3	5	4	6	8	1
8	6	1	9	7	2	4	3	5
4	5	3	8	1	6	7	2	9

INSANE - 64

6	5	8	9	7	3	4	2	1
1	7	9	5	2	4	8	3	6
2	3	4	8	1	6	5	7	9
3	6	7	2	8	5	9	1	4
4	2	1	6	3	9	7	5	8
8	9	5	7	4	1	2	6	3
5	8	6	3	9	2	1	4	7
9	1	2	4	6	7	3	8	5
7	4	3	1	5	8	6	9	2

INSANE - 65

3	5	8	9	7	2	1	6	4
9	1	4	8	3	6	2	7	5
7	6	2	1	4	5	9	8	3
8	7	1	6	5	3	4	9	2
4	2	6	7	9	1	3	5	8
5	3	9	4	2	8	7	1	6
1	9	3	5	8	4	6	2	7
2	8	7	3	6	9	5	4	1
6	4	5	2	1	7	8	3	9

INSANE - 66

8	5	9	6	1	4	7	2	3
3	6	7	2	9	8	1	4	5
4	1	2	3	7	5	8	9	6
1	7	4	9	2	3	6	5	8
9	8	6	5	4	1	2	3	7
5	2	3	8	6	7	4	1	9
7	3	1	4	5	6	9	8	2
2	4	8	7	3	9	5	6	1
6	9	5	1	8	2	3	7	4

INSANE - 67

5	7	2	1	6	3	4	9	8
3	6	1	4	8	9	7	5	2
9	8	4	7	2	5	1	3	6
7	3	6	2	9	8	5	1	4
8	4	5	3	1	6	2	7	9
2	1	9	5	7	4	6	8	3
6	5	3	9	4	7	8	2	1
1	9	8	6	5	2	3	4	7
4	2	7	8	3	1	9	6	5

INSANE - 68

9	3	1	2	4	6	5	7	8
4	6	8	3	5	7	9	1	2
2	5	7	9	1	8	4	3	6
8	7	9	5	2	4	3	6	1
1	4	5	8	6	3	7	2	9
6	2	3	7	9	1	8	5	4
3	9	2	6	8	5	1	4	7
5	1	6	4	7	9	2	8	3
7	8	4	1	3	2	6	9	5

INSANE - 69

2	4	1	3	7	5	6	9	8
9	3	8	2	1	6	7	5	4
5	7	6	9	8	4	3	1	2
8	1	4	6	3	7	9	2	5
3	9	7	8	5	2	4	6	1
6	2	5	1	4	9	8	3	7
1	8	2	4	9	3	5	7	6
7	6	9	5	2	8	1	4	3
4	5	3	7	6	1	2	8	9

INSANE - 70

4	2	9	3	5	1	8	7	6
3	7	6	8	2	9	5	4	1
8	5	1	6	4	7	9	3	2
9	4	2	5	1	8	3	6	7
6	8	5	2	7	3	1	9	4
1	3	7	9	6	4	2	8	5
7	1	8	4	9	5	6	2	3
5	6	3	7	8	2	4	1	9
2	9	4	1	3	6	7	5	8

INSANE - 71

5	6	7	1	9	8	3	2	4
4	9	2	3	7	6	5	8	1
3	1	8	5	2	4	9	7	6
8	5	1	7	6	9	2	4	3
2	4	9	8	5	3	6	1	7
7	3	6	2	4	1	8	5	9
1	7	3	6	8	5	4	9	2
9	2	5	4	3	7	1	6	8
6	8	4	9	1	2	7	3	5

INSANE - 72

1	8	3	9	2	4	6	5	7
9	6	7	3	1	5	8	2	4
5	4	2	8	6	7	9	1	3
6	2	4	7	8	3	1	9	5
8	3	1	2	5	9	7	4	6
7	5	9	6	4	1	3	8	2
2	7	8	4	9	6	5	3	1
3	9	5	1	7	2	4	6	8
4	1	6	5	3	8	2	7	9

INSANE - 73

7	3	5	8	9	2	4	1	6
8	1	9	4	5	6	3	2	7
4	2	6	3	1	7	9	8	5
9	8	1	5	7	3	6	4	2
2	4	3	6	8	9	5	7	1
6	5	7	2	4	1	8	9	3
5	9	2	1	6	8	7	3	4
3	6	8	7	2	4	1	5	9
1	7	4	9	3	5	2	6	8

INSANE - 74

8	2	3	9	4	5	7	1	6
9	1	5	6	8	7	3	2	4
4	6	7	3	2	1	9	8	5
7	8	2	5	1	4	6	9	3
5	3	4	2	6	9	1	7	8
1	9	6	8	7	3	4	5	2
3	5	8	1	9	6	2	4	7
2	7	1	4	3	8	5	6	9
6	4	9	7	5	2	8	3	1

INSANE - 75

6	7	4	3	5	1	8	9	2
5	2	8	7	4	9	6	1	3
3	9	1	6	2	8	7	5	4
2	8	3	9	1	5	4	7	6
4	6	5	8	7	3	9	2	1
7	1	9	4	6	2	3	8	5
1	4	7	5	8	6	2	3	9
8	3	2	1	9	4	5	6	7
9	5	6	2	3	7	1	4	8

INSANE - 76

5	8	4	6	1	7	2	3	9
6	7	2	5	9	3	4	1	8
3	9	1	4	2	8	6	5	7
8	3	7	9	5	4	1	2	6
4	2	9	8	6	1	5	7	3
1	6	5	7	3	2	9	8	4
7	1	3	2	4	9	8	6	5
9	5	8	1	7	6	3	4	2
2	4	6	3	8	5	7	9	1

INSANE - 77

6	8	3	4	7	5	1	2	9
1	7	2	9	3	8	5	4	6
4	9	5	1	2	6	3	8	7
9	1	8	5	6	7	4	3	2
3	2	4	8	1	9	7	6	5
5	6	7	3	4	2	8	9	1
8	3	9	2	5	1	6	7	4
2	5	6	7	8	4	9	1	3
7	4	1	6	9	3	2	5	8

INSANE - 78

5	9	1	4	3	6	2	7	8
6	7	8	5	2	9	4	3	1
2	3	4	1	8	7	9	6	5
3	8	6	9	1	2	5	4	7
1	2	9	7	5	4	3	8	6
4	5	7	3	6	8	1	2	9
9	1	2	8	7	3	6	5	4
7	6	5	2	4	1	8	9	3
8	4	3	6	9	5	7	1	2

INSANE - 79

5	6	8	4	9	2	7	3	1
7	9	4	3	8	1	2	5	6
1	2	3	5	6	7	8	9	4
8	7	6	9	1	4	5	2	3
2	3	1	7	5	6	4	8	9
4	5	9	2	3	8	1	6	7
3	8	2	1	4	9	6	7	5
9	1	7	6	2	5	3	4	8
6	4	5	8	7	3	9	1	2

INSANE - 80

1	3	9	6	4	8	5	2	7
4	5	7	1	2	9	6	3	8
6	2	8	7	3	5	9	1	4
5	9	2	4	8	3	7	6	1
3	7	1	2	9	6	4	8	5
8	4	6	5	7	1	2	9	3
9	6	5	8	1	4	3	7	2
7	8	3	9	5	2	1	4	6
2	1	4	3	6	7	8	5	9

INSANE - 81

5	9	1	2	3	6	8	4	7
7	6	8	5	4	1	3	2	9
2	4	3	8	7	9	1	5	6
3	8	2	1	6	7	4	9	5
9	7	6	3	5	4	2	1	8
1	5	4	9	2	8	7	6	3
4	1	9	6	8	3	5	7	2
6	3	5	7	1	2	9	8	4
8	2	7	4	9	5	6	3	1

INSANE - 82

3	7	9	8	2	1	4	5	6
4	1	2	5	7	6	8	9	3
6	8	5	4	9	3	2	7	1
2	6	8	7	4	9	3	1	5
7	5	1	3	6	2	9	4	8
9	4	3	1	8	5	6	2	7
8	2	4	6	5	7	1	3	9
1	9	7	2	3	8	5	6	4
5	3	6	9	1	4	7	8	2

INSANE - 83

4	5	9	8	1	6	7	2	3
3	8	1	7	9	2	4	6	5
2	7	6	4	3	5	8	9	1
8	4	3	5	2	7	9	1	6
7	6	2	9	8	1	5	3	4
1	9	5	3	6	4	2	8	7
9	3	7	1	4	8	6	5	2
5	2	8	6	7	3	1	4	9
6	1	4	2	5	9	3	7	8

INSANE - 84

9	8	5	1	7	2	6	4	3
1	4	3	6	9	8	2	5	7
7	6	2	5	3	4	1	8	9
6	3	4	7	5	1	9	2	8
5	7	1	8	2	9	3	6	4
2	9	8	4	6	3	7	1	5
4	5	7	3	1	6	8	9	2
3	2	6	9	8	5	4	7	1
8	1	9	2	4	7	5	3	6

INSANE - 85

1	4	3	2	7	9	8	6	5
5	8	9	4	3	6	2	1	7
2	6	7	5	8	1	3	4	9
7	2	1	6	5	8	9	3	4
3	5	6	1	9	4	7	8	2
8	9	4	3	2	7	6	5	1
4	3	5	7	6	2	1	9	8
9	1	2	8	4	3	5	7	6
6	7	8	9	1	5	4	2	3

INSANE - 86

6	1	8	7	9	2	4	3	5
3	2	5	8	1	4	7	9	6
7	4	9	3	5	6	2	8	1
4	6	2	1	3	9	5	7	8
5	7	1	6	2	8	9	4	3
8	9	3	4	7	5	6	1	2
9	3	4	5	6	1	8	2	7
1	8	6	2	4	7	3	5	9
2	5	7	9	8	3	1	6	4

INSANE - 87

7	1	5	2	4	3	6	9	8
2	3	6	8	1	9	7	4	5
4	8	9	6	7	5	3	2	1
1	9	8	5	3	4	2	6	7
3	4	2	7	8	6	1	5	9
5	6	7	9	2	1	8	3	4
8	5	4	3	6	7	9	1	2
6	7	1	4	9	2	5	8	3
9	2	3	1	5	8	4	7	6

INSANE - 88

7	9	8	4	1	6	5	3	2
6	2	3	7	9	5	1	8	4
1	4	5	3	8	2	7	9	6
3	6	4	9	5	8	2	1	7
5	8	9	2	7	1	4	6	3
2	1	7	6	4	3	9	5	8
9	7	6	1	3	4	8	2	5
4	5	2	8	6	9	3	7	1
8	3	1	5	2	7	6	4	9

INSANE - 89

4	8	6	9	1	3	2	5	7
1	9	7	8	2	5	4	3	6
5	3	2	7	4	6	8	9	1
2	4	5	6	7	9	1	8	3
9	7	1	3	8	4	5	6	2
8	6	3	2	5	1	7	4	9
7	1	9	5	6	8	3	2	4
3	2	8	4	9	7	6	1	5
6	5	4	1	3	2	9	7	8

INSANE - 90

4	5	6	9	8	3	1	7	2
3	9	1	2	6	7	4	8	5
7	2	8	1	5	4	9	6	3
8	3	2	5	7	1	6	9	4
5	1	9	3	4	6	8	2	7
6	7	4	8	9	2	3	5	1
9	6	7	4	3	5	2	1	8
2	8	3	7	1	9	5	4	6
1	4	5	6	2	8	7	3	9

INSANE - 91

8	6	4	9	3	7	5	1	2
5	3	7	4	1	2	9	8	6
2	1	9	5	8	6	3	4	7
4	5	8	3	7	1	2	6	9
3	9	1	6	2	4	7	5	8
6	7	2	8	5	9	1	3	4
9	2	6	1	4	5	8	7	3
1	4	3	7	9	8	6	2	5
7	8	5	2	6	3	4	9	1

INSANE - 92

5	8	9	3	4	1	7	2	6
6	3	4	9	7	2	1	8	5
2	7	1	5	8	6	3	4	9
8	2	6	1	3	5	4	9	7
7	4	5	2	9	8	6	3	1
9	1	3	7	6	4	2	5	8
3	5	2	8	1	7	9	6	4
4	9	7	6	5	3	8	1	2
1	6	8	4	2	9	5	7	3

INSANE - 93

4	1	3	6	8	2	9	5	7
2	8	9	1	5	7	6	4	3
5	7	6	3	9	4	2	1	8
6	9	8	7	2	1	5	3	4
7	5	4	9	6	3	8	2	1
1	3	2	8	4	5	7	9	6
8	2	5	4	1	6	3	7	9
3	6	1	2	7	9	4	8	5
9	4	7	5	3	8	1	6	2

INSANE - 94

6	3	7	5	8	9	2	4	1
9	5	4	6	2	1	8	3	7
8	1	2	7	4	3	5	9	6
1	9	8	3	6	4	7	2	5
2	6	5	8	9	7	4	1	3
7	4	3	2	1	5	9	6	8
4	2	6	1	5	8	3	7	9
3	8	1	9	7	2	6	5	4
5	7	9	4	3	6	1	8	2

INSANE - 95

1	4	7	2	5	8	9	6	3
6	9	8	4	3	1	5	2	7
3	2	5	6	9	7	8	4	1
8	7	1	9	4	2	3	5	6
4	3	9	5	8	6	1	7	2
2	5	6	1	7	3	4	8	9
7	6	3	8	1	5	2	9	4
5	1	4	7	2	9	6	3	8
9	8	2	3	6	4	7	1	5

INSANE - 96

2	3	4	8	5	1	9	7	6
6	7	8	9	3	4	2	1	5
1	5	9	6	7	2	4	8	3
7	1	6	5	4	3	8	2	9
8	2	5	7	1	9	3	6	4
9	4	3	2	6	8	1	5	7
3	9	7	1	8	5	6	4	2
5	8	2	4	9	6	7	3	1
4	6	1	3	2	7	5	9	8

INSANE - 97

5	9	3	8	7	1	2	4	6
4	6	7	2	9	3	5	1	8
1	2	8	5	6	4	9	7	3
3	1	9	7	4	5	8	6	2
7	8	2	9	1	6	3	5	4
6	4	5	3	8	2	1	9	7
8	3	6	1	5	7	4	2	9
2	7	1	4	3	9	6	8	5
9	5	4	6	2	8	7	3	1

INSANE - 98

2	1	4	8	9	5	3	7	6
9	6	8	4	7	3	5	1	2
7	3	5	6	1	2	9	4	8
8	5	7	1	2	6	4	9	3
6	2	9	5	3	4	7	8	1
3	4	1	9	8	7	6	2	5
4	7	6	2	5	8	1	3	9
5	9	2	3	4	1	8	6	7
1	8	3	7	6	9	2	5	4

INSANE - 99

4	6	9	3	2	7	8	1	5
8	3	7	6	5	1	2	9	4
5	1	2	8	4	9	7	6	3
1	8	5	4	9	3	6	7	2
9	4	3	7	6	2	5	8	1
7	2	6	5	1	8	3	4	9
2	9	8	1	3	6	4	5	7
6	5	1	2	7	4	9	3	8
3	7	4	9	8	5	1	2	6

INSANE - 100

3	6	7	5	1	4	2	9	8
9	8	5	7	3	2	1	6	4
1	2	4	8	9	6	5	7	3
2	7	3	9	6	1	8	4	5
5	1	8	4	7	3	6	2	9
4	9	6	2	5	8	7	3	1
6	4	2	3	8	5	9	1	7
8	3	9	1	2	7	4	5	6
7	5	1	6	4	9	3	8	2

INSANE - 101

9	1	3	2	7	5	8	6	4
5	4	6	1	3	8	9	7	2
7	2	8	6	9	4	5	1	3
3	8	9	4	6	7	1	2	5
1	5	4	9	8	2	7	3	6
2	6	7	3	5	1	4	9	8
4	3	5	7	2	9	6	8	1
8	7	2	5	1	6	3	4	9
6	9	1	8	4	3	2	5	7

INSANE - 102

8	7	6	9	3	1	4	2	5
4	2	5	6	7	8	1	9	3
1	3	9	5	2	4	6	8	7
5	1	4	7	6	2	8	3	9
9	8	7	3	4	5	2	6	1
3	6	2	1	8	9	7	5	4
6	5	3	2	1	7	9	4	8
2	4	1	8	9	3	5	7	6
7	9	8	4	5	6	3	1	2

INSANE - 103

5	8	3	9	2	6	1	4	7
9	1	7	3	5	4	6	2	8
4	2	6	1	7	8	5	9	3
3	6	4	8	9	7	2	5	1
8	5	9	4	1	2	7	3	6
1	7	2	5	6	3	9	8	4
7	3	1	2	4	5	8	6	9
6	4	5	7	8	9	3	1	2
2	9	8	6	3	1	4	7	5

INSANE - 104

1	6	9	4	3	2	5	7	8
5	8	3	7	1	6	2	9	4
2	7	4	8	5	9	1	6	3
6	4	5	9	8	3	7	2	1
7	1	2	6	4	5	3	8	9
9	3	8	1	2	7	4	5	6
8	5	7	3	6	1	9	4	2
4	9	1	2	7	8	6	3	5
3	2	6	5	9	4	8	1	7

INSANE - 105

8	9	3	2	1	6	7	5	4
1	4	5	9	7	8	3	6	2
2	7	6	5	4	3	9	8	1
9	3	7	4	6	2	8	1	5
4	5	2	1	8	7	6	3	9
6	8	1	3	9	5	4	2	7
5	2	4	8	3	9	1	7	6
3	6	9	7	2	1	5	4	8
7	1	8	6	5	4	2	9	3

INSANE - 106

2	9	4	1	3	5	8	6	7
8	5	1	9	7	6	4	3	2
6	7	3	8	4	2	1	5	9
5	8	6	4	9	1	7	2	3
3	4	9	2	8	7	5	1	6
7	1	2	6	5	3	9	8	4
1	6	7	5	2	4	3	9	8
9	3	5	7	6	8	2	4	1
4	2	8	3	1	9	6	7	5

INSANE - 107

5	6	8	2	9	4	7	1	3
3	1	9	7	8	5	2	6	4
7	2	4	6	3	1	5	9	8
6	8	2	5	7	3	9	4	1
1	5	3	8	4	9	6	2	7
4	9	7	1	2	6	3	8	5
2	4	1	3	6	7	8	5	9
9	7	6	4	5	8	1	3	2
8	3	5	9	1	2	4	7	6

INSANE - 108

6	7	9	5	8	3	1	2	4
2	8	1	7	9	4	5	3	6
4	5	3	6	1	2	8	9	7
7	1	5	2	4	9	3	6	8
8	6	2	3	7	5	4	1	9
3	9	4	1	6	8	7	5	2
5	4	6	9	3	7	2	8	1
9	2	8	4	5	1	6	7	3
1	3	7	8	2	6	9	4	5

INSANE - 109

6	5	3	7	8	4	2	9	1
4	1	8	5	2	9	3	7	6
9	2	7	6	1	3	8	5	4
5	9	4	2	3	1	7	6	8
3	7	6	9	5	8	4	1	2
1	8	2	4	7	6	5	3	9
2	6	9	3	4	7	1	8	5
8	3	5	1	9	2	6	4	7
7	4	1	8	6	5	9	2	3

INSANE - 110

3	2	7	9	8	6	5	4	1
8	1	9	7	5	4	6	2	3
6	5	4	3	1	2	8	9	7
1	6	3	4	7	5	2	8	9
4	8	2	1	6	9	3	7	5
9	7	5	8	2	3	4	1	6
5	4	8	6	9	1	7	3	2
2	3	1	5	4	7	9	6	8
7	9	6	2	3	8	1	5	4

INSANE - 111

5	9	2	4	3	8	1	7	6
8	3	1	7	5	6	4	9	2
4	7	6	1	2	9	3	5	8
2	8	4	6	7	5	9	3	1
3	6	5	2	9	1	7	8	4
9	1	7	3	8	4	2	6	5
1	2	8	9	6	7	5	4	3
7	5	3	8	4	2	6	1	9
6	4	9	5	1	3	8	2	7

INSANE - 112

8	5	6	2	9	7	1	3	4
2	4	3	5	6	1	7	8	9
7	1	9	3	4	8	2	6	5
1	2	4	7	5	6	3	9	8
3	6	5	1	8	9	4	2	7
9	7	8	4	3	2	6	5	1
4	9	2	8	1	3	5	7	6
6	3	1	9	7	5	8	4	2
5	8	7	6	2	4	9	1	3

INSANE - 113

6	7	4	5	9	8	3	2	1
5	9	8	1	2	3	7	4	6
1	2	3	4	6	7	9	8	5
8	5	1	9	7	4	6	3	2
7	3	9	6	5	2	8	1	4
2	4	6	8	3	1	5	9	7
9	6	2	3	1	5	4	7	8
3	8	7	2	4	6	1	5	9
4	1	5	7	8	9	2	6	3

INSANE - 114

1	4	9	2	5	8	3	7	6
7	3	8	1	6	4	2	9	5
6	5	2	7	9	3	1	8	4
8	6	4	5	1	9	7	3	2
5	9	3	4	2	7	6	1	8
2	7	1	8	3	6	5	4	9
3	1	6	9	8	5	4	2	7
4	8	5	3	7	2	9	6	1
9	2	7	6	4	1	8	5	3

INSANE - 115

3	4	1	5	6	7	2	8	9
7	6	8	1	9	2	3	5	4
5	2	9	4	3	8	1	7	6
8	1	4	2	5	3	6	9	7
2	5	3	9	7	6	8	4	1
6	9	7	8	4	1	5	2	3
1	7	5	3	2	9	4	6	8
4	3	6	7	8	5	9	1	2
9	8	2	6	1	4	7	3	5

INSANE - 116

3	7	4	6	9	2	8	5	1
8	5	6	1	7	3	9	4	2
2	1	9	5	8	4	7	3	6
6	9	1	2	3	5	4	8	7
7	8	5	4	1	9	2	6	3
4	2	3	7	6	8	5	1	9
5	6	2	9	4	1	3	7	8
9	3	7	8	5	6	1	2	4
1	4	8	3	2	7	6	9	5

INSANE - 117

3	8	2	9	1	5	6	4	7
7	1	4	8	2	6	9	3	5
9	5	6	3	4	7	8	2	1
6	2	7	1	9	3	4	5	8
4	9	1	6	5	8	3	7	2
8	3	5	4	7	2	1	9	6
5	6	3	2	8	4	7	1	9
2	4	9	7	6	1	5	8	3
1	7	8	5	3	9	2	6	4

INSANE - 118

9	8	2	6	4	5	7	1	3
1	4	7	3	2	9	6	8	5
3	5	6	1	8	7	9	2	4
8	6	5	7	1	4	2	3	9
7	1	4	9	3	2	5	6	8
2	9	3	5	6	8	4	7	1
5	3	8	4	7	6	1	9	2
4	7	1	2	9	3	8	5	6
6	2	9	8	5	1	3	4	7

INSANE - 119

2	8	7	6	5	1	9	3	4
5	1	3	4	8	9	2	6	7
4	6	9	7	2	3	1	8	5
8	7	2	3	4	6	5	9	1
3	4	6	9	1	5	7	2	8
1	9	5	8	7	2	3	4	6
6	2	4	5	9	7	8	1	3
9	5	8	1	3	4	6	7	2
7	3	1	2	6	8	4	5	9

INSANE - 120

6	5	8	9	7	3	4	2	1
1	7	2	5	8	4	9	3	6
3	4	9	1	2	6	8	7	5
5	2	3	6	4	1	7	9	8
9	8	1	3	5	7	2	6	4
4	6	7	8	9	2	5	1	3
8	9	6	7	1	5	3	4	2
7	1	4	2	3	8	6	5	9
2	3	5	4	6	9	1	8	7

INSANE - 121

7	3	5	4	1	8	2	9	6
9	4	1	2	3	6	8	7	5
2	6	8	7	9	5	3	1	4
3	7	9	5	6	2	4	8	1
4	5	2	3	8	1	9	6	7
1	8	6	9	7	4	5	3	2
8	1	4	6	2	9	7	5	3
6	2	3	8	5	7	1	4	9
5	9	7	1	4	3	6	2	8

INSANE - 122

9	1	3	2	7	4	6	8	5
6	2	4	5	9	8	3	1	7
8	7	5	6	1	3	2	4	9
4	6	8	7	5	1	9	3	2
2	5	1	4	3	9	7	6	8
7	3	9	8	2	6	4	5	1
1	8	7	3	4	2	5	9	6
5	4	6	9	8	7	1	2	3
3	9	2	1	6	5	8	7	4

INSANE - 123

7	9	2	8	5	3	4	1	6
4	8	3	6	9	1	5	2	7
6	5	1	4	7	2	9	3	8
2	1	6	5	8	9	3	7	4
9	4	5	1	3	7	8	6	2
8	3	7	2	4	6	1	5	9
1	2	8	9	6	5	7	4	3
3	6	9	7	1	4	2	8	5
5	7	4	3	2	8	6	9	1

INSANE - 124

6	9	4	8	2	5	3	1	7
7	1	3	9	6	4	5	2	8
8	2	5	7	3	1	9	6	4
4	5	9	1	8	2	6	7	3
3	7	8	4	5	6	2	9	1
2	6	1	3	7	9	8	4	5
5	4	7	6	9	3	1	8	2
1	3	6	2	4	8	7	5	9
9	8	2	5	1	7	4	3	6

INSANE - 125

1	4	5	3	7	9	6	2	8
7	9	8	4	2	6	5	3	1
2	3	6	1	8	5	7	9	4
8	7	4	6	9	2	3	1	5
6	1	9	8	5	3	4	7	2
5	2	3	7	4	1	9	8	6
4	6	1	9	3	8	2	5	7
9	8	2	5	6	7	1	4	3
3	5	7	2	1	4	8	6	9

INSANE - 126

4	9	5	7	1	2	6	3	8
1	6	8	4	9	3	7	2	5
3	7	2	8	5	6	9	4	1
7	3	9	2	4	8	5	1	6
2	5	1	6	3	9	8	7	4
8	4	6	5	7	1	3	9	2
5	1	7	9	8	4	2	6	3
6	8	4	3	2	7	1	5	9
9	2	3	1	6	5	4	8	7

INSANE - 127

1	4	7	6	2	3	8	9	5
9	6	2	7	5	8	3	1	4
3	8	5	9	4	1	7	2	6
2	9	1	3	8	6	4	5	7
7	5	6	2	9	4	1	8	3
4	3	8	5	1	7	2	6	9
5	2	3	8	7	9	6	4	1
8	7	4	1	6	5	9	3	2
6	1	9	4	3	2	5	7	8

INSANE - 128

6	9	4	7	8	2	5	3	1
7	1	5	3	4	9	8	2	6
2	8	3	1	5	6	4	9	7
8	6	9	5	3	4	7	1	2
5	2	7	6	9	1	3	4	8
3	4	1	8	2	7	6	5	9
9	3	8	2	7	5	1	6	4
4	5	6	9	1	8	2	7	3
1	7	2	4	6	3	9	8	5

INSANE - 129

5	9	4	1	2	3	7	8	6
1	7	3	6	9	8	5	4	2
8	6	2	7	4	5	1	9	3
4	1	7	2	8	9	3	6	5
9	2	8	5	3	6	4	7	1
6	3	5	4	7	1	8	2	9
2	5	9	8	1	4	6	3	7
3	8	1	9	6	7	2	5	4
7	4	6	3	5	2	9	1	8

INSANE - 130

8	5	6	9	1	3	4	2	7
4	3	9	5	7	2	8	1	6
7	1	2	8	6	4	3	5	9
5	7	8	3	2	1	9	6	4
2	9	1	6	4	8	5	7	3
3	6	4	7	5	9	1	8	2
1	8	3	2	9	7	6	4	5
9	2	5	4	8	6	7	3	1
6	4	7	1	3	5	2	9	8

INSANE - 131

2	1	7	5	3	4	9	8	6
5	8	3	9	1	6	2	7	4
9	6	4	8	2	7	5	1	3
8	9	1	7	6	3	4	5	2
7	3	2	4	8	5	1	6	9
4	5	6	1	9	2	7	3	8
3	7	5	2	4	8	6	9	1
6	2	9	3	7	1	8	4	5
1	4	8	6	5	9	3	2	7

INSANE - 132

4	8	1	5	6	7	9	3	2
6	9	5	2	8	3	4	7	1
2	3	7	9	1	4	5	8	6
3	7	8	6	4	9	2	1	5
1	5	4	7	2	8	6	9	3
9	2	6	1	3	5	7	4	8
5	1	9	8	7	6	3	2	4
8	6	3	4	9	2	1	5	7
7	4	2	3	5	1	8	6	9

INSANE - 133

4	2	7	9	8	5	3	6	1
9	8	3	1	2	6	4	5	7
5	6	1	4	7	3	8	2	9
6	7	5	2	9	8	1	4	3
2	4	8	6	3	1	7	9	5
1	3	9	7	5	4	6	8	2
7	9	6	8	1	2	5	3	4
3	1	4	5	6	9	2	7	8
8	5	2	3	4	7	9	1	6

INSANE - 134

5	6	9	1	7	8	4	2	3
8	3	4	5	2	9	1	7	6
1	7	2	4	6	3	9	8	5
9	5	8	6	1	4	7	3	2
2	4	7	9	3	5	6	1	8
6	1	3	2	8	7	5	9	4
7	9	5	8	4	2	3	6	1
3	2	6	7	5	1	8	4	9
4	8	1	3	9	6	2	5	7

INSANE - 135

5	2	1	8	4	3	7	9	6
7	6	8	2	9	5	4	1	3
4	3	9	7	1	6	2	8	5
8	5	6	9	2	4	3	7	1
2	9	4	1	3	7	5	6	8
3	1	7	5	6	8	9	4	2
9	4	5	6	8	2	1	3	7
6	7	3	4	5	1	8	2	9
1	8	2	3	7	9	6	5	4

INSANE - 136

6	4	8	9	5	1	7	2	3
5	7	2	6	3	4	1	9	8
9	3	1	7	2	8	4	6	5
8	6	7	2	9	3	5	1	4
2	9	3	4	1	5	8	7	6
1	5	4	8	7	6	2	3	9
3	8	6	1	4	2	9	5	7
4	1	9	5	6	7	3	8	2
7	2	5	3	8	9	6	4	1

INSANE - 137

2	4	3	8	5	1	9	7	6
8	5	6	4	9	7	3	1	2
7	1	9	3	6	2	5	4	8
4	8	7	1	2	5	6	9	3
9	2	1	7	3	6	8	5	4
6	3	5	9	8	4	7	2	1
3	7	2	6	4	9	1	8	5
1	6	4	5	7	8	2	3	9
5	9	8	2	1	3	4	6	7

INSANE - 138

2	9	5	4	6	3	1	8	7
3	8	4	9	7	1	2	5	6
7	6	1	5	8	2	4	9	3
4	1	9	6	2	7	8	3	5
5	2	6	3	4	8	9	7	1
8	7	3	1	9	5	6	2	4
1	4	2	7	3	9	5	6	8
9	5	7	8	1	6	3	4	2
6	3	8	2	5	4	7	1	9

INSANE - 139

9	1	7	8	3	4	2	6	5
2	6	3	5	7	1	8	4	9
4	8	5	2	6	9	1	7	3
1	3	6	7	5	8	9	2	4
5	9	2	6	4	3	7	8	1
8	7	4	1	9	2	3	5	6
6	2	9	3	8	5	4	1	7
3	5	8	4	1	7	6	9	2
7	4	1	9	2	6	5	3	8

INSANE - 140

2	3	1	5	7	6	4	9	8
5	9	4	1	8	2	3	7	6
6	7	8	3	4	9	5	1	2
4	8	6	7	2	1	9	3	5
7	2	5	8	9	3	1	6	4
9	1	3	6	5	4	2	8	7
3	4	7	9	6	5	8	2	1
1	6	2	4	3	8	7	5	9
8	5	9	2	1	7	6	4	3

INSANE - 141

2	3	1	6	8	7	4	5	9
9	6	7	3	5	4	1	8	2
8	4	5	2	9	1	6	3	7
5	7	3	8	4	2	9	6	1
6	2	9	5	1	3	7	4	8
1	8	4	9	7	6	5	2	3
3	9	2	1	6	5	8	7	4
7	5	8	4	3	9	2	1	6
4	1	6	7	2	8	3	9	5

INSANE - 142

1	5	7	4	3	2	6	9	8
3	6	4	9	7	8	2	1	5
9	2	8	5	6	1	4	7	3
4	8	1	6	2	3	7	5	9
5	7	3	1	4	9	8	2	6
6	9	2	8	5	7	3	4	1
8	1	6	7	9	4	5	3	2
7	3	9	2	8	5	1	6	4
2	4	5	3	1	6	9	8	7

INSANE - 143

9	5	7	6	8	2	4	1	3
4	6	2	1	5	3	7	9	8
1	3	8	7	4	9	2	6	5
5	9	3	8	6	4	1	7	2
6	7	4	2	1	5	8	3	9
2	8	1	9	3	7	6	5	4
3	1	9	4	7	8	5	2	6
8	2	6	5	9	1	3	4	7
7	4	5	3	2	6	9	8	1

INSANE - 144

8	4	6	5	9	7	1	3	2
3	1	7	6	4	2	5	9	8
5	9	2	8	1	3	7	4	6
6	8	3	7	2	9	4	5	1
4	7	9	1	6	5	8	2	3
1	2	5	3	8	4	9	6	7
2	6	4	9	7	8	3	1	5
7	5	1	4	3	6	2	8	9
9	3	8	2	5	1	6	7	4

INSANE - 145

1	3	6	9	2	7	8	5	4
8	2	7	6	4	5	3	9	1
4	9	5	8	1	3	6	2	7
7	1	9	4	5	6	2	8	3
3	8	4	1	9	2	7	6	5
6	5	2	7	3	8	1	4	9
2	6	3	5	7	9	4	1	8
9	4	8	3	6	1	5	7	2
5	7	1	2	8	4	9	3	6

INSANE - 146

7	1	9	8	4	6	5	3	2
2	5	6	9	3	1	8	4	7
4	3	8	5	7	2	6	9	1
1	9	5	2	6	3	7	8	4
6	8	2	4	5	7	9	1	3
3	7	4	1	8	9	2	5	6
5	4	3	6	2	8	1	7	9
8	6	1	7	9	4	3	2	5
9	2	7	3	1	5	4	6	8

INSANE - 147

5	6	4	9	2	1	7	8	3
1	2	7	8	3	4	6	9	5
8	3	9	7	6	5	2	4	1
4	8	6	3	7	2	5	1	9
3	7	1	4	5	9	8	2	6
2	9	5	6	1	8	4	3	7
7	5	2	1	8	3	9	6	4
9	1	8	5	4	6	3	7	2
6	4	3	2	9	7	1	5	8

INSANE - 148

2	4	7	1	3	8	9	5	6
1	9	8	7	6	5	4	2	3
6	3	5	9	2	4	1	7	8
3	6	4	5	8	9	7	1	2
7	2	1	3	4	6	8	9	5
5	8	9	2	1	7	6	3	4
8	1	3	6	7	2	5	4	9
4	5	2	8	9	1	3	6	7
9	7	6	4	5	3	2	8	1

INSANE - 149

4	5	3	9	8	7	1	6	2
7	9	1	3	6	2	4	5	8
6	8	2	5	4	1	7	3	9
9	1	8	4	5	3	2	7	6
5	7	6	8	2	9	3	1	4
3	2	4	1	7	6	9	8	5
1	4	5	7	9	8	6	2	3
2	3	9	6	1	5	8	4	7
8	6	7	2	3	4	5	9	1

INSANE - 150

6	1	7	9	2	3	8	5	4
4	3	8	7	6	5	1	9	2
9	2	5	1	4	8	6	3	7
2	5	1	3	8	4	7	6	9
8	6	3	2	9	7	5	4	1
7	9	4	6	5	1	2	8	3
3	4	6	5	1	2	9	7	8
1	8	9	4	7	6	3	2	5
5	7	2	8	3	9	4	1	6

INSANE - 151

6	1	3	4	8	2	5	9	7
7	8	5	6	9	1	4	3	2
2	9	4	7	3	5	6	1	8
3	2	1	8	4	7	9	5	6
4	6	7	1	5	9	8	2	3
8	5	9	2	6	3	1	7	4
5	3	8	9	7	6	2	4	1
9	4	2	3	1	8	7	6	5
1	7	6	5	2	4	3	8	9

INSANE - 152

9	8	1	3	5	4	6	7	2
6	2	5	7	8	9	1	3	4
3	4	7	6	2	1	8	5	9
4	1	6	9	3	8	7	2	5
8	5	9	1	7	2	3	4	6
7	3	2	4	6	5	9	8	1
1	7	4	5	9	3	2	6	8
2	9	3	8	4	6	5	1	7
5	6	8	2	1	7	4	9	3

INSANE - 153

6	9	8	3	4	5	2	1	7
3	1	2	7	8	9	5	4	6
7	5	4	2	1	6	8	3	9
1	6	7	9	5	8	3	2	4
2	8	3	1	7	4	9	6	5
9	4	5	6	2	3	1	7	8
5	7	6	8	3	2	4	9	1
4	3	9	5	6	1	7	8	2
8	2	1	4	9	7	6	5	3

INSANE - 154

6	7	9	4	3	8	5	1	2
5	4	3	6	1	2	8	9	7
1	2	8	9	7	5	3	4	6
2	8	4	3	9	1	6	7	5
7	3	6	2	5	4	9	8	1
9	1	5	7	8	6	4	2	3
4	6	1	8	2	3	7	5	9
3	5	7	1	4	9	2	6	8
8	9	2	5	6	7	1	3	4

INSANE - 155

2	8	4	7	3	9	5	1	6
5	7	9	2	1	6	8	3	4
1	3	6	8	4	5	2	7	9
7	6	3	9	5	4	1	2	8
4	2	5	3	8	1	6	9	7
8	9	1	6	7	2	3	4	5
6	4	8	1	9	3	7	5	2
9	1	7	5	2	8	4	6	3
3	5	2	4	6	7	9	8	1

INSANE - 156

9	2	7	4	6	1	8	5	3
4	5	3	9	2	8	6	7	1
1	8	6	3	5	7	9	2	4
2	7	5	1	9	6	3	4	8
3	4	1	8	7	2	5	9	6
6	9	8	5	4	3	7	1	2
7	6	9	2	8	4	1	3	5
5	1	2	6	3	9	4	8	7
8	3	4	7	1	5	2	6	9

INSANE - 157

6	2	7	3	8	1	9	5	4
5	8	1	7	9	4	2	3	6
9	4	3	6	2	5	7	8	1
3	6	5	1	4	2	8	7	9
8	7	9	5	6	3	1	4	2
2	1	4	8	7	9	5	6	3
1	9	6	4	5	7	3	2	8
7	3	8	2	1	6	4	9	5
4	5	2	9	3	8	6	1	7

INSANE - 158

3	4	2	7	1	8	5	6	9
9	6	8	4	3	5	1	7	2
5	7	1	2	6	9	8	3	4
7	8	6	3	2	4	9	1	5
4	3	5	1	9	7	2	8	6
2	1	9	5	8	6	7	4	3
8	5	7	9	4	3	6	2	1
6	2	4	8	5	1	3	9	7
1	9	3	6	7	2	4	5	8

INSANE - 159

3	5	2	4	8	6	7	1	9
4	9	6	3	1	7	8	5	2
8	7	1	5	9	2	3	4	6
2	6	8	7	4	3	5	9	1
7	4	3	1	5	9	6	2	8
9	1	5	6	2	8	4	7	3
1	2	4	8	6	5	9	3	7
5	8	7	9	3	1	2	6	4
6	3	9	2	7	4	1	8	5

INSANE - 160

4	9	3	8	5	6	2	7	1
8	6	2	7	3	1	4	5	9
7	1	5	9	4	2	8	6	3
9	7	8	1	2	3	6	4	5
5	4	1	6	8	9	7	3	2
3	2	6	5	7	4	1	9	8
1	5	7	4	9	8	3	2	6
6	3	9	2	1	7	5	8	4
2	8	4	3	6	5	9	1	7

INSANE - 161

1	3	4	5	8	2	9	6	7
9	6	8	1	3	7	2	4	5
2	7	5	4	6	9	8	3	1
8	5	6	3	4	1	7	2	9
7	2	1	6	9	8	3	5	4
4	9	3	7	2	5	1	8	6
5	1	2	8	7	4	6	9	3
3	4	9	2	1	6	5	7	8
6	8	7	9	5	3	4	1	2

INSANE - 162

2	6	7	8	4	5	9	1	3
3	4	5	9	7	1	6	8	2
1	8	9	6	3	2	5	4	7
5	7	6	1	9	4	3	2	8
8	3	1	7	2	6	4	9	5
4	9	2	3	5	8	7	6	1
7	2	3	4	8	9	1	5	6
6	5	4	2	1	7	8	3	9
9	1	8	5	6	3	2	7	4

INSANE - 163

1	5	8	7	4	2	6	9	3
7	3	6	8	9	1	5	2	4
4	2	9	6	3	5	7	1	8
8	1	7	4	5	6	9	3	2
5	9	4	1	2	3	8	6	7
3	6	2	9	7	8	1	4	5
6	7	5	3	1	4	2	8	9
9	4	1	2	8	7	3	5	6
2	8	3	5	6	9	4	7	1

INSANE - 164

7	9	8	2	3	1	4	5	6
3	1	2	4	6	5	9	8	7
5	4	6	7	9	8	3	1	2
9	6	5	1	8	4	7	2	3
2	3	4	6	5	7	1	9	8
8	7	1	9	2	3	6	4	5
6	5	7	8	4	9	2	3	1
1	8	9	3	7	2	5	6	4
4	2	3	5	1	6	8	7	9

INSANE - 165

3	8	6	5	4	2	1	7	9
9	4	1	6	8	7	5	2	3
2	5	7	1	3	9	6	4	8
5	6	2	8	7	1	9	3	4
7	9	4	2	6	3	8	1	5
1	3	8	9	5	4	2	6	7
6	7	5	3	2	8	4	9	1
4	2	9	7	1	5	3	8	6
8	1	3	4	9	6	7	5	2

INSANE - 166

4	5	6	2	8	9	7	1	3
9	1	3	7	6	5	4	8	2
7	8	2	4	1	3	5	6	9
8	3	7	5	4	6	2	9	1
2	4	5	3	9	1	8	7	6
1	6	9	8	7	2	3	4	5
6	2	4	9	5	7	1	3	8
3	7	1	6	2	8	9	5	4
5	9	8	1	3	4	6	2	7

INSANE - 167

6	4	2	7	8	1	9	5	3
9	7	3	6	4	5	8	1	2
5	8	1	9	3	2	4	7	6
8	6	5	4	7	3	1	2	9
3	9	4	1	2	8	7	6	5
1	2	7	5	9	6	3	4	8
7	1	6	3	5	9	2	8	4
2	5	9	8	1	4	6	3	7
4	3	8	2	6	7	5	9	1

INSANE - 168

2	8	3	5	9	1	4	6	7
4	9	6	7	2	8	5	3	1
7	1	5	4	6	3	9	2	8
1	6	2	8	4	7	3	5	9
5	7	8	6	3	9	1	4	2
3	4	9	2	1	5	8	7	6
9	3	4	1	7	2	6	8	5
8	2	1	3	5	6	7	9	4
6	5	7	9	8	4	2	1	3

INSANE - 169

8	7	2	5	3	1	6	4	9
3	5	6	2	4	9	1	7	8
1	9	4	8	6	7	5	3	2
6	1	5	7	2	8	3	9	4
2	3	7	6	9	4	8	5	1
4	8	9	1	5	3	2	6	7
5	4	1	3	7	2	9	8	6
7	2	3	9	8	6	4	1	5
9	6	8	4	1	5	7	2	3

INSANE - 170

9	8	4	5	7	3	6	2	1
7	2	1	9	4	6	8	3	5
5	6	3	1	8	2	4	7	9
3	4	9	7	2	8	5	1	6
6	1	8	3	9	5	2	4	7
2	7	5	4	6	1	9	8	3
4	5	7	8	3	9	1	6	2
8	9	6	2	1	7	3	5	4
1	3	2	6	5	4	7	9	8

INSANE - 171

6	2	8	4	9	5	7	3	1
9	1	5	7	2	3	4	8	6
3	7	4	1	8	6	9	5	2
7	9	2	8	3	4	1	6	5
8	4	1	5	6	9	2	7	3
5	6	3	2	1	7	8	9	4
2	5	6	9	7	1	3	4	8
4	8	7	3	5	2	6	1	9
1	3	9	6	4	8	5	2	7

INSANE - 172

4	3	7	2	6	9	8	1	5
8	2	6	7	5	1	4	3	9
5	9	1	3	8	4	7	2	6
1	5	3	6	9	7	2	8	4
2	4	9	8	3	5	1	6	7
7	6	8	1	4	2	9	5	3
3	8	2	4	7	6	5	9	1
9	1	4	5	2	3	6	7	8
6	7	5	9	1	8	3	4	2

INSANE - 173

5	1	4	2	6	7	3	8	9
9	7	3	5	1	8	2	6	4
8	6	2	3	4	9	1	7	5
6	2	5	7	8	4	9	1	3
3	8	1	6	9	5	4	2	7
7	4	9	1	2	3	8	5	6
1	3	8	9	5	6	7	4	2
2	9	6	4	7	1	5	3	8
4	5	7	8	3	2	6	9	1

INSANE - 174

9	1	4	5	3	2	8	7	6
8	7	2	1	9	6	5	3	4
3	6	5	4	7	8	2	9	1
2	4	8	9	6	3	7	1	5
7	5	9	8	2	1	4	6	3
1	3	6	7	4	5	9	8	2
5	2	7	6	1	9	3	4	8
6	9	3	2	8	4	1	5	7
4	8	1	3	5	7	6	2	9

INSANE - 175

2	4	7	8	9	6	1	3	5
1	5	3	7	2	4	6	8	9
6	8	9	3	5	1	2	7	4
9	1	8	5	4	3	7	2	6
3	7	4	1	6	2	9	5	8
5	6	2	9	7	8	3	4	1
7	3	5	6	8	9	4	1	2
8	2	6	4	1	7	5	9	3
4	9	1	2	3	5	8	6	7

INSANE - 176

5	4	8	2	9	7	3	1	6
9	2	3	5	6	1	8	4	7
1	6	7	8	3	4	2	5	9
2	1	6	7	5	8	9	3	4
3	5	4	6	2	9	1	7	8
8	7	9	4	1	3	5	6	2
4	3	1	9	7	2	6	8	5
7	9	5	3	8	6	4	2	1
6	8	2	1	4	5	7	9	3

INSANE - 177

5	8	9	2	6	7	4	1	3
3	6	7	5	1	4	2	9	8
2	1	4	9	8	3	7	6	5
8	7	6	1	5	9	3	2	4
4	2	1	8	3	6	9	5	7
9	5	3	4	7	2	6	8	1
7	3	5	6	2	1	8	4	9
1	9	2	7	4	8	5	3	6
6	4	8	3	9	5	1	7	2

INSANE - 178

6	3	5	8	9	7	2	1	4
7	4	9	3	1	2	5	6	8
8	2	1	4	6	5	7	3	9
1	6	4	2	5	9	8	7	3
2	7	8	6	4	3	1	9	5
9	5	3	1	7	8	4	2	6
5	8	6	9	2	1	3	4	7
4	1	7	5	3	6	9	8	2
3	9	2	7	8	4	6	5	1

INSANE - 179

1	9	3	5	6	4	8	2	7
4	2	8	7	1	3	6	5	9
6	7	5	8	2	9	4	3	1
5	4	2	6	3	1	9	7	8
3	6	7	4	9	8	5	1	2
9	8	1	2	7	5	3	6	4
2	1	9	3	8	6	7	4	5
7	3	4	9	5	2	1	8	6
8	5	6	1	4	7	2	9	3

INSANE - 180

7	9	5	8	3	2	1	6	4
1	3	6	9	4	7	2	8	5
8	2	4	1	6	5	7	9	3
5	4	7	3	1	6	8	2	9
9	1	2	5	7	8	4	3	6
6	8	3	2	9	4	5	7	1
3	6	8	7	5	1	9	4	2
4	7	1	6	2	9	3	5	8
2	5	9	4	8	3	6	1	7

INSANE - 181

9	6	1	7	2	8	3	5	4
5	8	4	3	1	6	9	7	2
2	3	7	4	9	5	8	1	6
4	5	2	9	8	1	6	3	7
7	1	6	2	3	4	5	8	9
8	9	3	6	5	7	4	2	1
1	7	9	5	4	3	2	6	8
3	2	8	1	6	9	7	4	5
6	4	5	8	7	2	1	9	3

INSANE - 182

1	4	5	3	8	9	6	2	7
6	3	9	1	7	2	8	5	4
2	7	8	6	5	4	9	3	1
9	5	7	8	3	1	4	6	2
4	8	2	5	9	6	1	7	3
3	6	1	4	2	7	5	8	9
8	1	6	2	4	3	7	9	5
7	2	4	9	6	5	3	1	8
5	9	3	7	1	8	2	4	6

INSANE - 183

3	2	8	5	6	9	7	1	4
7	6	9	1	8	4	3	2	5
5	4	1	3	7	2	9	6	8
2	8	5	9	4	3	1	7	6
6	1	4	2	5	7	8	9	3
9	3	7	6	1	8	5	4	2
8	5	2	4	9	1	6	3	7
1	7	3	8	2	6	4	5	9
4	9	6	7	3	5	2	8	1

INSANE - 184

1	5	6	8	7	9	4	3	2
9	7	8	2	3	4	1	5	6
2	4	3	1	6	5	9	8	7
6	9	2	4	1	3	8	7	5
8	3	7	9	5	6	2	1	4
5	1	4	7	8	2	3	6	9
7	2	5	3	9	8	6	4	1
3	6	9	5	4	1	7	2	8
4	8	1	6	2	7	5	9	3

INSANE - 185

4	6	1	9	5	2	7	8	3
5	9	3	8	1	7	2	4	6
2	7	8	3	6	4	9	5	1
9	2	7	6	4	8	3	1	5
1	4	5	7	9	3	6	2	8
3	8	6	1	2	5	4	9	7
7	5	4	2	8	6	1	3	9
8	3	9	4	7	1	5	6	2
6	1	2	5	3	9	8	7	4

INSANE - 186

3	1	9	6	2	8	4	5	7
6	4	8	3	7	5	9	2	1
7	5	2	4	9	1	6	3	8
4	6	7	9	5	2	8	1	3
2	3	5	8	1	6	7	4	9
9	8	1	7	3	4	5	6	2
8	7	4	1	6	3	2	9	5
5	9	3	2	4	7	1	8	6
1	2	6	5	8	9	3	7	4

INSANE - 187

6	4	8	2	1	7	5	3	9
3	1	2	5	6	9	4	7	8
5	9	7	4	3	8	1	2	6
9	2	6	8	7	4	3	5	1
7	5	3	1	2	6	9	8	4
1	8	4	9	5	3	2	6	7
2	3	9	6	8	1	7	4	5
8	7	1	3	4	5	6	9	2
4	6	5	7	9	2	8	1	3

INSANE - 188

2	9	7	3	8	5	4	1	6
8	3	5	1	6	4	2	7	9
4	1	6	9	2	7	5	3	8
5	6	1	2	3	8	7	9	4
3	4	8	6	7	9	1	5	2
9	7	2	5	4	1	6	8	3
7	5	3	4	9	6	8	2	1
1	2	4	8	5	3	9	6	7
6	8	9	7	1	2	3	4	5

INSANE - 189

1	9	4	8	3	7	6	5	2
5	8	3	6	9	2	4	7	1
7	2	6	5	4	1	3	8	9
9	6	8	4	5	3	2	1	7
2	5	7	1	8	6	9	3	4
4	3	1	7	2	9	8	6	5
8	1	2	9	6	5	7	4	3
3	4	5	2	7	8	1	9	6
6	7	9	3	1	4	5	2	8

INSANE - 190

6	8	7	5	1	2	9	3	4
9	4	5	6	8	3	1	2	7
1	2	3	9	7	4	5	8	6
7	1	8	3	2	9	6	4	5
4	9	2	1	5	6	3	7	8
5	3	6	7	4	8	2	1	9
3	6	1	8	9	7	4	5	2
2	7	9	4	3	5	8	6	1
8	5	4	2	6	1	7	9	3

INSANE - 191

4	7	5	6	9	2	3	1	8
6	8	3	4	1	7	5	2	9
9	1	2	5	8	3	4	6	7
5	3	6	7	4	9	1	8	2
2	9	1	3	6	8	7	4	5
8	4	7	1	2	5	9	3	6
1	2	9	8	5	4	6	7	3
7	5	4	2	3	6	8	9	1
3	6	8	9	7	1	2	5	4

INSANE - 192

7	5	4	8	3	6	1	9	2
1	6	9	7	4	2	8	5	3
8	3	2	5	9	1	7	6	4
9	1	3	2	8	7	6	4	5
2	4	8	3	6	5	9	1	7
5	7	6	9	1	4	2	3	8
6	9	5	4	2	8	3	7	1
4	2	1	6	7	3	5	8	9
3	8	7	1	5	9	4	2	6

INSANE - 193

2	7	3	5	4	6	1	9	8
9	8	5	3	7	1	6	4	2
1	6	4	8	2	9	3	7	5
5	2	6	9	1	4	8	3	7
3	9	8	6	5	7	2	1	4
7	4	1	2	3	8	5	6	9
4	3	7	1	8	2	9	5	6
8	1	9	4	6	5	7	2	3
6	5	2	7	9	3	4	8	1

INSANE - 194

8	3	6	7	2	5	4	1	9
4	7	9	1	3	8	2	5	6
5	1	2	9	4	6	3	7	8
6	2	4	3	9	1	7	8	5
3	9	7	5	8	2	6	4	1
1	8	5	6	7	4	9	2	3
7	5	8	2	6	9	1	3	4
2	6	1	4	5	3	8	9	7
9	4	3	8	1	7	5	6	2

INSANE - 195

6	1	3	5	2	4	8	7	9
2	5	4	9	7	8	1	6	3
8	9	7	6	1	3	5	4	2
1	3	6	8	4	5	9	2	7
7	2	9	3	6	1	4	5	8
5	4	8	2	9	7	6	3	1
3	6	2	4	8	9	7	1	5
9	7	5	1	3	6	2	8	4
4	8	1	7	5	2	3	9	6

INSANE - 196

6	3	4	7	9	2	5	1	8
2	1	8	4	6	5	9	3	7
5	7	9	3	1	8	4	6	2
1	9	7	8	5	4	6	2	3
3	2	6	9	7	1	8	4	5
8	4	5	6	2	3	7	9	1
7	6	3	1	8	9	2	5	4
9	5	1	2	4	7	3	8	6
4	8	2	5	3	6	1	7	9

INSANE - 197

5	9	1	8	4	3	2	7	6
3	7	8	5	2	6	1	9	4
6	2	4	1	9	7	3	5	8
2	5	3	9	1	8	6	4	7
1	6	9	2	7	4	5	8	3
4	8	7	3	6	5	9	2	1
8	4	2	6	5	1	7	3	9
7	1	5	4	3	9	8	6	2
9	3	6	7	8	2	4	1	5

INSANE - 198

6	8	1	9	5	4	3	2	7
4	9	2	7	1	3	5	8	6
3	7	5	6	8	2	4	9	1
1	4	3	2	7	6	9	5	8
2	6	9	5	3	8	7	1	4
8	5	7	4	9	1	6	3	2
9	2	8	3	4	7	1	6	5
7	3	6	1	2	5	8	4	9
5	1	4	8	6	9	2	7	3

INSANE - 199

9	4	3	8	5	6	2	7	1
1	5	2	4	7	9	8	3	6
7	8	6	1	2	3	5	9	4
3	9	7	2	4	5	6	1	8
6	2	8	7	9	1	3	4	5
4	1	5	6	3	8	7	2	9
2	6	1	9	8	7	4	5	3
5	7	9	3	6	4	1	8	2
8	3	4	5	1	2	9	6	7

INSANE - 200

7	1	5	8	4	6	9	3	2
6	8	9	3	2	5	7	1	4
3	4	2	1	7	9	8	6	5
2	3	8	9	6	7	4	5	1
5	6	7	4	1	2	3	8	9
1	9	4	5	8	3	6	2	7
4	2	3	6	9	1	5	7	8
8	7	6	2	5	4	1	9	3
9	5	1	7	3	8	2	4	6

INSANE - 201

2	9	4	6	8	3	1	5	7
7	6	3	9	5	1	8	2	4
8	1	5	7	4	2	3	9	6
6	7	9	2	1	8	4	3	5
3	4	2	5	7	6	9	1	8
1	5	8	4	3	9	6	7	2
5	8	1	3	2	4	7	6	9
4	2	6	1	9	7	5	8	3
9	3	7	8	6	5	2	4	1

INSANE - 202

8	4	1	7	6	3	9	5	2
3	5	2	4	8	9	6	7	1
6	9	7	2	1	5	3	4	8
2	8	5	9	7	4	1	6	3
9	7	6	1	3	8	4	2	5
4	1	3	5	2	6	7	8	9
1	3	4	8	5	7	2	9	6
5	6	9	3	4	2	8	1	7
7	2	8	6	9	1	5	3	4

INSANE - 203

5	3	1	6	7	8	9	4	2
6	8	7	4	2	9	5	1	3
4	9	2	5	3	1	8	7	6
9	6	8	3	4	5	7	2	1
3	7	5	1	9	2	4	6	8
1	2	4	8	6	7	3	9	5
8	5	6	9	1	4	2	3	7
7	1	9	2	5	3	6	8	4
2	4	3	7	8	6	1	5	9

INSANE - 204

9	2	5	6	4	1	8	3	7
3	6	8	9	7	5	4	1	2
1	7	4	2	8	3	5	9	6
6	8	1	7	5	9	2	4	3
5	9	3	4	2	8	7	6	1
7	4	2	1	3	6	9	8	5
2	3	6	5	9	4	1	7	8
4	1	7	8	6	2	3	5	9
8	5	9	3	1	7	6	2	4

Made in United States
Orlando, FL
17 December 2022

27076050R00070